# Intermediate Statistics FOR DUMMIES®

# Intermediate Statistics FOR DUMMIES®

by Deborah Rumsey, PhD

Wiley Publishing, Inc.

**Intermediate Statistics For Dummies®**
Published by
**Wiley Publishing, Inc.**
111 River St.
Hoboken, NJ 07030-5774
www.wiley.com



For general information on our other products and services, please contact our Customer Care Department within the U.S. at 800-762-2974, outside the U.S. at 317-572-3993, or fax 317-572-4002.

For technical support, please visit www.wiley.com/techsupport.

Wiley also publishes its books in a variety of electronic formats. Some content that appears in print may not be available in electronic books.

Library of Congress Control Number: 2006939467

ISBN: 978-0-470-04520-6

Manufactured in the United States of America

10 9 8 7 6 5 4 3 2 1

1B/SQ/QS/QX/IN

WILEY

# About the Author

**Deborah Rumsey** has a PhD in Statistics from The Ohio State University (1993). She is a Statistics Education Specialist/Auxiliary Faculty Member for the Department of Statistics. Dr. Rumsey has been given the distinction of being named a Fellow of the American Statistical Association. She has also won the Presidential Teaching Award from Kansas State University. She is the author of *Statistics For Dummies, Statistics Workbook For Dummies,* and *Probability For Dummies*. She has published numerous papers and given many professional presentations on the subject of statistics education. Her passions include being with her family, bird watching, solving Sudoku puzzles, getting more seat time on her Kubota tractor, and cheering the Ohio State Buckeyes on to another National Championship.

# Dedication

To my husband Eric: My sun rises and sets with you. To my son Clint: I love you up to the moon and back.

# Author's Acknowledgments

Thanks again to Kathy Cox for giving me the opportunity to write this book; to Chrissy Guthrie for her unwavering support and perfect chiseling and molding of my words and ideas; Phyllis Curtiss, Grand Valley State University, for a thorough technical view; and Sarah Westfall for great copy editing. Special thanks to Elizabeth Stasny for guidance and support from day one; and Joan Garfield for constant inspiration and encouragement.

## Publisher's Acknowledgments

We're proud of this book; please send us your comments through our Dummies online registration form located at www.dummies.com/register/.

Some of the people who helped bring this book to market include the following:

***Acquisitions, Editorial, and Media Development***

**Senior Project Editor:** Christina Guthrie

**Acquisitions Editors:** Kathy Cox, Lindsay Lefevere

**Copy Editor:** Sarah Westfall

**Technical Editor:** Phyllis Curtiss, PhD

**Editorial Manager:** Christine Meloy Beck

**Editorial Assistants:** Erin Calligan, Joe Niesen, David Lutton, Leeann Harney

**Cover Photo:** Ingram Publishing

**Cartoons:** Rich Tennant (www.the5thwave.com)

***Composition Services***

**Project Coordinator:** Jennifer Theriot

**Layout and Graphics:** Brooke Graczyk, Denny Hager, Joyce Haughey, Stephanie D. Jumper

**Anniversary Logo Design:** Richard Pacifico

**Proofreaders:** Cynthia Fields, Linda Quigley

**Indexer:** Dakota Indexing

---

**Publishing and Editorial for Consumer Dummies**

**Diane Graves Steele,** Vice President and Publisher, Consumer Dummies

**Joyce Pepple,** Acquisitions Director, Consumer Dummies

**Kristin A. Cocks,** Product Development Director, Consumer Dummies

**Michael Spring,** Vice President and Publisher, Travel

**Kelly Regan,** Editorial Director, Travel

**Publishing for Technology Dummies**

**Andy Cummings,** Vice President and Publisher, Dummies Technology/General User

**Composition Services**

**Gerry Fahey,** Vice President of Production Services

**Debbie Stailey,** Director of Composition Services

# Contents at a Glance

# Table of Contents

# Introduction

Ready to load your statistical toolbox with a new level of tools? *Intermediate Statistics For Dummies* picks up where *Statistics For Dummies* (or your introductory statistics course) leaves off, and keeps you moving along the road of statistical ideas and techniques in a positive step-by-step way.

The focus of intermediate statistics is on building and testing models based on data. You're trying to estimate, investigate, correlate, and congregate certain variables based on the information at hand. The process for doing this is two-fold. First you build a model that you think describes your situation (the model-building phase), and then you test your model, using the data you've collected (the data analysis phase).

The techniques presented in intermediate statistics are used even more heavily in medical and scientific studies than the introductory topics were. The reason is that most real-world studies have more complex problems to solve; they ask more questions and collect more data. Given that the results of these more complex studies are used to make decisions in a host of different areas (including medical science, biology, engineering, business, and politics to name a few) most anyone can benefit from reading this book. You can see applications that give you exposure to real problems and to the process of interpreting and understanding other people's results.

## About This Book

This book is designed for people who want to get into (or at least be able to understand and interpret) some of the more involved techniques in statistics, beyond medians and means, the Central Limit Theorem, and confidence intervals and hypothesis tests. (However, I do add some brief overviews of introductory statistics as needed, just to remind everyone of what was covered and get new readers up to speed.) The topics this time around are many flavors of regression (including simple, multiple, nonlinear, and logistic); ANOVA (one-way and two-way); Chi-square tests (for independence and goodness-of-fit); and nonparametric procedures.

I also include interpretation of computer output for data analysis purposes. I do show how to use the software to get the results, but I focus more on how to interpret the results found in the output. It's likely that more people will be interpreting this kind of information rather than doing the programming specifically. And because the equations and calculations can get too involved by hand, you often use a computer to get your results. I include instructions for using Minitab to conduct many of the calculations in this book. Most statistics teachers who cover these intermediate topics hold this philosophy as well. (What a relief!)

This book is different from the other intermediate statistics books in many ways, including the following:

- **Full explanations of intermediate statistical ideas**. Many statistics textbooks squeeze all the intermediate level topics at the very end of their huge introductory-level textbooks; as a result, these topics tend to get condensed and presented as if they were optional topics. But no worries; I take the time to clearly and fully explain all the information you need to survive and thrive.
- **Dissection of computer output.** Throughout the book, I present many examples that use statistical software to analyze the data. In each case, I present the computer output as well as an explanation of how I got the output and what it means.
- **An extensive number of examples.** I include several examples to cover the many different types of problems you will face.
- **Lots of tips, strategies, and warnings.** I share with you some of the trade secrets, based on my experience teaching and supporting students and grading their papers.
- **Nonlinear approach.** The setup of this book allows you to skip around in the book and still have easy access and understanding of any given topic.
- **Understandable language.** I try to keep things conversational to help you understand, remember, and put into practice statistical definitions, techniques, and processes.
- **Clear and concise step-by-step procedures.** In most chapters, you can find steps that intuitively explain how to work through intermediate statistics problems, and remember how to do it later on.

# Conventions Used in This Book

Throughout this book, I've used several conventions that I want you to be aware of:

- I indicate multiplication by using a times sign, indicated by a lowered asterisk, *.
- I also indicate the null and alternative hypotheses as Ho (for the null hypothesis) and Ha (for the alternative hypothesis).

- The statistical software package I use and display throughout the book is Minitab 14, but I simply refer to it as *Minitab.*
- Whenever I introduce a new term, I *italicize* it.
- Keywords and numbered steps appear in **boldface.**
- Web sites and e-mail addresses appear in `monofont`.

## What You're Not to Read

At times I get into some of the more technical details of formulas and procedures for those individuals who may need to know (or just really want to). These minutiae are marked with a Technical Stuff icon. I also include sidebars as an aside to the essential text, usually in the form of a real-life statistics example or some bonus info you may find interesting. You can feel free to skip those icons and sidebars because you won't miss any of the main information you need (but by reading it, you may just be able to impress your stat professor with your above-and-beyond knowledge of intermediate statistics!).

## Foolish Assumptions

Because this books deals with *intermediate* statistics, I assume you have had one previous course in introductory statistics under your belt (or have at least read *Statistics For Dummies* [Wiley]), with topics taking you up through the Central Limit Theorem and perhaps an introduction to confidence intervals and hypothesis tests (although I review these concepts briefly in Chapter 3). Prior experience with simple linear regression isn't necessary. Only college algebra is needed for the mathematics details. Some experience using statistical software is a plus but not required.

As a student, you may be covering these topics in one of two ways: either at the tail end of your introductory statistics course (perhaps in a hurried way, but in some way nonetheless); or through a two-course sequence in statistics in which the topics in this book are the focus of the second course. If so, this book provides you just the information you need to do well in those courses.

You may simply be interested in intermediate statistics from an everyday point of view or want to add to your understanding of studies and statistical results presented in the media. If this is you, you can find plenty of real-world examples and applications of these statistical techniques in action as well as cautions for interpreting them.

## How This Book Is Organized

This book is organized into five major parts that explore the main topic areas in intermediate statistics, along with one bonus part that offers a series of quick top-ten references for you to use. Each part contains chapters that break down the part's major objective into understandable pieces.

### Part I: Data Analysis and Model-Building Basics

This part goes over the big ideas of descriptive and inferential statistics and simple linear regression in the context of model building and decision making. Some material from introductory statistics receives a quick review. I also present you with the typical jargon of intermediate statistics.

### Part II: Making Predictions by Using Regression

Here, you can review and extend the ideas of simple linear regression to that of using more than one predictor variable. This part presents techniques for dealing with data that follows a curve (nonlinear models) and models for yes or no data used to make predictions about whether or not an event will happen (logistic regression). It includes all you need to know about conditions, diagnostics, model building, data-analysis techniques, and interpreting results.

### Part III: Comparing Many Means with ANOVA

You may want to compare the means of more than two populations. In this case, you use analysis of variance (ANOVA). I discuss the basic conditions required, the *F*-test, one-way and two-way ANOVA, and multiple comparisons. The final goal of these analyses is to show whether the means of the given populations are different and if so, which ones are higher or lower than the rest.

## Part IV: Building Strong Connections with Chi-Square Tests

This part deals with the Chi-square distribution and how you can use it to model and test qualitative (categorical) data. You see how to test for independence of two categorical variables using a Chi-square test. (No more speculations just by looking at the data in a two-way table!) You also see how to use Chi-square to test how well a model for categorical data fits.

## Part V: Rebels without a Distribution: Nonparametric Statistics

You can look at techniques used in situations where you can't (or don't want to) assume your data comes from a population with a certain distribution. For example, when your population isn't normal (the condition required by most other methods in intermediate statistics).

## Part VI: The Part of Tens

Reading this part can give you an edge in two major areas that go beyond the formulas and techniques of intermediate statistics. Those areas are starting the problem right (knowing what type of problem it is and how to attack it) and ending the problem right (knowing what kinds of conclusions you can and can't make).

You also find an appendix at the back of the book that contains all the tables you need to understand and complete the calculations used in this book.

# Icons Used in This Book

I use icons in this book to draw your attention to certain features that occur on a regular basis. Think of them as road signs that you encounter on a trip. Some signs tell you about shortcuts, but others offer more information that you may need; some signs alert you to possible warnings, while others leave you with something to remember.

When you see this icon, it means I'm explaining how to carry out that particular data analysis using Minitab. I also explain the information you get in the computer output so you can interpret your results.

I use this icon to reinforce certain ideas that are critical for success in intermediate statistics, such as things I think are important to go over as you prepare for an exam.

This icon points out exciting and perhaps surprising situations where intermediate statistics is being used in the real world.

When you see this icon, you can skip over it if you don't want to get into the nitty-gritty details. They exist mainly for people who have a special interest or obligation to know more about the more technical aspects of the statistical issues.

Tips refer to helpful hints, ideas, or shortcuts that you can use to save time, or alternative ways to think about a particular concept.

I use warning icons to help you stay away from common misconceptions and pitfalls you can face when dealing with intermediate statistics ideas and techniques.

# Where to Go from Here

This book is written in a nonlinear way, so you can start anywhere and still be able to understand what's happening. However, I can make some recommendations to those who are interested in knowing where to start.

If you're thoroughly familiar with the ideas of hypothesis testing and simple linear regression, start with Chapter 5 (multiple regression). Use Chapter 1 if you need a reference for the jargon that statisticians use in intermediate statistics.

If you have covered all topics up through the various types of regression (simple, multiple, nonlinear, and logistic) or a subset of those as your professor deemed important, proceed to Chapter 9, the basics of analysis of variance (ANOVA).

Chapter 14 is the place to begin if you want to tackle qualitative (categorical) variables before hitting the quantitative stuff. You can work with the Chi-square test there.

Nonparametric statistics are presented starting with Chapter 16. This area is a hot topic in today's statistics courses, yet one that doesn't seem to get as much space in textbooks as it should. Start here if you want the full details on the most common nonparametric procedures.

# Part I
# Data Analysis and Model-Building Basics

The 5th Wave By Rich Tennant

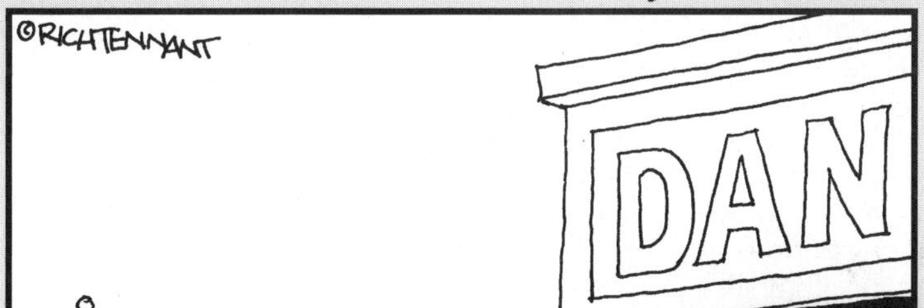

"Is it just me or did the whole '50% satisfaction' statistic seem a little unimpressive?"

## In this part . . .

To get everyone on the same page moving from introductory to intermediate statistics, I go over the basics of data analysis, important terminology, the main goals and concepts of model building, tips for choosing appropriate statistics to fit the job, and a review of the most heavily referred to items from introductory statistics. You also get a head start on making and looking at some basic computer output.

# Chapter 1

# Beyond Number Crunching: The Art and Science of Data Analysis

### *In This Chapter*

- Realizing your role as a data analyst
- Avoiding statistical faux pas
- Delving into the jargon of intermediate statistics

Because you're reading this book, you're likely familiar with the basics of statistics. You're now ready to take it up a notch. That next level involves using what you know, picking up a few more tools and techniques at the intermediate level, and finally putting it all to use to help you answer more realistic questions by using real data.

In statistical terms, you're ready to enter the world of the *data analyst.* This world's an exciting one, with many options to explore and many tools available. But, as you may have guessed, you have to navigate this world very carefully, choosing the right methods for each situation. In this book, you can see that I'm including the underlying theories and ideas behind the methods where necessary to help you make good decisions — and not just get into the point-and-click mode that today's software packages offer.

In this chapter, you review the terms involved in statistics as they pertain to data analysis at the intermediate level. You get a glimpse of the impact that your results can have by seeing what these analysis techniques can do. You also gain insight into some of the common misuses of data analysis and their effects.

## Data Analysis: It's Not Just for Statisticians Anymore

It used to be that statisticians were the only ones who really analyzed data. The reason for this is because the only computer programs that were available then were very complicated to use, requiring a great deal of knowledge about statistics to set up and carry out. The calculations were tedious and at times unpredictable and required a thorough understanding of the theories and methods behind the calculations to get correct and reliable answers.

Today, anyone who wants to analyze data can do it easily. Many user-friendly statistical software packages are made expressly for that purpose — Microsoft Excel, Minitab, SAS, and SPSS, just to name a few. Free online programs are even available, such as Stat Crunch, to help you do just what it says — crunch your numbers and get an answer. As you see in this section, the modern easy-to-use statistical packages are good in some ways, and not-so-good in other ways.

The most important idea when applying statistical techniques to analyze data is to know what's going on behind the number crunching, so you (not the computer) are in control of the analysis. That's why knowledge of intermediate statistics is so critical.

## Remembering the old days

In the old days, in order to determine whether methods gave different results, you had to write a computer program to do it, using code that you had to take a class to learn. You had to type in your data in a specific way that the computer program demanded, and you had to submit your program to a mainframe computer and wait for the printer to print out your results. This method was time consuming and a general all-around pain.

I remember the day in college when I reached bottom. I was just learning to write those sophisticated programs you needed to do the simplest analysis. No matter how hard I tried to write the perfect program, the computer kept spitting my work back at me without doing my analysis, noting error after error in the way I typed the commands. The last straw came when I gave my program to the computer for the umpteenth time: At the end of the printout, the computer told me on the very last line: "Error #34410: Too many errors."

Now, don't get the idea that your author doesn't know what she's doing. I had all the statistical methods right; I just wasn't very good at writing computer programs. So for anyone out there who's ever been frustrated by a computer, I feel your pain, and I try to minimize your troubles throughout this book.

Enough lamenting about having to walk to school uphill both ways in the snow with plastic bags on my feet instead of boots. The point is, statistical software packages have undergone an incredible evolution in the last 10 to 15 years, to the point where you can now enter your data quickly and easily in almost any format. Moreover, the choices for data analysis are well organized and listed in pull-down menus. Now almost anyone (even me) can quickly see how to find the necessary procedure and tell the computer what to do. The results come instantly and successfully, and you can cut and paste them into a word-processing document without blinking an eye. For example, comparing the weight loss for people on different weight-loss programs now takes less than three clicks of the mouse to perform, which is great news for folks like me.

Many very useful and efficient statistical software packages exist, including SAS, SPSS, Data Desk, Stat Crunch, MS Excel, and Minitab, and each one has its own pros and cons (and its own users and protesters). My software of choice, and the one I reference throughout this book, is Minitab, because it's very easy to use, the results are correct, the output is very clear and professional looking, and the software's loaded with all the data-analysis techniques that are used in intermediate statistics as well as in this book. While a site license for Minitab can be expensive, the downloadable student version is available for rent for only a few bucks a semester.

## The downside of today's statistical software

You may be wondering where the downside is in all of this. Is it too good to be true that what was once a tedious, complicated process for analyzing data has now become as easy as checking your e-mail on your cell phone? Yes and no. Yes, it's too good to be true that the software practically does everything for you — if you don't pay attention to what the programs are really doing. Yes, it's too good to be true if you don't understand that conditions need to be checked in every situation before an analysis should be applied. Yes, it's too good to be true if you take all the results as complete and utter gospel (as too many statistician wannabees do).

Bottom line: Today's software packages are too good to be true if you don't have a clear and thorough understanding of the intermediate level of statistics that lie underneath them.

Here's the good news, though. By reading this book, you gain the understanding you need to set you up for success. You get enough of the underlying intermediate statistical concepts to be empowered, but not be dangerous. You find out what conditions need to be checked on the data before applying an analysis and how to check them. You get a good feel for which analyses to use to answer your question (and which ones can cause you trouble), and you become aware of the kinds of results you can expect. Most importantly, you discover what's possible and appropriate to conclude from your analysis and what limitations and caveats you need to make.

# Rule #1: Look Before You Crunch

Many people don't realize that statistical software can't tell you when to use and not to use a certain statistical technique. You have to determine that on your own. As a result, people think they're doing their analyses correctly, but they can end up making all kinds of mistakes. Statistical software packages are centered on mathematical formulas, and mathematical formulas aren't smart enough to know how you're applying them or to warn you when you're doing something wrong (that's where this book comes in).

In this section, I give some examples of some of the major situations where innocent data analyses can go wrong and why it's important to know what's happening behind the scenes from a statistical standpoint before you start crunching numbers.

## Nothing (even a straight line) lasts forever

After you get a statistical equation, or *model,* that tries to explain or predict some random phenomena, you need to specify for what values the equation applies and for what values the equation doesn't apply. Equations don't know when they work and when they don't; it's up to the data analyst to determine that. This idea is the same for applying the results of any data analysis that you do.

Bill Prediction is a statistics student, studying the affect of study time on exam score. Based on his experience, and that of a few friends, Bill comes up with the equation $y = 10x + 30$, where $y$ represents the test score you get if you study a certain number of hours $(x)$. This equation is Bill's model for predicting exam score using study time. Notice that this model is the equation of a straight line with a $y$-intercept of 30 and a slope of 10.

So Bill predicts, using this model, that if you don't study at all, you'll get a 30 on the exam (plugging $x = 0$ into the equation and solving for $y$; this point represents the $y$-intercept of the line). And he predicts, using this model, that if you study for five hours, you'll get an exam score of $y = 10 * 5 + 30 = 80$. So, the point (5, 80) is also on this line. (I won't talk in detail at this point about how well Bill's model does at predicting exam score, but you can just say he's got some work to do on this and leave it at that for now.)

I'm sure you would agree that because $x$ is the amount of study time, that $x$ can never be a number less than zero. If you plug a negative number in for $x$, say $x = -10$, you get $y = 10 * -10 + 30 = -70$, which makes no sense. The worst possible score, according to Bill's model, is 30, which occurs when $x$ equals 0.

And, you can't study a negative number of hours, so a negative number for $x$ itself isn't even possible.

On the other side of the coin, $x$ probably isn't a number in the two-digit range (10 or more). Why is this? Say someone did study ten hours for this exam. Plugging in 10 for $x$ in Bill's equation, you get $y = 10 * 10 + 30$, which equals 130. Remember, $y$ is the predicted exam score. Because most exams are out of 100 possible points, a score of 130 isn't possible. (I'm all for extra credit on exams, but 30 points of extra credit is too much, even for me.)

The point is that there are limits on the values of $x$ that make sense in this equation. However, the equation itself, $y = 10x + 30$, doesn't know that, and if you graph this line, it'll go on forever in both the positive and negative directions (see Figure 1-1).

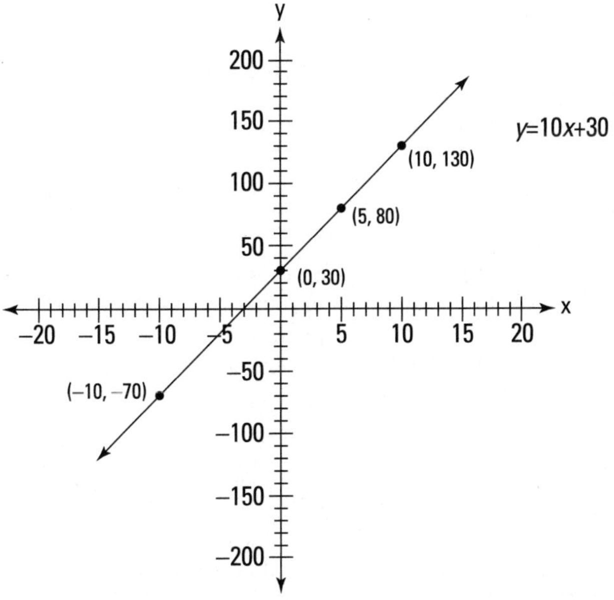


**Figure 1-1:** The line $y = 10x + 30$, for all possible values of $x$.

# Data snooping isn't cool

WARNING!

Statisticians have come up with a saying that you may have heard of: "Figures don't lie. Liars figure." Make sure that you find out about all the analyses that were performed on a data set, not just the ones reported as being statistically significant.

Suppose Bill Prediction tries to apply his simple model (from the preceding section) to predict exam scores for his whole class, based on their reported amounts of study time, and he finds out that his results fall flat. He figures out that he needs more information, so he tries to uncover what other factors help determine exam score on a statistics test besides study time. Bill measures everything from soup to nuts. His set of possible variables includes study time, GPA, previous experience in statistics, math grades in high school, attitudes toward statistics, whether you listen to classical music while studying, shoe size, whether you chew gum during the exam, and even what your favorite color is (after all, you never know, he figures). For good measure, he includes 11 other variables, for a total of 20 possible factors that he thinks may relate to exam score.

Bill starts out by looking for relationships between each of these variables and exam score, so he does 20 correlations. (*Correlation* is a measure of the linear relationship between two variables; see the section on correlation later in this chapter). He finds out that four variables have a statistically significant relationship with exam score (that means the results are supposed to be correct with a 95 percent chance — but only if he collected the data properly and did the analysis correctly).

The variables that Bill found to be related to exam score were study time, math grades in high school, GPA, and whether the person chews gum during the exam. It turns out that his new model fits pretty well (by criteria I discuss in Chapter 5 on multiple linear regression models). Bill now thinks he's scored a home run and has answered that all-elusive question: How can I do better on my statistics test?

But as they said in *Apollo 13,* "Houston, we have a problem." By looking at all possible correlations between his 20 variables and exam score, Bill is actually doing 20 separate statistical analyses. Under typical conditions (I describe these conditions in Chapter 3), each statistical analysis has a 5 percent chance of being wrong just by chance (this value of 5 percent is called the *significance level* of the test).

Because 5 percent of 20 analyses is equal to one, you can expect that when you do 20 statistical analyses, one of them will give the wrong result, just by chance, over the long term. I bet you can guess which one of Bill's correlations likely came out wrong in this case. Of course, study time has *nothing* to do with exam score, and gum-chewing is the answer to all of our problems, right? (If that were the case, all statisticians would be out of business and working for chewing-gum companies instead.)

What Bill is doing is called *data snooping* in the data-analysis business. Bill looks around until he finds something, and then he believes the result. This strategy is dangerous, but one that's done all too often in the real world. One of the reasons data snooping is running rampant today is because everyone and his brother is out there collecting data and analyzing it — and everyone wants to find something. They're using statistical software that allows them to just point and click to do as many analyses as they want, without any warning about what statisticians call the *overall error rate* (that is, the probability of making an error due to chance during any step of the entire analysis, not just the probability of making an error due to chance on any single analysis).

## No (data) fishing allowed

Redoing analyses in different ways to try to get the results you want is called *data fishing* in the statistics business, and folks in the stat biz consider it to be a major no-no (however, people unfortunately do it all too often in the name of research).

For example, Ellen Go-getter is convinced that dissolving sugar in the water helps cut flowers last longer. She performs an experiment to prove her hypothesis. She cuts two dozen roses and puts one rose in each vase. She fills each vase with 3 cups of water, but in 12 of the vases she adds 1 tablespoon of sugar (the other 12 vases constitute the control group, meaning that Ellen doesn't apply any new treatment to them to show what happens if she adds nothing). In the next sections, you follow Ellen through her experiment, keeping an eye on the statistical analyses that pop up along the way.

### Examining Ellen's data

Ellen counts how many days the flowers still look nice and uses the same criteria for each flower. After ten days, all the flowers have withered to the point where they need to be thrown away, so the experiment is over. You can see Ellen's data in Table 1-1.

**Table 1-1 Ellen's Data: Days Roses Lasted in Sugar Water versus Regular Water (Control Group)**

| *Observation* | *Days Lasted: Water Only* | *Days Lasted: Sugar Water* |
|---|---|---|
| 1 | 3 | 5 |
| 2 | 3 | 5 |
| 3 | 4 | 5 |
| 4 | 4 | 4 |
| 5 | 4 | 4 |
| 6 | 4 | 4 |
| 7 | 3 | 3 |

*(continued)*

**Table 1-1 *(continued)***

| *Observation* | *Days Lasted: Water Only* | *Days Lasted: Sugar Water* |
|---|---|---|
| 8 | 3 | 4 |
| 9 | 2 | 3 |
| 10 | 4 | 3 |
| 11 | 4 | 5 |
| 12 | 4 | 5 |

### *Setting the hypothesis*

Ellen wants to compare the two methods, water and sugar, to see whether the roses that had sugar added lasted longer than the regular water group. She needs to conduct a hypothesis test whose null hypothesis is Ho: There is no difference in days lasted for sugar group versus control group. Her alternative hypothesis, which she hopes to show, is Ha: The roses in the sugar group lasted longer than the control group. She figures a two-sample *t*-test is in order here. (I discuss hypothesis tests in Chapter 3.)

### *Checking the conditions*

Ellen has taken a few statistics classes before and knows that before she plunges into an analysis, she needs to check the proper conditions. For a comparison of two groups, she has to plot the data from each group on a *histogram* (a bar graph showing the number of days the flowers lasted, organized into groupings in numerical order versus the number of flowers that lasted each number of days). According to what she knows about a two-sample *t*-test, the data in each group has to have a normal distribution before she starts. That is, the data has to have a bell-shaped curve when you look at the histogram. Ellen plots the data in histograms for the two groups and gets the following results (see Figures 1-2a and 1-2b).

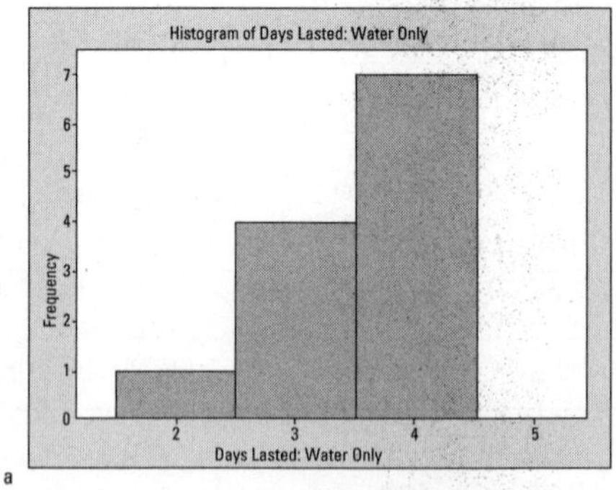


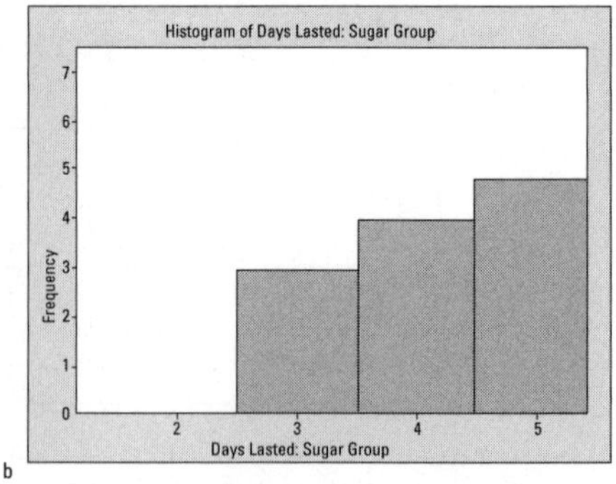


**Figure 1-2:** Histograms showing number of days roses lasted, using water only versus sugar added.

### Getting the bad news

As you can see in Figures 1-2a and 1-2b, Ellen's data doesn't follow the typical bell-shaped curve. One of the problems is her data only takes on values that are positive whole numbers, so numbers like 1.2, 2.3, and the like aren't possible. (Normal distributions are supposed to have many possible values.) The other problem is that the data has no values outside the typical two-, three-, four-, or five-day range, so the histogram doesn't have a chance to take on a bell shape. Perhaps more data would have curbed this problem. At any rate, Ellen knows that the conditions for a two-sample *t*-test aren't met here; namely that the data doesn't have a normal distribution and is, in fact, *skewed* (meaning set off to one side or the other).

### Going nonparametric

Undaunted by this turn of events, Ellen employs a nonparametric test of her data, which is the right thing to do. Statisticians use *nonparametric statistics* in situations where the assumptions of the typical analyses aren't met (like not having a normal distribution). However, nonparametric stats often give more conservative (albeit more accurate) results than the typical (parametric) procedures you're used to using. (I discuss nonparametrics a bit more in the last section of this chapter. Nonparametric procedures are discussed in full detail in Chapters 16–19.)

Because Ellen's data doesn't have a normal distribution or even a *symmetric distribution* (meaning one that looks the same on each side when you split it down the middle), the mean (or average) isn't a good measure of the center of the data, so a two-sample *t*-test isn't possible. As an alternative, she can test whether the two histograms are the same or not, if she compares the histograms of the two populations in question (all roses given water, versus all roses given sugar water).

Because she's comparing two groups, Ellen uses a Wilcoxon Rank Sum test, also known as the Mann-Whitney test (see Chapter 19). The Wilcoxon Rank Sum test checks whether two populations have the same distribution (meaning whether the two histograms look the same) versus one of the populations shifting to the right or left. Ellen's theory is that the sugar group lasts longer, so she tests Ho: Sugar group and control group have the same distribution versus Ha: Sugar group is shifted to the right of the control group.

### Ellen strikes out

To cut to the chase, the Wilcoxon Rank Sum test unfortunately fails to reject Ellen's null hypothesis. She didn't prove what she wanted to confirm by her experiment. Not enough roses in the sugar group lasted longer than those roses in the control group. You can see the underlying reason for this result by comparing the medians of the two groups. When you find the median of each of the data sets in Table 1-1, you get the value of 4 in each case. Because the medians of the two data sets are equal, it's unlikely that Ellen can find a statistically significant result by using this test.

### Breaking the rules

According to the rules that all good statisticians live by, Ellen's story should end there. She may still be convinced that sugar indeed helps roses last longer. She may use sugar with her roses for the rest of time and tell her friends to use it too. But, she isn't allowed to say that sugar water gives statistically different results than water alone; her analysis failed to show that.

But remember, Ellen's last name is Go-getter, so she's out to get those results. She knows that nonparametric tests usually give more conservative results than regular tests, and despite the fact that the conditions aren't met, she decides to analyze her data again, this time using the two-sample *t*-test.

Putting her data into a two-sample *t*-test takes only two more clicks of the mouse, and Ellen's results give her a *p*-value of 0.043. Using the usual significance level used for hypothesis tests, 0.050, her *p*-value is less than this number, so she can reject Ho. (In a two-sample *t*-test, Ho is that there's no difference in the means of the two groups. And her Ha in this case is that the mean of the sugar group is larger than the mean of the control group.) So Ellen gleefully cheers herself on for getting the results she wanted and decides there's no harm in trying a different analysis when all else fails.

### Seeing the error of Ellen's ways

But again, "Houston. . ." — you know the rest. Ellen's problem is that she cheated her way to getting a result that's incorrect. She knew that the conditions for the two-sample *t*-test weren't met, but when the correct analysis failed to get the results she wanted, she found an analysis that did. The trouble is, the results of the two-sample *t*-test are bogus.

Now it may not be a life-and-death situation whether your roses actually do last a little bit longer on sugar or not. (Incidentally, the gardening crowd says they don't, and that sugar in fact can encourage the growth of stem-clogging bacteria so the flower can't take in water.) But imagine a situation where doctors are trying to test to see whether a certain medication helps people get over an illness faster or whether some procedure helps cancer patients live longer. Now you're talking about results with a very serious impact.

REMEMBER

Using the wrong data analysis for the sake of getting the results you desire results in two major problems:

- You mislead your audience into thinking that your hypothesis is actually correct, which it may not be.
- Sooner or later someone is going to try to replicate those results and will find out that they can't be replicated. This discovery will result in a loss of your credibility *big time.* And unfortunately, you mislead many people in the meantime.

# Getting the Big Picture: An Overview of Intermediate Statistics

Because of the dangers and lingering effects of using the wrong techniques in the wrong situation to analyze data to answer questions, knowing what's happening behind the scenes of any data analysis and staying within the rules of well-chosen techniques and appropriate practices is very important. In other words, it's crucial for you to take your knowledge of statistics to the next level.

Intermediate statistics is an extension of introductory statistics, so the jargon follows suit and the techniques build on what you already know. If you've been able to grasp the ideas from the first course, you'll find no trouble with the terminology for intermediate statistics. If you're still unsure about some of the terms from introductory statistics, you can consult your textbook from your first course or see my other book, *Statistics For Dummies* (Wiley), for a complete rundown.

In this section, you get an introduction to the terminology you use in intermediate statistics, and you get a broad overview of the techniques that statisticians use for the purpose of analyzing data and the big picture behind them.

## Population parameter

REMEMBER

A *parameter* is a number that summarizes the population (the entire group you're interested in investigating). Examples of parameters include the mean of a population, the median of a population, or the proportion of the population that falls into a certain category.

Suppose you want to determine the average length of a cell-phone call among teenagers (ages 13 to 18). You're not interested in making any comparisons; you just want to make a good guesstimate as to what the average time is. So you want to estimate a population parameter (such as the mean or average). The population is all cell-phone users between the ages of 13 and 18 years old. The parameter is the average length of a phone call this population makes.

## Sample statistic

You normally can't study every member of an entire population (how would you like to measure and record the length of every single cell-phone call made by all teenagers?). So you can't determine population parameters exactly; you can only estimate them. But all is not lost; by taking a sample (a subset of individuals) from the population and studying them, you can come up with a good guess (estimate) of the population parameter, if you play your cards right. A subset of this population is called a *sample.* A *sample statistic* is a single number that summarizes that subset of the population.

For example, in the cell-phone scenario, you select a sample of teenagers and measure the length of their cell-phone calls over a period of time (or look at their cell-phone records if you can gain access legally). You take the average of the cell-phone call lengths. For example, the average length of 100 cell-phone calls may be 12.2 minutes — this average is a statistic. This particular statistic is called the *sample mean,* because it's the average value from your sample data.

You can also find a statistic called the *sample proportion* (the proportion of individuals in the sample that have a certain characteristic — for example, the percentage of female teens who use cell phones). Many different statistics are available (which you probably picked up in intro stats) to study different characteristics of a sample, such as the median, variance, and standard deviation.

## Confidence interval

A *confidence interval* is a range of values that provides reasonable estimates for a population parameter. A confidence interval is based on a sample and the statistics that come from that sample. The main reason you want to provide a range of possible values rather than a single number is that sample results vary from sample to sample.

For example, say you want to estimate the percentage of people who eat chocolate. According to the Simmons Research Bureau, 78 percent of adults reported eating chocolate, and of those, 18 percent admitted to eating sweets frequently. What's missing in these results? These numbers are only a single sample of people, and those sample results are guaranteed to vary from sample to sample. You need some measure of how much you can expect those results to move if you were to repeat the study.

This expected movement in your statistic is measured by the *margin of error,* which reflects a certain number of standard deviations of your statistic you add and subtract to have a certain confidence in your results (see Chapter 3 for more on margin of error). If the chocolate-eater results were based on 1,000 people, the margin of error would be approximately 3 percent, meaning the actual percentage of people who eat chocolate in the entire population is expected to be 78 percent, plus or minus 3 percent. In other words, it's somewhere between 75 percent and 81 percent. Now if you only base these results on a sample of 100 people, the margin of error balloons to 10 percent, meaning the percentage of chocolate eaters can only be reported to be between 68 and 88 percent. Notice how much wider the interval becomes when a smaller sample size is used. This result confirms that more data means more precision in your results (provided the data is collected properly).

## Hypothesis test

A *hypothesis test* is a statistical procedure that you use to test an existing claim about the population, using your data. The claim is noted by Ho (the null hypothesis). If your data support the claim, you fail to reject Ho. If your data don't support the claim, you reject Ho and conclude an alternative hypothesis, Ha. The reason most people conduct a hypothesis test is not to merely show that their data support an existing claim, but rather to show that the existing claim is false, in favor of the alternative hypothesis.

The Pew Research Center studied the percentage of people who go to ESPN for their sports news. Their statistics, based on a survey of about 1,000 people, found that in 2000, 23 percent of people said they go to ESPN; while in 2004, only 20 percent reported going to ESPN. The question is this: Does this 3-percent reduction in viewers from 2000 to 2004 represent a significant trend that ESPN should worry about?

To test these differences formally, you can set up a hypothesis test. You set up your null hypothesis as the result you have to believe without your study, Ho = no difference exists between 2000 and 2004 data for ESPN viewership. Your alternative hypothesis (Ha) is that a difference is there.

In very general terms, here's what's happening with a hypothesis test. You have the sample data, and you find the statistics that are relevant. In this case, you have two sample percentages, one for 2000 and one for 2004. You take the difference between the two samples (3 percent), and divide it by the standard error for the difference. The standard error measures how much the difference in the statistics is expected to change from sample to sample. In this case, the standard error comes to about 1.8 percent (for specific calculations see Chapter 3).

Taking the difference in the statistics (3 percent = 0.03) divided by the standard error (1.8 percent = 0.018) gives you the value of 1.67 (called the *test statistic*). This value represents the difference between the two statistics, in terms of number of standard errors. This result has a universal interpretation. Roughly speaking, if your test statistic falls between –2.00 and +2.00, that means the results you found don't differ enough to get excited about, because 95 percent of the time, this outcome happens just by chance. (And this example falls right into that situation.) After you take the variability of the sample results into account, the difference in these particular samples doesn't transfer over to the populations they represent. So, because you can't reject Ho, you have to say the percentage of viewers of ESPN in the entire population probably didn't change from 2000 to 2004.

REMEMBER

Because you have a 95 percent confidence level, this test uses a significance level ($\alpha$ level) of 1 – 0.95 = 0.05 or 5 percent. This percentage measures how likely your results would have been just by chance.

The trouble is that people often just report the sample statistics and give no regard to the expected amount of change with a new sample. This disregard leads to big mistakes in the conclusions (more on hypothesis testing in Chapter 3).

## Analysis of variance (ANOVA)

ANOVA is the acronym for *analysis of variance.* You use ANOVA in situations where you want to compare the means of more than two populations. For example, you want to compare the lifetime of four brands of tires, in number of miles. You take a random sample of 50 tires from each group, for a total of 200 tires, and set up an experiment to compare the lifetime of each tire, and record it. You have four means and four standard deviations now, one for each data set. But you have different types of variability in your data, each measured by using various sums of squares. (Remember from your intro stats that the variance of a data set is the total of all the squared distances between the data and the mean, all divided by $n - 1$.)

One of the types of variability in your data is called the variability *between* treatments (also known as *SST,* the treatment sums of squares). SST measures the variation in the average lifetimes of each brand of tire, compared to the overall average lifetime. If SST is large, you have a chance that there's a difference in lifetimes due to the treatment (in this case, the brand of tire).

Next, you have the variability *within* the treatments (also known as *SSE,* the error sums of squares). SSE measures the overall average amount of variability of the tire lifetimes within each particular brand (after all, not all tires are created equal, even if they're of the same brand). If SSE is large, you have so much variability within the tire brands themselves, that it will be harder to see any real difference between the brands, even if it actually exists.

And finally, you have the *total* overall variability in the data values if you just put them all together into one big data set. This variability is known as *SSTO,* the total sums of squares. ANOVA splits up the total variability (SSTO) into the between-groups variability (SST) plus the within-groups variability (SSE).

Then, to test for differences in average lifetime for the four brands of tires, you compare the mean sums of squares for treatments (MST) to the mean sums of squares for error (MSE) in a ratio called the *F-statistic.* If this ratio is large, then the variability between the brands is more than the variability within the brands, giving evidence that not all the means are the same for the different tire brands. If the *F*-statistic is small, that means not enough difference was between the treatment means, compared to the general variability within the treatments themselves. In this case, you can't say that the means are different for the groups. (I give you the full scoop on ANOVA in Chapters 9 and 10.)

## Multiple comparisons

Suppose you conduct ANOVA, and you find a difference in the average lifetimes of the four brands of tire (see preceding section). Your next questions would probably be, which brands are different, and how different are they? To answer these questions, you use multiple-comparison procedures.

A *multiple-comparison procedure* is a statistical technique that compares means to each other and finds out which ones are different and which ones aren't. You're then able to put the groups in order, from those with the largest mean to those with the smallest mean, realizing that sometimes two or more groups were too close to tell and so you put them in the same group.

Suppose you compare the exam scores of four different classes (call them class one, class two, class three, and class four), and your ANOVA procedure finds out that not all the means were the same. That means the *F*-statistic is large. Next, you use multiple-comparison procedures in order to make separate comparisons and figure out which classes were about the same and which ones were different, and come up with an ordering of the classes. It may be, for example, that class four was statistically higher than all the others; classes one and two were statistically equivalent, but both were lower than class four. And class one was in a group all by itself at the bottom. The ordering is: class four (highest average), classes two and three (tied for second highest), and class one (the lowest average).

Never take that second step to compare the means of the groups if the ANOVA procedure doesn't find any significant results during the first step. (See Chapter 11 for more information.)

Many different multiple-comparison procedures exist to compare individual means and come up with an ordering in the event that your *F*-statistic does find that some difference exists. Some of the multiple-comparison procedures include Tukey's test, LSD, and pairwise *t*-tests. (While these tests' names may cause you to raise an eyebrow, don't worry. They're legitimate statistical tests.) Some procedures are better than others, depending on the conditions and your goal as a data analyst. I discuss multiple-comparison procedures in detail in Chapter 11.

## Interaction effects

An *interaction effect* in statistics operates the same way that it does in the world of medicine. Sometimes if you take two different medicines at the same time, the combined effect is much different than if you take the two individual medications separately.

REMEMBER

Interaction effects come up when you have a model that includes two or more variables, and you're using those variables to explain differences or to make comparisons regarding some outcome. When you have two or more variables in a model, you can't automatically study the effect of each variable separately; you also have to take into account the way those variables interact in terms of the outcome. In other words, you have to examine whether or not an interaction effect is present.

For example, suppose medical researchers are studying a new drug for depression and want to know how this drug affects the change in blood pressure for a low dose versus a high dose of the drug. They also compare the effects for children versus adults. In total, the model being studied has one response variable, an increase in blood pressure, and two factors that may possibly explain changes in the outcome, namely age group (adults versus children) and dosage level (low versus high). It could be that dosage level affects the blood pressure of adults differently than the blood pressure of children. This type of model is called a *two-way ANOVA model,* with a possible interaction effect between the two factors (age group and dosage level). See Chapter 11 for more.

One of the first things statisticians do when they have a two-way ANOVA is to plot the mean outcomes for each group they're comparing and look for patterns. This is called an *interaction plot.* One interaction plot for the drug-study scenario is in Figure 1-3.

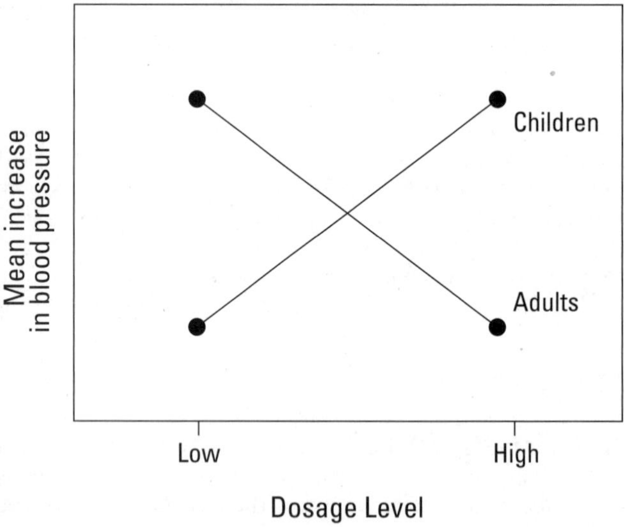

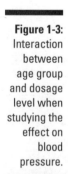

**Figure 1-3:** Interaction between age group and dosage level when studying the effect on blood pressure.

As you can see by Figure 1-3, the lines cross. If you look at the line representing children, you can see that the mean increase in blood pressure is low for the low dose of the drug, but then for the high dose of the drug; the increase in blood pressure goes way up. Alternatively, the reaction is the exact opposite for adults; on the low dose, the mean increase in blood pressure is very high, but for the high dose, the increase is very low. If the doctors neglected to study children as well as adults, the results of this study could be extremely damaging to children if doctors applied the rules for adults to children. This example shows that interaction effects are very important to look at.

Figure 1-4 shows the situation where you have no interaction effect for this drug. The lines are parallel, which tells you that the mean blood pressure increases more on a higher dosage of the drug for both adults and children. Because the line for the adults is higher up than the line for children, that means that overall, the increase in blood pressure is more for adults than the increase in blood pressure for children, no matter what the dosage level is.

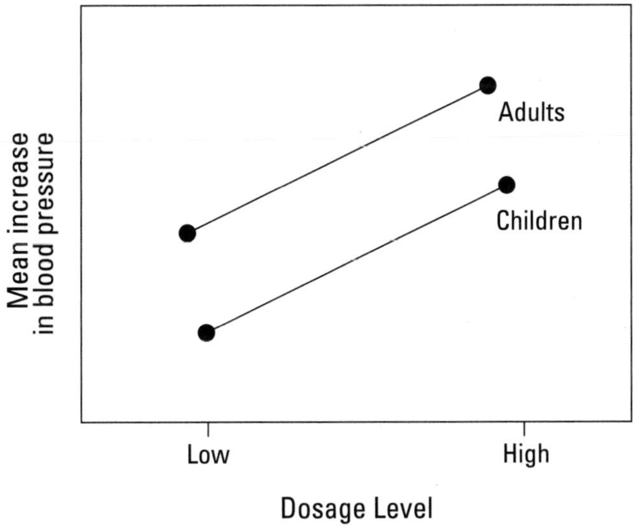


**Figure 1-4:** No interaction between age group and dosage level when studying the effect on blood pressure.

# Correlation

The term *correlation* is often misused. Statistically speaking, the correlation measures the strength and direction of the linear relationship between two quantitative variables (variables that represent counts or measurements only).

You aren't supposed to use the word *correlation* to talk about relationships of any other kind. For example, it's wrong to say that a correlation exists between eye color and hair color. While these variables may be related in some way, they're not quantitative variables, so you can't discuss their relationship in terms of a correlation. (In this case, you would use the term *association;* in Chapter 14, you see how to test for association of two categorical variables.)

The long and short of correlation is the following: *Correlation* is a number between –1.0 and +1.0. Positive one indicates a perfect positive relationship; in other words, as you increase one variable, the other one increases in perfect sync. On the other side of the coin, a correlation that is –1.0 indicates a perfect negative relationship between the variables. As one variable increases, the other one decreases in perfect sync. A correlation of zero indicates that you found no linear relationship at all between the variables. Most correlations in the real world aren't exactly +1.0, –1.0, or 0 — they fall somewhere in between. The closer to +1.0 or –1.0, the stronger the relationship is; the closer to 0, the weaker the relationship is.

Figure 1-5 shows an example of a plot showing the number of coffees sold at football games in Buffalo, New York, as well as the air temperature (in Fahrenheit) at each game. This data set seems to follow a downhill straight line fairly well, indicating a negative correlation. When you calculate the correlation, you get the value of –0.741. This value says that coffees sold has a fairly strong negative relationship with the temperature of the football game. This makes sense, because on days when the temperature is low, people will get cold and want more coffee. On days when the temperature is higher, people will tend to drink less coffee and perhaps tend more toward soft drinks, which are cold. I discuss correlation further, as it applies to model building, in Chapter 4.

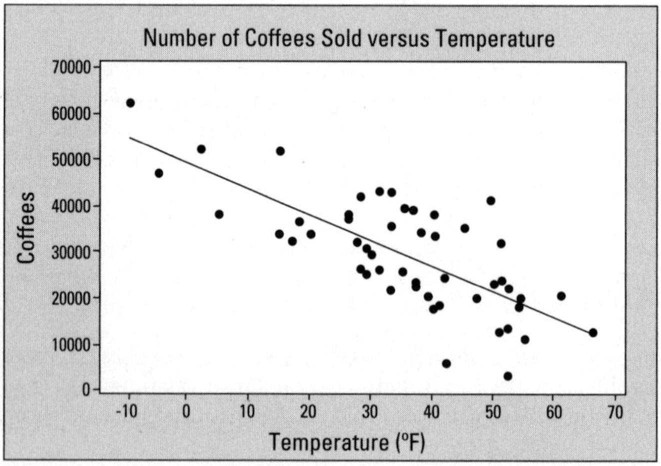


**Figure 1-5:** Coffees sold at various air temperatures on football game day.

## Linear regression

After you've determined that two variables have a fairly strong linear relationship, you may want to try to make predictions for one variable based on the value of the other variable. For example, if you know that a fairly strong negative linear relationship exists between coffees sold and the air temperature at a football game, you may want to use this information to predict how much coffee is needed for a game, just by knowing the temperature. This method of finding the best-fitting line is called *linear regression.*

In the coffees and temperature example (see Figure 1-5), the best-fitting line has the equation $y = 49{,}337 - 554 * x$, where $x$ is temperature and $y$ is the number of coffees sold. So when the temperature $(x)$ is zero degrees, you can expect to sell around 49,337 coffees (this is how you interpret the $y$-intercept of the line). To interpret the slope of this line, think of –554 as –554 divided by one and use the old rise-over-run idea using coffees and degrees of temperature. Applied here, it means that for every one degree increase in temperature, you can expect the coffee sales to decrease by 554. You can use this line to make predictions for reasonable values of the temperature $(x)$. For example, if the temperature is a cold 20-degrees Fahrenheit, you can predict that the number of coffees sold will be around $49{,}337 - 554 * 20 = 38{,}257$.

When you use only one variable to predict the response, the method of regression is called *simple linear regression.* (I review the basics of simple linear regression in Chapter 4. But many other types of regression are out there, many of which I discuss in this book.)

Most researchers use more than one variable to predict a response; this technique is called *multiple linear regression.* (Check out Chapter 5 for the details about multiple linear regression.) Multiple linear regression has many issues of its own because some variables you can use in the model may be related to each other, making overlapping contributions to the response. That possibility of overlapping makes their individual contributions hard to track. You also have to watch for interaction effects when using more than one variable to predict a response.

Simple and multiple linear regression assume that the response variable (the one being studied) is quantitative in nature (that is, it measures or counts something). However, you may be interested in making predictions about a variable that has only two outcomes: yes or no. For example, whether or not a certain horse will win a race; whether a baby will be a girl or a boy; or whether or not a tropical storm is going to make landfall. These situations require a different kind of regression called *logistic regression.* (I discuss logistic regression in Chapter 8.)

Finally, you may be interested in building a model for which a straight line doesn't fit. For example, you may want to predict miles per gallon, using the speed of the car. While high speeds get low miles per gallon, low speeds can get low miles per gallon as well. So the relationship between speed and miles per gallon actually follows that of a *parabola* (an upside-down bowl, in this case). This kind of relationship is called a *quadratic relationship.* More generally speaking, relationships that don't follow a straight line are called *nonlinear relationships,* and the technique you use to handle these situations is called (no surprise) *nonlinear regression.* I get into the meat of this technique in detail in Chapter 7.

## Chi-square tests

Correlation and regression techniques all assume that the variable being studied in most detail (the response variable) is quantitative. That is, the variable measures or counts something. However, you can run into many situations where the data being studied isn't quantitative, but rather qualitative. In other words, the data themselves represent categories, not measurements or counts.

For example, suppose you want to compare the views of the president by political affiliation. Say that in this particular year, the president is a Republican, and you select a random sample of 150 Republicans, 150 Democrats, and 150 Independents to find out their views on the president. The data may look like Table 1-2.

**Table 1-2 Views on a (Republican) President by Political Affiliation**

| | *Approve* | *Neutral* | *Disapprove* |
|---|---|---|---|
| Republican | 100 | 40 | 10 |
| Democrat | 40 | 10 | 100 |
| Independent | 50 | 50 | 50 |

In looking at how the numbers appear across the columns for various rows in Table 1-2, you may suspect that something is up. It appears that Republicans tend to approve of the president, while Democrats tend to disapprove, and Independents are split down the middle. (So much for the spirit of bipartisanship. . . .)

Now does this association you found in the data set for this sample of 450 people carry over to the entire population? In order to answer this question, you need to conduct a hypothesis test. And not just any hypothesis test — a *Chi-square test for independence.* You're testing to see whether the two qualitative variables, political affiliation and views on the president, are related or not. If they are related, the variables are deemed not independent; if they are unrelated, the variables are independent.

A Chi-square test basically does the following: It figures out the number of values that you expect to see in each cell of the table if the variables are independent (these values are brilliantly called the *expected cell counts*). The Chi-square test then compares these expected cell counts to what you actually saw in the data (called the *observed cell counts*) and compares them to each other in a Chi-square statistic (see Chapter 14).

If the Chi-square test statistic is large, you're likely to find an association between the two variables, because the total differences are large between the observed and expected cell counts. In other words, the variables are not independent, and you can look at the observed cell counts to discuss the relationship you see. If the Chi-square test statistic is small, then you can't conclude you've found a relationship, and the two variables are independent.

In the case of political affiliation and views on the president, the Chi-square test statistic is huge, and you conclude a relationship is there somewhere. You can say that in the population, Republicans tend to support the president, Democrats tend to oppose the president, and the Independents are split down the middle. (You can find the details of how to find the expected counts and conduct the Chi-square test in Chapter 14.)

You can also use the Chi-square test to see whether your theory about what percent of each group falls into a certain category is true or not. For example, can you guess what percentage of M&Ms fall into each color category? More on these Chi-square variations, as well as the M&Ms question, in Chapter 15.

## Nonparametrics

*Nonparametrics* is an entire area of statistics that provides analysis techniques to use when the conditions for the more traditional and commonly used methods aren't met. For example, in order to use a *t*-test, the data needs to be collected from a population that has a normal distribution (that is, it has to have a bell-shaped curve). In order to do a hypothesis test for two means, the data from each group must be from its own normal population. In fact, most all of the commonly used data-analysis procedures have conditions that must be met in order to use them.

The trouble with these requirements is that many times people forget or just don't bother to check those conditions, and if the conditions are actually not met, the entire analysis goes out the window, and the researcher doesn't even know it. Or, someone finds out that the conditions aren't being met, yet she still goes ahead and uses the procedures anyway (for more on this faux pas, see the section in this chapter "No [data] fishing allowed").

While many of the traditional methods are what statisticians call *robust,* with respect to violations of their conditions (that's fancy terminology for the fact that they're pretty forgiving), you can only push the window so far. Proceeding to use a statistical procedure that isn't appropriate causes a great deal of trouble with respect to the correctness of the conclusions and the credibility of the researcher.

Have no fear, nonparametrics comes to your rescue. If the conditions aren't met for a data-analysis procedure that you want to do, chances are that an equivalent nonparametric procedure is waiting in the wings. And the good news is that they're generally pretty tame, in terms of formulas, and most statistical software packages can do them just as easily as the regular (parametric) procedures.

Conditions aren't checked automatically by statistical software packages, before doing a data analysis. It's up to the user to check any and all appropriate conditions, and if they're seriously violated, to take another course of action. Many times a nonparametric procedure is just the ticket. For much more information on different nonparametric procedures, see Chapters 16 through 19.

## Chapter 2

# Sorting through Statistical Techniques

### In This Chapter

- Deciphering the difference between qualitative and quantitative variables
- Choosing appropriate statistical techniques for the task at hand
- Evaluating bias and precision levels
- Interpreting the results properly

One of the most critical elements of statistics and data analysis is the ability to choose the right statistical technique for each job. Carpenters and mechanics know the importance of having the right tool when they need it and the problems that can occur if they use wrong tool. They also know that the right tool helps to increase their odds of getting the results they want the first time around, using the "work smarter not harder" approach.

In this chapter, you look at the some of the major statistical analysis techniques from the point of view of the mechanics and carpenters — knowing what each statistical tool is meant to do, how to use it, and when to use it. You also zoom in on mistakes some number crunchers make in applying the wrong analysis or doing too many analyses. Knowing how to spot these problems can help you avoid making the same mistakes, but it also helps you to steer your way through the ocean of statistics that may await you in your job and in everyday life.

If many of the ideas you find in this chapter seem like a foreign language to you and you feel like you need more background information, don't fret. Before continuing on in this chapter, head to your nearest intro stats book or check out another one of my books, *Statistics For Dummies* (Wiley).

# Qualitative versus Quantitative Variables in Statistical Analysis

After you've collected all the data you need from your sample, you want to organize it, summarize it, and analyze it. Before plunging the data in to do all the number crunching though, you need to first identify the type of data you're dealing with. The type of data you have points you to the proper types of graphs, statistics, and analyses you're able to use.

Before I begin, here's an important piece of jargon: Statisticians call any quantity or characteristic you measure on an individual a *variable;* the data collected on a variable is expected to vary from person to person (hence the creative name).

The two major types of variables are the following:

- **Qualitative:** A qualitative variable classifies the individual based on categories. For example, political affiliation may be classified into four categories: Democrat, Republican, Independent, and other; gender as a variable takes on two possible categories: male and female. A person may be categorized as a female Republican, which means that, regarding the gender variable, she falls into the female category, and regarding the political affiliation variable, she falls into the Republican category. Another name for a qualitative variable is a *categorical variable.*
- **Quantitative:** A quantitative variable measures or counts a quantifiable characteristic, such as height, weight, number of children you have, your GPA in college, or the number of hours of sleep you got last night. The quantitative variable value represents a quantity (count) or a measurement and has numerical meaning. That is, you can add, subtract, multiply, or divide the values of a quantitative variable, and the results make sense as numbers. This characteristic isn't true of qualitative variables, which can take on numerical values only as placeholders.

Because the two types of variables represent such different types of data, it makes sense that each type has its own set of statistics. Qualitative variables, such as gender, are somewhat limited in terms of the statistics that can be performed on them. For example, suppose you have a sample of 500 classmates classified by gender — 180 of them are male, and 320 are female. How can you summarize this information? You already have the total number in each category (this statistic is called the *frequency*). You're off to a good start, but frequencies are hard to interpret because you find yourself trying to compare them to a total in your mind in order to get a proper comparison. In the previous example, you may be thinking "One hundred and eighty males out of what? Let's see, it's out of 500. Hmmm . . . what percentage is that? I can't think."

The next step is to find a means to relate these numbers to each other in an easy way. You can do this by using what is called a relative frequency. The *relative frequency* is the percentage of data that falls into a specific category of a qualitative variable. You can find a category's relative frequency by dividing the frequency by the sample total (500, using this example) and multiplying by 100. In this case, you have $\frac{180}{500} = 0.36 * 100 = 36$ percent males and $\frac{320}{500} = 0.64 * 100 = 64$ percent females.

You can also express the relative frequency as a proportion in each group by leaving the result in decimal form and not multiplying by 100. This statistic is called the *sample proportion*. If you continue with the same example, the sample proportion of males is 0.36, and the sample proportion of females is 0.64.

You mainly summarize qualitative variables by using two statistics — the number in each category (frequency) and the percentage (relative frequency) in each category.

# Statistics for Qualitative Variables

The types of statistics done on qualitative data may seem to be limited; however, the wide variety of analyses you can perform using frequencies and relative frequencies offers answers to an extensive range of possible questions you may want to explore.

In this section, you see that the proportion in each group is the number-one statistic for summarizing qualitative data. Beyond that, you see how you can use proportions to estimate, compare, and look for relationships between the groups that compose the qualitative data.

## Comparing proportions

Researchers, the media, and even everyday folk like you and me love to compare groups (whether you like to admit it or not). For example, what proportion of Democrats support oil drilling in Alaska, compared to Republicans? What percentage of women watch college football versus men? What proportion of readers of *Intermediate Statistics For Dummies* pass their stats exams with flying colors, compared to nonreaders? To answer these questions, you need to compare the sample proportions using a hypothesis test for two proportions (see Chapter 3 or your intro stat textbook).

Suppose you've collected data on a random sample of 1,000 United States voters. You may want to compare the proportion of female voters to the proportion of male voters and find out whether they're equal. Suppose in your sample you find that the proportion of females is 0.53, and the proportion of males is 0.47. So for this sample of 1,000 people, you have a higher proportion of females than males. But here's the big question: Are these sample proportions different enough to say that the entire population of U.S. voters has more females in it than males? After all, sample results vary from sample to sample. The answer to this question requires comparing the sample proportions by using a hypothesis test for two proportions. I demonstrate and expand on this technique in Chapter 3.

## Estimating a proportion

You can also use relative frequencies (check out the section "Qualitative versus Quantitative Variables in Statistical Analysis") to make estimates about a single population proportion.

Say, for example, you want to know what proportion of females in the United States are Democrats. According to a sample of 29,839 female voters from the U.S. conducted by the Pew Research Foundation in 2003, the percentage of female Democrats was 36. Now because the Pew researchers based these results on only a sample of the population and not on the entire population, these results may vary from sample to sample. The amount of variability is measured by the *margin of error* (the amount that you add and subtract from your sample statistic), which for this sample is only about 0.5 percent. (To find out how to calculate margin of error, explore Chapter 3.) That means that the estimated percentage of female Democrats in the U.S. voting population is estimated to be somewhere between 35.5 percent and 36.5 percent.

The margin of error, combined with the sample proportion, forms what statisticians call a confidence interval for the population proportion. Recall from intro stats that a *confidence interval* is a range of likely values for a population parameter, formed by taking the sample statistic plus or minus the margin of error. (For more on confidence intervals, see Chapter 3.)

## Looking for relationships between qualitative variables

Suppose you want to know whether two qualitative variables are related (for example, is gender related to political affiliation?). Answering this question requires putting the sample data into a two-way table (using rows and columns to represent the two variables), and analyzing the data by using a Chi-square test (see Chapter 14). By following this process, you can determine whether two categorical variables are independent (unrelated) or whether a relationship exists between them. If you find a relationship, you can use percentages to describe it.

Table 2-1 shows an example of data organized in a two-way table. The data was collected by the Pew Research Foundation.

**Table 2-1 Gender and Political Affiliation for 56,735 U.S. Voters**

| *Gender* | *Republican* | *Democrat* | *Other* |
|---|---|---|---|
| Males | 32% | 27% | 41% |
| Females | 29% | 36% | 35% |

Notice that the percentage of male Republicans in the sample is 32 and the percentage of female Republicans in the sample is 29. These percentages are quite close in relative terms. However, the percentage of female Democrats seems much higher than the percentage of male Democrats (36 percent versus 27 percent); also, the percentage of males in the "Other" category is quite a bit higher than the percentage of females in the "Other" category (41 percent versus 35 percent). These large differences in the percentages indicates that gender and political affiliation are related in the sample. But do these trends carry over to the population of all U.S. voters? This question requires a hypothesis test to answer. The particular hypothesis test you need in this situation is a Chi-square test, which I discuss in detail in Chapter 14.

To make a two-way table from a data set by using Minitab, first enter the data in two columns, where column one is the row variable (continuing with the previous example, this variable would be gender) and column two is the column variable (in this case, political affiliation). For example, suppose the first person is a male Democrat. In row one of Minitab, enter *M* (for male) in column one and *D* (Democrat) in column two. Then go to Stat>Tables>Cross Tabulation and Chi-square. Highlight column one and click Select to enter this variable in the For Rows line. Highlight column two and click Select to enter this variable in the For Columns line. Click on OK.

People often use the word *correlation* to discuss relationships between variables, but in the statistical world, you can use correlation only to discuss the relationship between two quantitative (numerical) variables, not two qualitative (categorical) variables. Correlation measures how closely the relationship between two quantitative variables, such as height and weight, follows a straight line and tells you the direction of that line as well. In total, for any two quantitative variables, *x* and *y*, the correlation measures the strength and direction of their linear relationship. As one increases, what does the other one do?

Because qualitative variables don't have a numerical order to them, they don't increase or decrease in value. For example, just because male = 1 and female = 2 doesn't mean that a female is worth twice a male. (Although some women may want to disagree.) These numbers represent categories, not values. Therefore, you can't use the word *correlation* to describe the relationship between, say, gender and political affiliation. The appropriate term to describe the relationships of qualitative variables is *association*. You can say that political affiliation is associated with gender, and explain how. (For full details on association, see Chapter 13. For more information on correlation, see Chapter 4.)

## Building models to make predictions

You can also build models to predict the value of a qualitative variable based on other related information. In this case, building models is more than a lot of little plastic pieces and some irritatingly sticky glue. When you build a model, you look for variables that help explain, estimate, or predict some response you're interested in (the variables that do this are called *explanatory variables*). You sort through the explanatory variables and figure out which ones do the best job of predicting the response, and you put them together into a type of equation like $y = 2x + 4$ where $x$ = shoe size and $y$ = length of your calf. That equation is a *model*.

For example, what if you want to know which factors or variables can help you predict someone's political affiliation? Is a woman without children more likely to be a Republican or a Democrat? What about a middle-aged man who proclaims Hinduism as his religion? In order for you to compare these complex relationships, you must build a model to evaluate each group's impact on political affiliation (or some other qualitative variable). This kind of model building is explored more in-depth in Chapter 8, where I discuss the topic of logistic regression.

TIP

Logistic regression builds models to predict the outcome of a qualitative variable, such as political affiliation. If you want to make predictions about a quantitative variable, such as income, you need to use the standard type of regression (check out Chapters 4 and 5).

In 2003, the Pew Research Foundation studied the following variables in terms of their relationship with political affiliation: gender, race, state of residence, income level, age, education, religion, marital status, and whether or not you have children. While you can do individual Chi-square analyses to examine possible connections between each of these variables and political affiliation separately, you can't find out which combinations of these variables increase the likelihood of someone being a Democrat, Republican, or other.

For example, the Foundation found that women are more likely to be Democrats than men, but age is also a factor. Younger people tend to be more inclined to be Republican, and older people lean toward being Democrat. However, if you look at the combination of gender and age, you can see mixed results; males who are older are more likely than young females to be Democrat rather than Republican, for example. This kind of result is called an *interaction effect* between gender and age group. An interaction effect occurs when certain combinations of variables produce different results than other combinations. The only way to look for these kinds of more-complex relationships is to do model building, which allows you to examine the combinations of variables and their impact on political affiliation. The Pew Foundation was able to make conclusions about the United States population based on its model linking political affiliation, age and gender, as well as their interactions.

# Statistics for Quantitative Variables

Quantitative variables, unlike qualitative variables, have a wider range of statistics that you can do, depending on what questions you want to ask. The main reason for this wider range is that *quantitative data* are numbers that represent measurements or counts, so it makes sense that you can order, add or subtract, and multiply or divide them — and the results all have numerical meaning. Examining quantitative date opens up a whole world of possibilities for analysis. In this section, I present the major data-analysis techniques for quantitative data. I further expand each technique in later chapters of this book.

## Making comparisons

Suppose you want to look at income (a quantitative variable) and how it relates to a qualitative variable, such as gender or region of the country. Your first question may be: Do males still make more money than females? In this case, you can compare the mean incomes of two populations — males and females. This assessment requires a hypothesis test of two means (oftentimes called a *t*-test for independent samples). I present more information on this technique in Chapter 3.

When comparing the means of *more* than two groups, don't simply look at all the possible *t*-tests that you can do on the pairs of means, because you have to control for an overall error rate in your analysis. Too many analyses can result in errors — adding up to disaster. For example, if you conduct 100 hypothesis tests, each one with a 5 percent error rate, then 5 of those 100 tests give wrong results on average, just by chance.

If you want to compare the average wage in different regions of the country (the East, the Midwest, the South, and the West, for example), this comparison requires a more sophisticated analysis, because you're looking at four groups rather than just two. The procedure you can use to compare more than two means is called *analysis of variance* (ANOVA), and I discuss this method in detail in Chapters 9 and 10.

## Finding connections

Suppose you're an avid golfer and you want to figure out how much time you should spend on your putting game. The question is this: Is the number of putts related to your total score? If the answer is yes, then spending time on your putting game makes sense. If not, then you can slack off on it a bit. Both of these variables are quantitative variables, and you're looking for a connection between them. You collect data on 100 rounds of golf played by golfers at your favorite course over a weekend. Table 2-2 shows the first few lines of your data set.

**Table 2-2 First Ten Golf Scores (ordered)**

| *Number of Putts* | *Total Score* |
|---|---|
| 23 | 76 |
| 27 | 80 |
| 28 | 80 |
| 29 | 80 |
| 30 | 80 |
| 29 | 82 |

| *Number of Putts* | *Total Score* |
|---|---|
| 30 | 83 |
| 31 | 83 |
| 33 | 83 |
| 26 | 84 |

The first step in looking for a connection between putts and total scores (or any other quantitative variables) is to make what is called a *scatterplot* of the data, which graphs your data set in two-dimensional space by using an *x* and *y* plane. You can take a look at the scatterplot of the golf data in Figure 2-1. Here, *x* represents the number of putts, and *y* represents the total score. For example, the point in the lower-left corner of the graph represents someone who had only 23 putts and a total score of 75. (For instructions on making a scatterplot by using Minitab, see Chapter 4.)

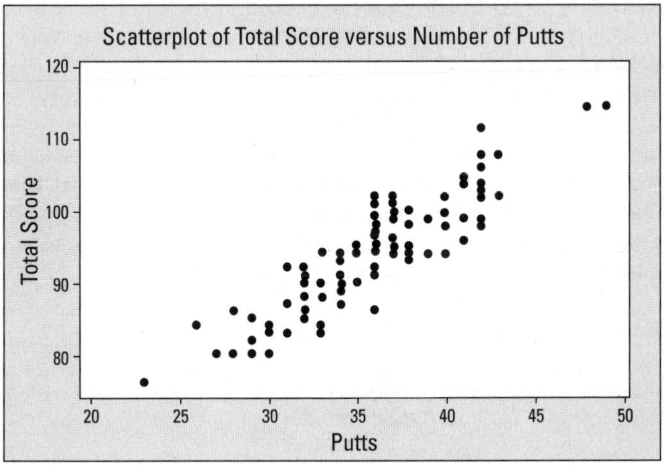


**Figure 2-1:** A scatterplot is a two-dimensional graph you can use to look for relationships in data.

According to Figure 2-1, it appears that as the number of putts increases, so does total score. The relationship seems pretty strong — the number of putts plays a big part in determining the total score.

Now you need a measure of how strong the relationship is between *x* and *y* and whether it goes uphill or downhill. Correlation is the number that measures how close the points follow a straight line. Correlation is always between –1.0 and +1.0, and the more closely the points follow a straight line, the closer the correlation is to –1.0 or +1.0. A positive correlation means that as $x$ increases on the $x$-axis, $y$ also increases on the $y$-axis. Statisticians call this type of relationship an *uphill relationship.* A negative correlation means that as $x$ increases on the $x$-axis, $y$ goes down. Statisticians call this type of relationship — you guessed it — a *downhill relationship.*

For the golf data set, the correlation is 0.896 = 0.90, which is extremely high as correlations go. This strong correlation (close to +1.0) is a good thing because it means number of putts can do a great job of predicting total score. Because the sign of the correlation is positive, it means as you increase number of putts, your total score increases (an uphill relationship). For instructions on calculating a correlation in Minitab, see Chapter 4.

## Making predictions

If you want to predict some response variable *(y)* using one explanatory variable *(x),* and you want to use a straight line to do it, you can use *simple linear regression* (see Chapter 4 for all the fine points on this topic). Linear regression finds the best-fitting line that cuts through the data set, called the *regression line.* After you get the regression line, you can plug in a value of $x$ and get your prediction for $y$. (For instructions on using Minitab to find the best-fitting line for your data, see Chapter 4.)

To use the golf example from the previous section, suppose you want to predict the total score you can get for a certain number of putts. In this case, you want to calculate the linear regression line. By using the data set shown in Table 2-2, and running a regression analysis, the computer tells you that the best line to use to predict total score using number of putts is the following:

Total score = 39.6 + 1.52 * Number of putts

So if you have 35 putts in an 18-hole golf course, your total score is predicted to be about 39.6 + 1.52 * 35 = 92.8, or 93. (Not bad for 18 holes!)

Notice that the slope of the regression line tells you what you really want to know — how much does your total score increase with every additional putt? In other words, how much damage is done when you miss the hole on your first, or second, or third putt? The slope of the regression line for the golf data set is 1.52. Because the slope of a line is the ratio of the change in $y$ (total score) to the change in $x$ (number of putts) this means that every additional putt you need results in an overall increase in total score by 1.52. Maybe that's why Tiger Woods spends so much time on his short game.

Don't try to predict $y$ for $x$-values that fall outside the range of where the data was collected; you have no guarantee that the line still works outside of that range, or that it will even make sense. For the golf example, you can't say that if $x$ (the number of putts) = 0 the total score would be $39.6 + 1.52 * 0 = 39.6$ (unless you just call it good after your ball hits the green). This mistake is called *extrapolation.*

You can discover more about simple linear regression, and expansions on it, in Chapters 4 and 5.

# Avoiding Bias

Bias is the bane of a statistician's existence; it's easy to create and very hard to deal with, if not impossible in most situations. The statistical definition of *bias* is the systematic overestimation or underestimation of the actual value. In language the rest of us can understand, it means that the results are always off by a certain amount in a certain direction. For example, a bathroom scale may always report a weight that's five pounds more than it should be (I'm convinced this is true of my doctor's office scale); this consistent adding of five points to every outcome represents a systematic overestimation of the actual weight.

The most important idea when dealing with bias is prevention, or at least minimizing it. Bias is like weeds in a garden: After they're present, they're very hard to deal with, and it's always better to eliminate them from the start. In this section, you see ways bias can creep into a data set, or even into a statistic, and what you can do about it.

## Looking at bias through statistical glasses

Bias can show up in a data set a variety of different ways. Here are some of the most common ways bias can creep into your data:

- **Selecting the sample from the population:** Bias occurs when you leave some intended groups out of the process, and/or give certain groups too much weight.

  For example, TV surveys (the ones where they ask you to phone in your opinion) are biased because no one has selected a prior sample of people to represent the population — people call in on their own. When people participate in a survey on their own, they're more likely to have stronger opinions than those who don't choose to participate. Such samples are called *self-selected samples* and are typically very biased.

- **Designing the data-collection instrument:** Poorly designed instruments (including surveys) can result in inconsistent or even incorrect data.

  For example, a survey question's wording plays a large role in whether or not results are biased. A leading question can make people feel like they should answer a certain way. For example: "Don't you think that the president should be allowed to have a line-item veto to prevent government spending waste?" Who would feel they should say *no* to that?
- **Collecting the data:** In this case, bias can infiltrate the results if someone makes errors in the recording of the data or if interviewers deviate from the script.
- **Deciding how and when the data is collected:** The time and place you collect data can affect whether your results are biased. For example, if you conduct a telephone survey during the middle of the day, people who work from nine to five aren't able to participate. Depending on the issue, the timing of this survey could lead to biased results.

TIP

Bias can creep into a data set very easily. The best way to deal with bias is to avoid it in the first place. You can do this in two major ways:

- **Use a random process to select the sample from the population.** The only way a sample is truly random is if every single member of the population has an equal chance of being selected. Self-selected samples aren't random.
- **Make sure that the data is collected in a fair and consistent way.** Be sure to use neutral question wording and time the survey properly.

## Settling the variance controversy: The battle of n–1 versus n

Not all statistical formulas are free of bias. In other words, some statistics have good characteristics (like offering great precision) and some not-so-good characteristics (like not giving the best possible result in all situations). Statisticians definitely prefer statistics that are both precise and unbiased, and the techniques you find in this book have both qualities. However, precise and unbiased statistics doesn't always happen naturally; sometimes the basic idea requires a little tweaking to get a statistic that actually meets the standards of the statistical powers that be (of which I am not one). The classic example of this need to fine-tune is the formula for the variance of a data set, which I describe in the following section.

### The problem

Statistics textbooks sometimes show two formulas for the variance of a data set. One formula shown for the variance is $s^2 = \frac{\sum_{i=1}^{n}(x_i - \bar{x})^2}{n}$, where $n$ is the sample size, the values of $x$ are the data values, and the sample mean (or the average of all the values of the data set) is $\bar{x} = \frac{\sum_{i=1}^{n} x_i}{n}$. This formula for variance, you may note, contains an $n$ all by itself in the denominator. The fact that the denominator is $n$ and not $n - 1$ makes a teacher's job of explaining variance a whole lot easier, because it represents the average squared distance from the mean. In this case, the values being squared are the differences between the data values and their mean. You get the average of these squared values by summing them up and dividing by $n$, the sample size.

However, this version of a formula for variance, as it's written, is biased. That means in a statistical sense, you know that in the long term, the results are always off by a very small amount from their target value. If you take repeated samples, find the variance, and do this over and over, the results on average are a little smaller than they should be. (Statisticians can prove this, but you don't have to worry about that. I'm sure you have better things to do.)

### The solution

Because statisticians prefer results being correct to results that can be more easily explained, they decided to do something about this bias problem in the formula for the sample variance. A group of stat big wigs figured out that dividing by $n$ was the problem, and if you divide by $n - 1$ rather than $n$, you can get answers that are right on target. That's how the following commonly used formula for sample variance came into being:

$$s^2 = \frac{\sum_{i=1}^{n}(x_i - \bar{x})}{n - 1}$$

Notice that an $n - 1$ rather than an $n$ is now in the denominator. However, trying to explain why the formula isn't dividing by $n$ does tend to open up a can of worms for statistics professors (and explains why biased statistics are a topic left for the intermediate-level students, like you!).

Because statistics can be biased too, in terms of the results they create through their formulas alone, it's always a good idea to check with a statistician or someone else in the know whether a particular statistic is unbiased before you use it.

## Don't put all your data into one basket!

An animal science researcher came to me one time with a data set he was so proud of. He was studying cows and the variables involved in helping determine their longevity. He came in with a super-mega data set that contained over 100,000 observations. He was thinking "Wow, this is gonna be great! I've been collecting this data for years and years, and I can finally have it analyzed. There's got to be loads of information I can get out of this. The papers I'll write, the talks I'll be invited to give . . . the raise I'll get!" He turned his precious data over to me with an expectant smile and sparkling eyes.

But after looking at his data for a few minutes I made a terrible realization — all of his data came from exactly one cow. With no other cows to compare with and a sample size of just one, he had no way to even measure how much those results would vary if he wanted to apply them to another cow. His results were so biased toward that one animal that I couldn't do anything with the data. After I summed up the courage to tell him, it took a while to peel him off the floor. The moral of the story, I suppose, is to find a statistician and check out your big plans with her before you go down a cow path like this guy did.

# Getting Good Precision

*Precision* is the amount of movement you expect to have in your sample results if you repeat your entire study again with a new sample. *Low precision* means that you expect your sample results to move a lot (not a good thing). *High precision* means you expect your sample results to remain fairly close in the repeated samples (a good thing). In this section, you find out what precision does and doesn't measure, and you see how to measure the precision of a statistic in general terms.

## Understanding precision from a statistical point of view

You may think that precision means the level of correctness you have in your statistical results. But precision only measures the *level of consistency* in the results from sample to sample. Your results can be consistently correct or consistently incorrect.

For example, a field-goal kicker on a football team may consistently kick the ball two feet to the right of the goalposts every single time. Even though he's consistent, he never gets to score, because his results are systematically off by the same amount each time. In other words, his results are biased, even though they're precise.

A statistic can be precise with or without bias, and vice versa. The best situation is when your results are both precise (consistent) as well as unbiased (on target). That goal is what statisticians always strive for. How often does it happen? You can have a lot of control of the precision part by simply taking a larger sample. However, the goal of completely unbiased results is rarely achieved, but that doesn't stop statisticians from trying. And you do have ways to minimize it (keep reading).

## Measuring precision with margin of error

You can measure precision by the margin of error. The *margin of error* is the amount that you expect your statistical results to change from one sample to the next. While you always hope, and may even assume, that statistical results shouldn't change much with another sample, that's not always the case. It's like a commercial that tries to sell a weight-loss product by showing a person who lost 50 pounds in a single weekend; then in small letters at the bottom of the screen, you see the words "results will vary." Before you report or try to interpret any statistical results, you need to have some measurement of how much those results are expected to vary from sample to sample.

The following sections show how to calculate the precision of your statistic and how to come up with a margin of error.

### Calculating precision

The exact formulas for margin of error differ depending on the type of data that you're analyzing; however, they all contain two major components:

- Confidence coefficient
- Standard error of the statistic

The general structure of a formula for margin of error is the following, where standard error is the standard deviation of the population divided by the square root of the sample size (you can see all the details on margin of error in Chapter 3):

Margin of error = ± Confidence coefficient * Standard error

The big idea is that the confidence coefficient tells you the number of standard errors you're willing to add and subtract in order to have a certain level of confidence in your results. If you want to be more confident in your results, you add or subtract more standard errors. If you don't have to be as confident, you don't have to add or subtract as many standard errors. Typically, you add and subtract about two standard errors if you want to be 95 percent confident and three standard errors if you want to be more than 99 percent confident. This rule of thumb follows a statistical result called the *Empirical Rule,* also known as the *68-95-99.7 Rule.*

The *standard error* is the average amount of movement in the statistic you're using. It's a function of two quantities:

- **Sample size:** Sample size is perhaps the most important factor in controlling margin of error. The sample size is in the denominator of the standard error, meaning that as your sample size increases, the standard error goes down, and that's why the margin of error goes down.

  This result makes sense, because having a larger sample means having more information in your analysis, which should lead to greater precision. As the sample size decreases, the margin of error goes up, because you have less information to work with and that makes for less-precise results.

- **Standard deviation in the population:** Standard deviation is close to the average distance from the mean. If the population you took your sample from has a large amount of variability, the standard deviation is large, and the margin of error for your statistic goes up (because standard deviation is in the numerator of the margin of error). If the population is more homogeneous, your sample results are more homogeneous as well, and the margin of error goes down (because the standard error gets smaller).

STATISTICS IN ACTION

## Up close and personal: Survey results

The Gallup Organization states its survey results in a universal, statistically correct format. Using a specific example from a recent survey it conducted, you can see the language it uses to report its results:

> "These results are based on telephone interviews with a randomly selected national sample of 1,002 adults, aged 18 years and older, conducted June 9–11, 2006. For results based on this sample, one can say with 95% confidence that the maximum error attributable to sampling and other random effects is ±3 percentage points. In addition to sampling error, question wording and practical difficulties in conducting surveys can introduce error or bias into the findings of public opinion polls."

The first sentence of the quote refers to how the Gallup Organization collected the data, as well as the size of the sample. As you can guess, precision is related to the sample size, as seen in the section "Calculating precision."

The second sentence of the quote refers to the precision measurement: How much did Gallup expect these sample results to vary? The fact that Gallup is 95 percent confident means that if this process were repeated a large number of times, in 5 percent of the cases the results would be wrong, just by chance. This inconsistency occurs if the sample selected for the analysis doesn't represent the population — not due to biased reasons, but due to chance alone (more on this in Chapter 3).

(Check out the section "Bias not included" to get the info on why the third sentence is included in this quote.)

For more details on how to calculate margin of error in various statistical techniques, see Chapter 3.

### Interpreting margin of error

Finding the margin of error is one thing — figuring out what it means is a whole other ball o' wax. But don't fear; it's actually not so bad. To interpret the margin of error, just think of it as the amount of play you allow in your results to cover most of the other samples you could have taken.

Suppose you're trying to estimate the proportion of people in the population who support a certain issue, and you want to be 95 percent confident in your results. You sample 1,002 individuals and find that 65 percent support the issue. The margin of error for this survey turns out to be plus or minus 3 percentage points (you can find the details of this calculation in Chapter 3). That result means that you can expect the sample proportion of 65 percent to change by as much as 3 percentage points either way if you took a different sample of 1,002 individuals. In other words, you believe the actual population proportion is somewhere between 65 – 3 = 62 percent and 65 + 3 = 68 percent. That's the best you can say.

### Bias not included!

Realizing that the margin of error measures the consistency (precision) of a statistic only, not its level of bias is extremely important. In other words, a margin of error can appear on paper to be very small yet actually be way off target because of bias in the data that was collected. (In the nearby sidebar, you can see that Gallup discusses margin of error and bias separately.)

WARNING!

Any reported margin of error was calculated on the basis of having zero bias in the data. However, this assumption is rarely true. Before interpreting any margin of error, check first to be sure that the sampling process and the data-collection process don't contain any obvious sources of bias. If a great deal of bias exists, you should ignore the results, or take them with a great deal of skepticism.

## Making Conclusions and Knowing Your Limitations

The most important goal of any data analyst is to remain focused on the big picture — the question that you or someone else is asking — and make sure that the data analysis used is appropriate and comprehensive enough to answer that question correctly and fairly.

Here are some tips for analyzing data and interpreting the results, in terms of the statistical procedures and techniques that you may use — at school, in your job, and in everyday life. These tips are implemented and reinforced throughout this book:

- **Be sure that the research question being asked is clear and definitive.** Some researchers don't want to be pinned down on any particular set of questions because they have the intent of mining the data (looking for any relationship they can find, and then stating their results after the fact). This can lead to overanalyzing the data, making the results subject to skepticism by statisticians.
- **Double-check that you clearly understand the type of data being collected.** Is the data qualitative or quantitative? The type of data used drives the approach that you take in the analysis.
- **Make sure that the statistical technique you use is designed to answer the research question.** If you want to make comparisons between two groups and your data is quantitative, use a hypothesis test for two means. If you want to compare five groups, use analysis of variance (ANOVA). You can use this book as a resource to help you determine the technique you need.
- **Look for the limitations of the data analysis.** For example, if the researcher wants to know whether negative political ads affect the population of voters, and she bases her study on a group of college students, you can find severe limitations here. For starters, student reactions to negative ads don't necessarily carry over to all voters in the population. And even if the population were limited to all student voters, the students from this particular class don't represent all students. In this case, it's best to limit the conclusions to college students in that class (which no researcher would ever want to do). Ultimately what needs to be done is design the study so the sample contains a representation of the intended population of all voters in the first place (a much more difficult task, but well worth it).

One of the hardest parts of my job as a statistical consultant is dealing with analyses after the design was already done — and done incorrectly. It's much better to put in a little work to get a good design together first, and then the analysis will take care of itself.

# Chapter 3

# Building Confidence and Testing Models

### *In This Chapter*

- Utilizing confidence intervals to estimate parameters
- Testing models by using hypothesis tests
- Finding the probability of getting it right and getting it wrong
- Discovering power in a large sample size

One of the major goals in statistics is to use the information you collect from a sample in order to get a better idea of what's going on in the entire population you're studying (because populations are generally large and exact info is often unknown). The most common items to study are the mean of the population, the proportion of the population that has a certain characteristic, or a comparison of the means or proportions from two different populations. These unknown values that summarize the population are called *population parameters.* Researchers typically either want to get a handle on what those parameters are, or they want to test a hypothesis about the population parameters. In introductory statistics, you typically go over confidence intervals and hypothesis tests for one and two population means and one and two population proportions. Your instructor hopefully emphasized that no matter which parameters you're trying to estimate or test, the general process is the same. If not, don't worry; that's what this chapter's all about.

The most important idea you can gain from this chapter is that intermediate statistics focuses on building and testing models. You're typically faced with some random phenomena, and you're trying to build a model that explains or predicts that phenomena. The situation is more complex than it was in intro stats, where you used one variable to predict another variable in simple linear regression. Intermediate statistics takes it up a notch to using many variables to predict another one. But as long as you keep the big picture of how the process works in your mind, you'll be okay.

It all comes down in the end to testing hypotheses to see whether certain models fit, and if they do, to using confidence intervals to estimate certain values in the population or to make predictions based on the model that you built.

This chapter reviews the basic concepts of confidence intervals and hypothesis tests, including the probabilities of making errors by chance. I also discuss how statisticians measure the ability of a statistical procedure to do a good job — of detecting a real difference in the populations, for example. Hang on — you're in for quite a ride.

# Estimating Parameters by Using Confidence Intervals

Confidence intervals are a statistician's way of covering themselves when it comes to estimating a population parameter. For example, instead of just giving a one-number guess as to what the average household income is in the United States, a statistician would give a range of likely values for this number. Statisticians do this for two reasons:

- All good statisticians know sample results vary from sample to sample, so a one-number estimate isn't any good.
- Statisticians have developed some awfully nice formulas you can use to give a range of likely values, so why not use them?

In this section, you get the general formula for a confidence interval, including the margin of error, and a good look at the common approach to building confidence intervals. I also discuss interpretation and the chance of making an error.

## Getting the basics: The general form of a confidence interval

The big idea of a confidence interval is coming up with a range of likely values for a population parameter. The *confidence level* represents the chance that if you repeated your sample-taking over and over, you'd get a range of likely values that actually contains the actual population parameter. In other words, it's the long-term chance of being correct.

The general formula for a confidence interval is the following:

Confidence interval = Sample statistic ± Margin of error

The confidence interval has a certain level of precision (measured by the margin of error). Precision calculates how close you expect your results to be to the truth.

For example, you want to know the average amount of time a student at Ohio State University spends listening to music per day, using an MP3 player. The average time for the entire population of OSU students that are MP3-player users is the parameter you're looking for. Certain that you can't ask every student who uses an MP3 player at OSU this question, you take a random sample of students and find the average from there.

Suppose the average time a student uses an MP3 player per day to listen to music based on a random sample of 1,000 OSU students is 2.5 hours, and the standard deviation is 0.5 hours. Is it right to say that the population of all OSU-student MP3-player owners use their players an average of 2.5 hours per day for music listening? No. You hope and may assume that the average for the whole population is close to 2.5, but it probably isn't exact. After all, you're only sampling a tiny fraction of the 60,000 member population of all OSU students. The fact is that sample results vary from sample to sample.

What's the solution to this problem? The solution is to not only report the average from your sample, but along with it, report some measure of how much you expect that sample average to vary from one sample to the next, with a certain level of confidence. You want to cover your bases, so to speak (at least most of the time). The number that you use to represent this level of precision in your results is called the *margin of error.* You take your sample average and add and subtract the margin of error (to get that plus-or-minus factor going), which gives you a confidence interval for the average time all OSU students use their MP3 players.

## Finding the confidence interval for a population mean

The sample statistic part of the confidence-interval formula is fairly straightforward. If you want to estimate the population mean, you use the sample mean. If you want to estimate the population proportion, use the sample proportion. If you want to find the difference of two population means, take two samples, find their sample means, and subtract them.

In the case of the population mean, you use the sample mean to estimate it. The sample mean has a standard error of $\frac{\sigma}{\sqrt{n}}$. In this formula, you can see the population standard deviation ($\sigma$) and the sample size *($n$)*.

If you think about it though, why would you know the standard deviation of the population, $\sigma$, when you don't even know the mean (recall that the mean is what you're trying to estimate)? To handle this additional unknown, do what statisticians always do — estimate it and move on. So you estimate $\sigma$, the population standard deviation, using (what else?) the standard deviation of the sample, denoted by $s$. So you replace $\sigma$ by $s$ in the formula for the standard error of the mean.

To estimate the population mean by using a confidence interval when $\sigma$ is unknown, you use the formula $\bar{x} \pm t_{n-1}\left(\frac{s}{\sqrt{n}}\right)$. This formula contains the sample standard deviation *($s$)*, the sample size *($n$)*, and a $t$-value representing how many standard errors you want to add and subtract to get the confidence you need. To get the margin of error for the mean, you see the standard error, $\frac{s}{\sqrt{n}}$, is being multiplied by a factor of $t$. Notice that $t$ has $n - 1$ as a subscript to indicate which of the myriad $t$-distributions you use for your confidence interval. The $n - 1$ is called *degrees of freedom,* where $n$ is the sample size.

The value of $t$ in this case represents the number of standard errors you add and subtract to or from the sample mean to get the confidence you want. If you want to be 95 percent confident, for example, you add and subtract about two of those standard errors. If you want to be 99.7 percent confident, you add or subtract about three of them. (Table A-1 in the Appendix presents the $t$-distribution from which you can find $t$-values for any confidence level you want.)

**TIP**

If you do know the population standard deviation for some reason, you would certainly use it. In that case, you use the corresponding number from the $Z$-distribution (standard normal distribution) in the confidence interval formula. (The $Z$-distribution from your intro stat book can give you the numbers you need.) Or if you know $\sigma$ and have a large sample size, you can simply use the bottom line of the $t$-distribution, because a $t$-distribution with a large number of degrees of freedom gives very similar values to the $Z$-distribution.

For the MP3 player example from the preceding section, a random sample of 1,000 OSU students spends an average of 2.5 hours using their MP3 players to listen to music. The standard deviation is 0.5 hours. Plugging this information into the formula for a confidence interval, you get $2.5 \pm 1.96\left(\frac{0.5}{\sqrt{1{,}000}}\right) = 2.5 \pm 0.03$ hours. You can conclude that OSU MP3-player owners spent an average of between 2.47 and 2.53 hours listening to music on their players. (The value for $t$ in this example came from the last line of Table A-1 in the Appendix, because this line represents the situation where $n$ is large.)

## What changes the margin of error?

What do you need to know in order to come up with a margin of error? Margin of error, in general, depends on three elements:

- The standard deviation of the population, $\sigma$ (or an estimate of it, denoted by $s$, the sample standard deviation)
- The sample size, $n$
- The level of confidence you need

You can see these elements in action in the following formula for margin of error of the sample mean: $t_{n-1} * \frac{s}{\sqrt{n}}$. Here I assume that $\sigma$ isn't known; $t_{n-1}$ represents the value on the *t*-distribution (Table A-1 in the Appendix) with $n - 1$ degrees of freedom.

Each of these three elements has a major role in determining how large the margin of error will be when you estimate the mean of a population. At times it may seem that different elements work against each other (and they do!), but you can find ways around that. In the following sections, I show how each of the elements of the margin of error formula work separately and together to affect the size of the margin of error.

### The population standard deviation's affect on margin of error

The standard deviation of the population is typically combined with the sample size in the margin of error formula, with the population standard deviation on top of the fraction, and $n$ in the bottom. (In this case, the standard error of the population, $\sigma$, is estimated by the standard deviation of the sample, $s$, because $\sigma$ is typically unknown.)

This combination of standard deviation of the population and sample size is known as the *standard error* of your statistic. It measures how much the sample statistic deviates from its mean in the long term.

How does the standard deviation of the population ($\sigma$) affect margin of error? As the standard deviation of the population (or its estimate, $s$) gets larger, the margin of error increases, so your range of likely values is wider. That's why you typically see the population standard deviation in the numerator of margin of error formulas. The formula for the margin of error for one population is an example of this.

Suppose you have two gas stations, one on a busy corner (gas station #1) and one farther off the main drag (gas station #2). You want to estimate the average time between customers at each station. At the busy gas station #1, customers are constantly using the gas pumps, so you basically have no time between customers, and that model holds day after day. At gas station #2, customers sometimes come all at once, and sometimes you don't see a single person for an hour or more. So the time between customers varies quite a bit.

For which gas station would it be easier to estimate the overall average time between customers as a whole? Gas station #1 has much more consistency, which represents a smaller standard deviation of times between customers. Gas station #2 has much more heterogeneity of times between customers, so that one is harder to get a handle on. That means $\sigma$ for gas station #1 is smaller than $\sigma$ for gas station #2.

### *Sample size and margin of error*

Sample size affects margin of error in a very intuitive way. Suppose you're trying to estimate the average number of pets per household in your city. Which sample size would give you better information: 10 homes or 100 homes? You'd agree that 100 homes would give more precise information (as long as the data on those 100 homes was collected properly).

If you have more data to base your conclusions on, and that data is collected properly, your results will be more precise. Precision is measured by margin of error; so as the sample size increases, the margin of error of your estimate goes down. That's why you typically see an $n$ (sample size) in the denominator of margin of error formulas. In the formula for the margin of error of the sample mean, you can see $n$ in the denominator.

WARNING!

Bigger is only better in terms of sample size if the data is collected properly. That is, you should find no bias in the way the members of the sample were selected or in the way the data was collected on those subjects. If the quality of the data can't be maintained with a larger sample size, it does no good to have it.

### *Confidence level and margin of error*

The amount of confidence you need to have differs from problem to problem. Suppose you're estimating the mean weight that an elevator can hold. You would want to be pretty confident about your results, right? But, if you wanted to estimate the percentage of females that may come to your party on Saturday night, you may not need to be so confident (despite the desperation you see in your single buddies' eyes). For each problem at hand, you have to address how confident you need to be in your results over the long term, and, of course, more confidence comes with a price in the margin of error formula. This level of confidence in your results over the long term is reflected in a number called the confidence level, reported as a percentage. In general, more confidence requires a wider range of likely values. Ninety-five percent is the most common confidence level statisticians use.

Every margin of error is interpreted as plus or minus a certain number of standard errors. The number of standard errors added and subtracted is determined by the confidence level. If you need more confidence, you add and subtract more standard errors. If you need less confidence, you add and subtract fewer standard errors. The number that represents how many standard errors to add and subtract is different from situation to situation. For one population mean, you use a value on the *t*-distribution, represented by $t_{n-1}$, where $n$ is the sample size. See Table A-1 in the Appendix.

Here's an example. Suppose you have a sample size of 20, and you want to estimate the mean of a population. The number of standard errors you add and subtract is represented by $t_{n-1}$, which in this case is $t_{19}$. Suppose your confidence level is 90 percent. To find the value of $t$, you look at row 19 in the *t*-distribution table (Table A-1 in the Appendix). The table uses the area to the right, so that area in this case is 0.05. (You get this value because 90 percent is within the confidence interval, so 10 percent is outside of it. Half of that 10 percent lies above the confidence interval, and the other half lies below it.) So look at row 19 and the column headed by the value 0.05. You get the value of $t = 1.73$. So to be 90 percent confident with a sample size of 20, you need to add and subtract 1.73 standard errors.

Now suppose you want to be 95 percent confident in your results, with the same sample size of $n = 20$. The area above the interval is now half of 5 percent, which is 2.5 percent or 0.025. Row 19 and column 0.025 in Table A-1 gives you the value of $t_{19} = 2.09$. Notice that this value of $t$ is larger than the value of $t$ for 90 percent confidence, because in order to be more confident, you need to go out more standard deviations on the *t*-distribution table to cover more possible results.

### *Large confidence, narrow intervals — just the right size*

A narrow confidence interval is much more desirable than a wide one. For example, if you said that you think the average cost of a new home is $150,000 plus or minus $100,000, that wouldn't be helpful at all because this makes your estimate anywhere between $50,000 and $250,000. (Who has an extra hundred grand to throw around?) But you *have* to be 99 percent confident, so your statistician has to add and subtract more standard errors to get there, which makes the interval that much wider (a downer). She tells you to be happy with 95 percent confidence, but no!

Wait, don't panic — you can have your cake and eat it too! If you know you want to have a high level of confidence, but you don't want a wide confidence interval, just increase your sample size to meet that level of confidence. The effect of sample size and the effect of confidence level cancel each other out, so you can have a precise (narrow) confidence interval and a high level of confidence at the same time. It all depends on sample size, something you can control (up to the size of your pocketbook of course).

For example, say the standard deviation of the house prices from a previous study is $s = \$15{,}000$, and you want to be 95 percent confident in your estimate of average house price. Using a large sample size, your value of $t$ (from the last row of Table A-1 in the Appendix) would be 1.96. With a sample of 100 homes, your margin of error would be plus or minus 1.96 times $15,000 divided by the square root of 100, which comes out to $2,940. If this is too large for you but you still want 95 percent confidence, crank up your value of *n*. If you sample 500 homes, the margin of error decreases to plus or minus 1.96 times $15,000 divided by the square root of 500, which brings you down to $1,314.81.

You can actually use a formula to find the sample size you need to meet a desired margin of error. That formula is $n = \left(\frac{t_{n-1} s}{MOE}\right)^2$, where MOE is the desired margin of error (as a proportion), $s$ is the sample standard deviation, and $t$ is the value on the $t$-distribution that corresponds with the confidence level you want. (You can use the last line of Table A-1 in the Appendix, which will work fine, assuming that your sample size is fairly beyond 30.)

## Interpreting a confidence interval

Interpreting a confidence interval involves a couple of subtle but important issues, which I discuss in this section. The big idea is that a *confidence interval* presents a range of likely values for the population parameter, based on your sample. It includes this range because your sample results are going to vary, and you want to address that. A 95 percent confidence interval, for example, provides a range of likely values for the parameter such that the parameter is included in the interval 95 percent of the time in the long term.

A 95 percent confidence interval doesn't mean that your particular confidence interval has a 95 percent chance of capturing the actual value of the parameter; after the sample has been taken, it's either in the interval or it isn't. A confidence interval represents the long-term chances of capturing the actual value of the population parameter over many different samples.

Suppose a polling organization wants to estimate the percentage of people in the United States who drive a car with more than 100,000 miles on it, and it wants to be 95 percent confident in its results. The organization takes a random sample of 1,200 people and finds that 420 of them (35 percent) drive a much-driven car.

The meaty part of the interpretation lies in the confidence level — in this case, the 95 percent. Because the organization took a sample of 1,200 people in the U.S., asked each of them whether his or her car has more than 100,000 miles on it and made a confidence interval out of it, the polling organization is, in essence, accounting for all of the other samples out there that it could have gotten by building in the margin of error (±3 percent). The organization wants to cover its bases on 95 percent of those other situations, and the ±3 percent satisfies that.

Another way of thinking about the confidence interval is to say that if the organization sampled 1,200 people over and over again and made a confidence interval from its results each time, 95 percent of those confidence intervals would be right. (You just have to hope that yours is one of those right results.)

TIP

Using stat notation, you can write confidence levels as $1 - \alpha$. So if you want 95 percent confidence, you write it as $1 - 0.05$. The number that $\alpha$ represents is the chance that your confidence interval is one of the wrong ones. This number, $\alpha$, is also related to the chance of making a certain kind of error with a hypothesis test, which I explain in the hypothesis-testing section.

# Setting Up and Testing Models

A *model* is an equation that attempts to describe how a population behaves. It can be a claim that's made about a population parameter; for example, a shipping company might say that its packages are on time 95 percent of the time, or a campus official claims that 75 percent of students live off campus. It is important to test these models to see whether they actually hold up in the population, which you can do by using hypothesis tests.

In this section, you see the big ideas of hypothesis testing that are the basis for the data-analysis techniques in this book. You review and expand on the concepts involved in a hypothesis test, including the hypotheses, the test statistic, and the *p*-value.

## What do Ho and Ha represent — really?

The big idea here is that you set up a hypothesis test to see whether your model fits the population, based on your data. In the intro stat course, you tested simple hypotheses — like whether the population mean is equal to ten. At the intermediate statistics level, you get to look at much more sophisticated and relevant models that involve several variables and/or several different populations in a variety of situations. The good news, though, is that the basic ideas from intro stats apply here as well. (If you need a brief refresher before barreling through this section, feel free to flip through your intro stats book or check out my other book *Statistics For Dummies* [Wiley].)

You use a hypothesis test in situations where you have a certain model in mind, and you want to see whether that model fits your data. Your model may be one that just revolves around the population mean (testing whether that mean is equal to ten, for example). Your model may be testing the slope of a regression line (whether or not it's zero, for example, with zero meaning you find no relationship between *x* and *y*). You may be trying to use several different variables to predict the marketability of a product, and you believe a model using customer age, price, and shelf location can help predict it, so you need to run one or more hypothesis tests to see whether that model works. (This process is called multiple regression; more info on this in Chapter 5.)

A hypothesis test is made up of two hypotheses:

- **The null hypothesis (Ho):** Ho symbolizes the current situation — the one that everyone assumed was true until you got involved.
- **The alternative hypothesis (Ha):** Ha represents the alternative model that you want to consider. It stands for the researcher's hypothesis, and the burden of proof lies on the researcher to prove it.

Ho is the model that's on trial. If you get enough evidence against it, you conclude Ha, which is the model you're claiming is the right one. If you don't get enough evidence against Ho, then you can't say that your model (Ha) is the right one.

## Gathering your evidence into a test statistic

A *test statistic* is the statistic from your sample, standardized so you can look it up on a table, basically. While each hypothesis test is a little different, the main thought is the same. For whatever model you're trying to test, you come up with a statistic that you use to test that model. Take that statistic, standardize it (take the statistic minus its expected value from Ho and divide all that by the standard error). Then look up your test statistic on a table to see where it stands. That table may be the *t*-table (Table A-1 in the Appendix), it may be the Chi-square table (Table A-3 in the Appendix), or it may be a different table. The type of test you need to you use on your data dictates which table you use.

In the case of testing a hypothesis for a population mean, $\mu$, you use the sample mean, $\bar{x}$, as your statistic. To standardize it, you take $\bar{x}$ and convert it to a value of $t$ by using the formula $t_{n-1} = \frac{\bar{x} - \mu_0}{\frac{s}{\sqrt{n}}}$, where $\mu_0$ is the value in Ho. This value is your test statistic. You compare your test statistic to the *t*-distribution (check out Table A-1 in the Appendix).

## Determining strength of evidence with a p-value

If you want to know whether your data has the brawn to stand up against Ho, you want to figure out the *p*-value and compare it to a prespecified cutoff, $\alpha$ (typically 0.05). The *p-value* is a measure of the strength of your evidence against Ho. You can calculate the *p*-value by doing the following:

1. **Calculate the test statistic.** See the preceding section for more info on this.
2. **Look up the test statistic on the appropriate table (such as the *t*-table, A-1 in the Appendix).**
3. **Find the percentage of values on the table that fall beyond your test statistic.** This percentage is the *p*-value.

Suppose you're conducting a hypothesis test and have already decided you will reject Ho at level $\alpha = 0.05$. You collect your data and find the test statistic (see preceding section). If your test statistic is extremely high or extremely low compared to other values on the table (whatever that table is), then you reject Ho.

For example, say the cutoff value for rejecting Ho at a level $\alpha = 0.05$ is 1.645, where you're testing for the mean of one population. If you get a test statistic of 1.7, you reject Ho. If you get a test statistic of 2.7, you *really* reject Ho. That is, you have more evidence against Ho with a test statistic of 2.7 than with a test statistic of 1.7. The two *p*-values of 1.7 and 2.7 are what statisticians call *marginally significant* and *highly significant* results respectively, to use proper terms.

REMEMBER

Your friend, $\alpha$, is the cutoff for your *p*-value — and the star of this chapter. ($\alpha$ is typically set at 0.05 — sometimes 0.10.) If your *p*-value is less than your predetermined value of $\alpha$, reject Ho, because you have sufficient evidence against it. If your *p*-value is greater than or equal to $\alpha$, you can't reject Ho.

For example, if your *p*-value is 0.002, then your test statistic is so far away from Ho that the chance of getting this result only by chance is only 2 out of 1,000. So, you conclude that Ho is very likely to be false. However, if your *p*-value turns out to be 0.30, then this result happens 30 percent of the time anyway, so you see no red flags there, and you can't reject Ho. You don't have enough evidence against it. It doesn't mean Ho is true, but you don't have the evidence to say it's false — a subtle, but important, difference.

When I compare the *p*-value to the $\alpha$ (the cutoff value), I like to think of a football analogy, assuming that Ho is "the opposing team can't make a touchdown." The burden is on the other team to show enough evidence to reject

Ho. Now, imagine that their running back makes a touchdown by pushing the ball just barely over the goal line, so close that his team needs to have a referee review the film before calling it a touchdown. This situation is equivalent to rejecting Ho with a *p*-value just below your prespecified value of $\alpha = 0.05$. In this case, the *p*-value is close to the borderline, say 0.045. But, if their team makes a touchdown by catching a pass deep in the end zone, no one has any doubt about the result because the ball was obviously past the goal line, which is equivalent to the *p*-value being very small, say something like 0.001. The opposing team's showing a lot of evidence against Ho (and your team could be in a lot of trouble).

## Deconstructing Type I and Type II errors

Any technique you use in statistics to make a conclusion about a population based on a sample of data has the chance of making an error. The errors I am talking about, Type I and Type II errors, are due to random chance.

For example, you could flip a fair coin ten times and get all heads, making you think that the coin isn't fair at all. This thinking would result in an error, because the coin actually was fair, but the data just wasn't confirming that due to chance. On the other hand, another coin may be unfair, and, just by chance, you flip it ten times and get exactly five heads, which makes you think that particular coin is equally balanced and doesn't present any problem. (This tells you strange things can happen, especially when the sample size is small.)

The way you set up your test can help to reduce these kinds of errors, but they are always out there. As a data analyst, you need to know how to measure and understand the impact of the errors that can occur with a hypothesis test and what you can do to possibly make those errors smaller. In the following sections, I show you how you can do just that.

### Making false alarms with Type I errors

A Type I error represents the situation where the coin was actually fair (using the example from the preceding section), but your data led you to conclude that it wasn't, just by chance. I think of a Type I error as a false alarm: You blew the whistle when you shouldn't have.

TECHNICAL STUFF

To include a definition that makes all those stat experts happy, a Type I error is the conditional probability of rejecting Ho, given that Ho is true.

The chance of making a Type I error is equal to $\alpha$, which is predetermined before you begin collecting your data. This $\alpha$ is the same $\alpha$ that represents the chance of missing the boat in a confidence interval. It makes some sense that these two probabilities are both equal, because the probability of rejecting Ho when you shouldn't (Type I error) is the same as the chance that the true population parameter falls out of the range of likely values when it shouldn't. That chance is $\alpha$.

Say someone claims that the mean time to deliver packages for a company is 3.0 days on average (so Ho is $\mu = 3.0$), but you believe it's not equal to that (so Ha is $\mu \neq 3.0$). Your alpha level is 0.05, and because you have a two-sided test, this means you have 0.025 on each side. Your sample of 100 packages has a mean of 3.5 days with a standard deviation of 1.5 days. You find the test statistic $t_{n-1} = \frac{\bar{x} - \mu_0}{\frac{s}{\sqrt{n}}} = \frac{3.5 - 3.0}{\frac{1.5}{\sqrt{100}}}$, which equals 3.33. This value falls beyond 1.96 (the value on the last row and the 0.025 column of the *t*-distribution, Table A-1 in the Appendix). So you don't think 3.0 is a likely value for the mean time of delivery, over all possible packages, and you reject Ho. Your data led you to that decision and you stick to it.

But suppose your sample just by chance contained some longer than normal delivery times, and that in reality, the company's claim is right. You just made a Type I error. You made a false alarm about the company's claim.

To reduce the chance of a Type I error, reduce your value of $\alpha$. However I wouldn't recommend reducing $\alpha$ too far. On the positive side, this reduction makes it harder to reject Ho, because you need more evidence in your data to do so. On the negative side, by reducing your chance of a Type I error, you increase the chance of another type of error — the Type II error. To tackle Type II errors, keep reading!

### Missing an opportunity with a Type II error

A Type II error represents the situation where (continuing with the coin example) the coin was actually unfair, but your data didn't have enough evidence to catch it, just by chance. You can think of a Type II error as a missed opportunity — you didn't blow the whistle when you should have. In statistical terms, a Type II error is the conditional probability of not rejecting Ho, given that Ho is false. I call it a missed opportunity, because you were supposed to be able to find a problem with Ho and reject it, but you didn't.

The chance of making a Type II error depends on a couple of things:

- **Sample size:** If you have more data, you're less likely to miss something that's going on. For example, if a coin actually is unfair (and you don't know it), flipping the coin only ten times may not reveal the problem, because results can go all over the place when the sample size is small. But if you flip the coin 1,000 times, you have a good chance of seeing a pattern that favors heads over tails or vice versa.

- **Actual value of the parameter:** A Type II error is also related to how big the problem is that you're trying to uncover. For example, suppose a company claims that the average delivery time for packages is 3.5 days. If the actual average delivery time is 5 days, you won't have a very hard time detecting that with your sample (even a small sample). Evidence will mount up fast for rejecting Ho, which is exactly what you're supposed to do in this situation. But if the actual average delivery time is 4.0 days, you have to do more work to actually detect the problem. Note that you never do know the actual value of a parameter, but you want to protect yourself against the different possibilities, which is why you consider them.

TIP

To reduce the chance of a Type II error, take a larger sample size. A greater sample size makes it easier to reject Ho, but increases the chance of a Type I error. Type I and Type II errors sit on opposite ends of a seesaw — as one goes up, the other goes down. To try to meet in the middle, choose a large sample size (the bigger, the better; see Figures 3-1 and 3-2) and a small $\alpha$ level (0.05 or less) for your hypothesis test.

## Getting empowered by the power of a hypothesis test

Type II errors (see preceding section) show the downside of a hypothesis test. Statisticians, despite what many may think, actually try to look on the bright side once in a while, and this case is one of those times. Instead of looking at the chance of *missing* a difference from Ho that actually is there, you can look at the chance of *detecting* a difference that really is there. This detection is called the *power of a hypothesis test.*

REMEMBER

The power of a hypothesis test is one minus the probability of making a Type II error. So *power* is a number between 0 and 1 that represents the chance that you rejected Ho when Ho was false. (You can even sing about it "If Ho is false and you know it, clap your hands. . . .") Remember that power (just like Type II errors) depends on two elements: the sample size and the actual value of the parameter (see the preceding section for a description of these elements).

In the following sections, you discover what *power* means in statistics (not being one of the big wigs, mind you); you also find out how to quantify power by using a power curve.

### *Quantifying power with a power curve*

The specific calculations for the power of a hypothesis test are beyond the scope of this book (so, take that sigh of relief), but computer programs and graphs are available online to show you what the power is for different hypothesis tests and various sample sizes (just type "power curve for the [blah blah blah] test" into an Internet search engine). These graphs are called *power curves* for a hypothesis test. A power curve is a special kind of graph. It gives you an idea of how much of a difference from Ho you can detect with the sample size that you have. Because the precision of your test statistic increases as your sample size increases, sample size is directly related to power. But it also depends on how much of a difference from Ho you're trying to detect. For example, if a package delivery company claims that its packages arrive in 2 days or less, do you want to blow the whistle if it's actually 2.1 days? Or wait until it's 3 days? You need a much larger sample size to detect the 2.1-days situation versus the 3-days situation just because of the precision level needed.

In Figure 3-1, you can see the power curve for a particular test of Ho: $\mu = 0$ versus Ha: $\mu > 0$. You can assume that $\sigma$ (the standard deviation of the population) is equal to two (I give you this value in each problem) and doesn't change. I set the sample size at ten throughout.

The horizontal *(x)* axis on the power curve shows a range of actual values of $\mu$. For example, you hypothesize that $\mu$ is equal to 0, but it may actually be 0.5, 1.0, 2.0, 3.0, or any other possible value. If $\mu$ equals 0, then Ho is true, and the chance of detecting this (rejecting Ho) is equal to 0.05, the set value of $\alpha$. You work from that baseline. So, on the graph in Figure 3-1, when $x = 0$, you get a $y$-value of 0.05.

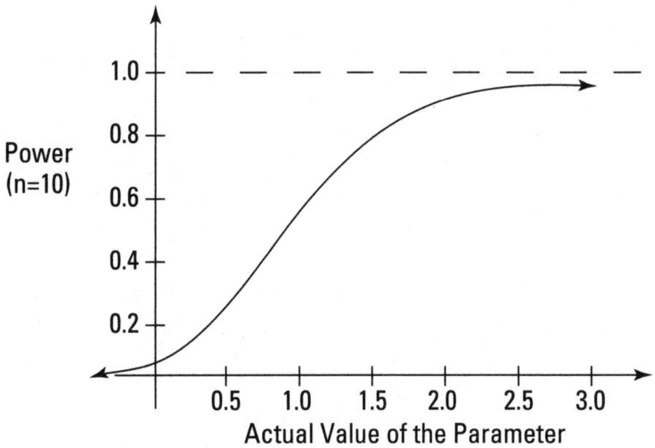


**Figure 3-1:** Power curve for Ho: $\mu = 0$ versus Ha: $\mu > 0$, for $n = 10$ and $\sigma = 2$.

Suppose that μ is actually 0.5, not 0, as you hypothesized. A computer tells you that the chance of rejecting Ho (what you're supposed to do here) is 0.197 = 0.20, which is the power. So, you have about a 20 percent chance of detecting this difference with a sample size of ten. As you move to the right, away from zero on the horizontal *(x)* axis, you can see that the power goes up, and the *y*-values get closer and closer to 1.0.

For example, if the actual value of μ is 1.0, the difference from 0 is easier to detect than if it's 0.50. In fact, the power at 1.0 is equal to 0.475 = 0.48, so you have almost a 50 percent chance of catching the difference from Ho in this case. And as the values of the mean increase, the power gets closer and closer to 1.0. Power never reaches 1.0, because statistics can never prove anything with 100 percent accuracy. But you can get close to 1.0 if the actual value is far enough from your hypothesis.

### *Controlling the sample size*

You don't have any control over what the actual value of the parameter is, though, because that number is unknown. So what do you have control over? The sample size. As the sample size increases, it becomes easier to detect a real difference from Ho.

Figure 3-2 shows the power curve with the same numbers as Figure 3-1, except for the sample size *(n)*, which is 100 instead of 10. Notice that the curve increases much more quickly and approaches 1.0 when the actual mean is 1.0, compared to your hypothesis of 0. You want to see this kind of curve — one that moves up quickly toward the value of 1.0, while the actual values of the parameter increase on the *x*-axis.

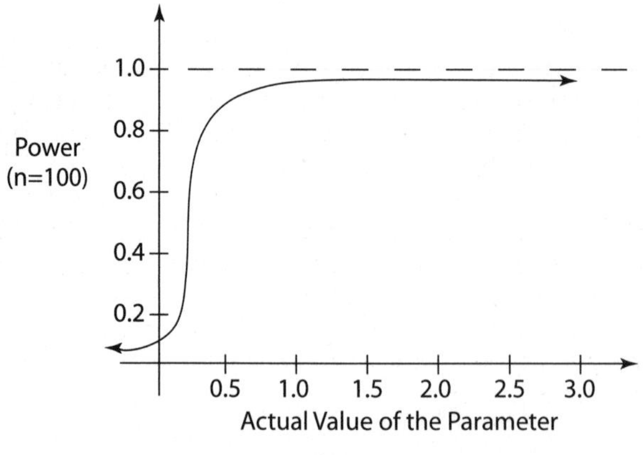

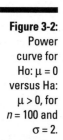

**Figure 3-2:** Power curve for Ho: $\mu = 0$ versus Ha: $\mu > 0$, for $n = 100$ and $\sigma = 2$.

If you compare the power of your test when $\mu$ is 1.0 for the $n = 10$ situation (in Figure 3-1) versus the $n = 100$ situation (in Figure 3-2), you see that the power increases from 0.475 to more than 0.999. Table 3-1 shows the different values of power for the $n = 10$ case versus the $n = 100$ case, when you test Ho: $\mu = 0$ versus Ha: $\mu > 0$, assuming a value of $\sigma = 2$.

**Table 3-1 Comparing the Values of Power for $n = 10$ versus $n = 100$ (Ho is $\mu = 0$)**

| *Actual Value of $\mu$* | *Power when n = 10* | *Power when n = 100* |
|---|---|---|
| 0.00 | 0.050 = 0.05 | 0.050 = 0.05 |
| 0.50 | 0.197 = 0.20 | 0.804 = 0.81 |
| 1.00 | 0.475 = 0.48 | approx. 1.0 |
| 1.50 | 0.766 = 0.77 | approx. 1.0 |
| 2.00 | 0.935 = 0.94 | approx. 1.0 |
| 3.00 | 0.999 = approx. 1.0 | approx. 1.0 |

REMEMBER

You can find power curves for a variety of hypothesis tests under many different scenarios. Each has the same general look and feel to it: starting at the value of $\alpha$ when Ho is true, increasing in an *S*-shape as you move from left to right on the *x*-axis, and finally approaching the value of 1.0 at some point. Power curves with large sample sizes approach 1.0 faster than power curves with low sample sizes.

WARNING!

You can have too much power. For example, if you make the power curve for $n = 10{,}000$ and compare it to Figures 3-1 and 3-2, you can find that it's practically at 1.0 already for any number other than 0.0 for the mean. In other words, the actual mean could be 0.05 and with your hypothesis Ho: $\mu = 0.00$, you would reject Ho, because of the huge sample size you've got. If you zoom in enough, you can always detect something, even if that something makes no practical difference. If the sample size is incredibly large, it can inflate power to the point where you can detect differences from Ho that are smaller than you really want, from a practical standpoint. Beware of surveys and experiments that have what appears to be an excessive sample size — for example, in the tens of thousands. They may be reporting "statistically significant" results that don't mean diddly.

## Power in manufacturing

The power of a test plays a role in the manufacturing process. Manufacturers often have very strict specifications regarding the size, weight, and/or quality of their products. During the manufacturing process, manufacturers want to be able to detect deviations from these specifications, even small ones, so they must think about how much of a difference from Ho they want to detect, and then figure out the sample size they need in order to detect that difference when it appears. For example, if the candy bar is supposed to weight 2.0 ounces, the manufacturer may want to blow the whistle if the actual average weight shifts to, say, 2.5 ounces. Statisticians can work backwards in calculating the power and find the sample size they need to know to stop the process.

Medical scientists also think about power when they set up their studies (called clinical trials). Suppose they're checking to see whether an antidepressant adversely affects blood pressure (as a side effect of taking the drug). Scientists need to be able to detect small differences in blood pressure, because for some patients, any change in blood pressure is important to note.

# Part II

# Making Predictions By Using Regression

The 5th Wave By Rich Tennant

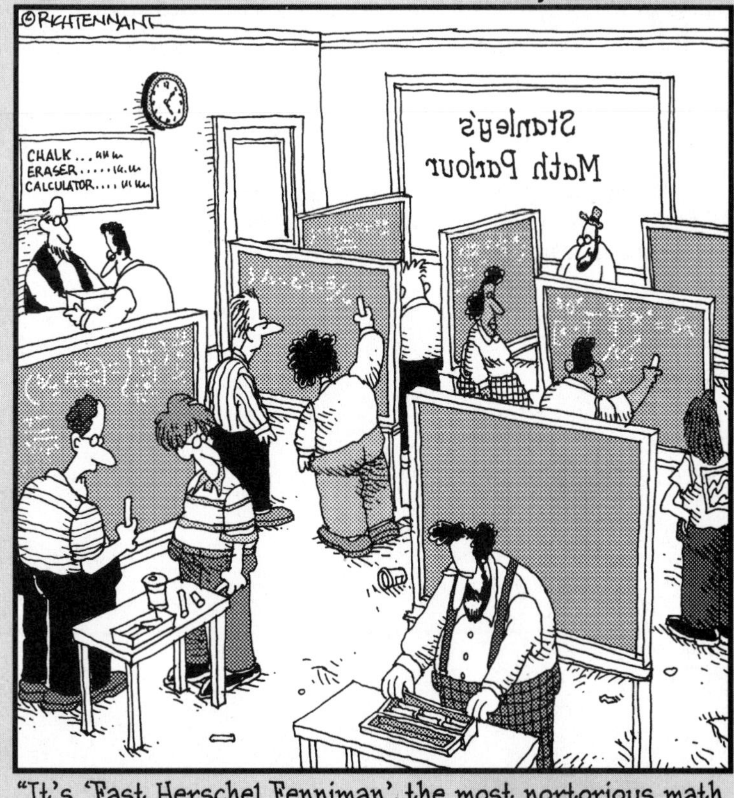


"It's 'Fast Herschel Fenniman', the most nortorious math hustler of all time. If he asks if you'd like to run some regression models with him, just walk away."

## In this part . . .

You really get into the modeling process, using various pieces of known info to predict one elusive variable. (Sounds sneaky? In a way, it is . . . ) This part goes way beyond using one variable to predict another, beyond simple linear regression to multiple, nonlinear, and even logistic regression. These methods can solve more complex problems, so they lend themselves to many real-world applications.

# Chapter 4

# Getting in Line with Simple Linear Regression

### *In This Chapter*

- Using scatterplots and correlation coefficients to examine relationships
- Building a simple linear regression model to estimate $y$ from $x$
- Testing how well the model fits
- Interpreting the results and making good predictions

Looking for relationships and making predictions is one of the staples of data analysis. Everyone wants to answer questions like "Can I predict how many units I'll sell if I spend $x$ amount of advertising dollars?"; or "Does drinking more diet cola really relate to more weight gain?"; or "Do children's backpacks seem to be getting heavier each year in school, or is it just me?"

Linear regression tries to find relationships between two or more variables and comes up with a model that tries to describe that relationship, much like the way the line $y = 2x + 3$ explains the relationship between $x$ and $y$. But unlike math where functions like $y = 2x + 3$ tell the entire story about the two variables, in statistics, things don't come out that perfectly; some variability and error is involved (that's what makes it fun!).

This chapter is partly a review of the concepts of simple linear regression presented in an intro stats book. But the fun doesn't stop there. I expand on the ideas you learned about regression in your intro stat course and set you up for some of the other types of regression models you see in Chapters 5 through 8.

In this chapter, you see how to build a simple linear regression model that examines the relationship between two variables. You also see how simple linear regression works from a model-building standpoint.

## Exploring Relationships with Scatterplots and Correlations

Before looking ahead to predicting a value of $y$ by using a value of $x$, you need to first establish that you have a legitimate reason to do so by using a straight line, and you also need to feel confident that using a line to make that prediction will actually work well. In order to achieve both of these important steps, you need to first plot the data in a pairwise fashion so you can visually look for a relationship; then you need to somehow quantify that relationship in terms of how well those points follow a line. In this section, you do just that, using scatterplots and correlations.

Here's a perfect example of a situation where simple linear regression is useful: In 2004, the California State Board of Education wrote a report entitled "Textbook Weight in California: Analysis and Recommendations." In this report, they discussed the great concern over the weight of the textbooks in student's backpacks, and the problems it presents for students. They conducted a study where they weighed a variety of textbooks from each of four core areas studied in grades 1 through 12 (reading, math, science, and history — where's statistics?) over a range of textbook brands and found the average total weight for all four books for each grade.

The California Board of Education consulted pediatricians and chiropractors, who recommended that the weight of a student's backpack should not exceed 15 percent of his body weight. From there, the Board hypothesized that the total weight of the textbooks in these four areas increases for each grade level and wanted to see whether they could find a relationship between the average child's weight in each grade and the weight of his books. So along with the average weight of the four core-area textbooks for each grade, they also recorded the average weight for the students in that grade. Their results are shown in Table 4-1.

**Table 4-1 Average Textbook Weight and Student Weight (Grades 1–12)**

| *Grade* | *Average Student Wt. (lbs.)* | *Average Textbook Wt. (lbs.)* |
|---|---|---|
| 1 | 48.50 | 8.00 |
| 2 | 54.50 | 9.44 |
| 3 | 61.25 | 10.08 |
| 4 | 69.00 | 11.81 |
| 5 | 74.50 | 12.28 |

| Grade | Average Student Wt. (lbs.) | Average Textbook Wt. (lbs.) |
|---|---|---|
| 6 | 85.00 | 13.61 |
| 7 | 89.00 | 15.13 |
| 8 | 99.00 | 15.47 |
| 9 | 112.00 | 17.36 |
| 10 | 123.00 | 18.07 |
| 11 | 134.00 | 20.79 |
| 12 | 142.00 | 16.06 |

In this section, you begin exploring whether or not a relationship exists between these two quantitative variables. You start by displaying the pairs of data using a two-dimensional scatterplot to look for a possible pattern, and you quantify the strength and direction of that pattern using the correlation coefficient.

Data analysts should never make any conclusions about a relationship between $x$ and $y$ based solely on either the correlation or the scatterplot alone; the two elements need to be examined together. It is possible (but of course not a good idea) to manipulate graphs to look better or worse than they really are just by changing the scales on the axes. Because of this, statisticians never go with the scatterplot alone to determine whether or not a linear relationship exists between $x$ and $y$. A correlation without a scatterplot is dangerous too, because the relationship between $x$ and $y$ may be very strong, but just not linear.

## Using scatterplots to explore relationships

In order to explore a possible relationship between two variables, such as textbook weight and student weight, you first plot the data in a special graph called a *scatterplot.* A scatterplot is a two-dimensional graph that displays pairs of data, one pair per observation in the $(x, y)$ format. Figure 4-1 shows a scatterplot of the textbook weight data from Table 4-1.

You can see that the relationship appears to follow the straight line that's included on the graph, except possibly for the last point, where textbook weight is 16.06 pounds and student weight is 142 pounds (for grade 12). This point appears to be an *outlier* — it's the only point that doesn't fall into the pattern. So overall, an uphill, or *positive* linear relationship appears to exist between textbook weight and student weight; as student weight increases, so does textbook weight.

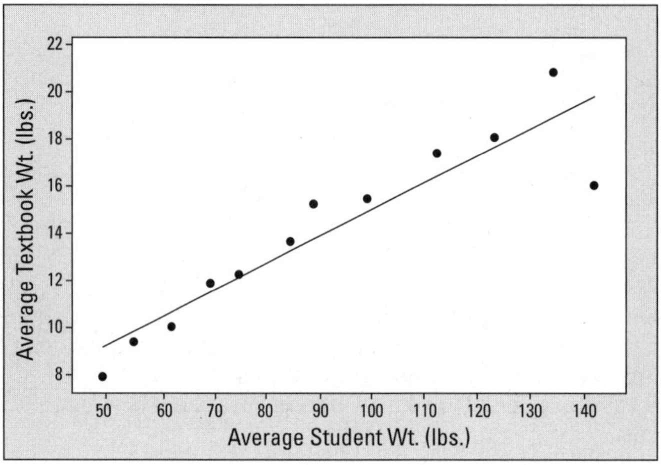


**Figure 4-1:** Scatterplot of average student weight versus average textbook weight in grades 1–12.

To make a scatterplot in Minitab, enter the data in columns one and two of the spreadsheet. Go to Graphs>Scatterplot. Click Simple and then OK. Highlight the response variable *(y)* in the left-hand box, and click Select. This variable shows up as the *y* variable in the scatterplot. Click on the explanatory *(x)* variable in the left-hand box and click Select. It shows up in the *x* variable box. Click OK, and you get the scatterplot.

## Collating the information by using the correlation coefficient

After you've displayed the data using a scatterplot (see preceding section), the next step is to find a statistic that quantifies the relationship somehow. The *correlation coefficient* (also known as *Pearson's correlation coefficient*) measures the strength and direction of the linear relationship between two quantitative variables *x* and *y*. It's a number between –1 and +1 that's unit-free; that means if you change from pounds to ounces, the correlation coefficient doesn't change. (What a messed-up world it would be if this wasn't the case!)

Statistical software packages, such as Minitab, refer to the correlation coefficient as Pearson's correlation coefficient. (Don't worry — they're the same!)

If the relationship between *x* and *y* is uphill, or positive (as *x* increases so does *y*), the correlation is a positive number. If the relationship is downhill, or negative (as *x* increases, *y* gets smaller), then the correlation is negative. If the correlation is zero, you can find no linear relationship between *x* and *y*. (It may be that a different relationship exists, such as a curve; see Chapter 7 for more on this.)

If the value of the correlation is +1 or –1, this value indicates that the points fall in a perfect, straight line. If the correlation is close to +1 or –1, this correlation value signifies a strong relationship. If the correlation is closer to +0.5 or –0.5, these values show a moderate relationship. A value close to 0 signifies a weak relationship or no linear relationship at all.

You can calculate the correlation coefficient by using a formula involving the standard deviation of *x*, the standard deviation of *y*, and the covariance of *x* and *y*, which measures how *x* and *y* move together, in relation to their means. However, the formula isn't the focus here (you can find it in your intro stats text or in my other book *Statistics For Dummies* [Wiley]); it's the concept that's important. Any computer package can calculate the correlation coefficient for you with a simple click of the mouse.

To have Minitab calculate a correlation for you, go to Stat>Basic Statistics>Correlation. Highlight the variables you want correlations for and click Select. Then click on OK.

The correlation for the textbook weight example is (can you guess before looking at it?) 0.926, which is very close to 1.0. This correlation means that a very strong linear relationship is present between average textbook weight and average student weight for grades 1 through 12, and that relationship is positive and linear (follows a straight line). This correlation is confirmed by the scatterplot shown in Figure 4-1.

# Building a Simple Linear Regression Model

After you have a handle on which *x* variables may be related to *y* in a linear way, you go about the business of finding that straight line that best fits the data. You find the slope and *y*-intercept, put them together to make a line, and you use the equation of that line to make predictions for *y*. All of this is part of building a simple linear regression model.

In this section, you set the foundation for regression models in general (including those you can find in Chapters 5 through 8). You plot the data, come up with a model that you think makes sense, assess how well it fits, and use it to guesstimate the value of *y* given another variable, *x*.

## Finding the best-fitting line to model your data

After you've established that $x$ and $y$ have a strong linear relationship, as evidenced by both the scatterplot and the correlation coefficient (see the previous sections), you're ready to build a model that estimates $y$ using $x$. In the textbook-weight case, you want to estimate average textbook weight using average student weight.

The most basic of all the regression models is the *simple linear regression model* that comes in the general form of $y = a + bx$. Here $a$ represents the $y$-intercept of the line; $b$ represents the slope.

A straight line that's used in simple linear regression is just one of an entire family of models (or functions) that statisticians use to express relationships between variables. A *model* is just a general name for a function that you can use to estimate or guess what outcome will occur if you have some given information about related items.

To find the right model for your data, the idea is to scour all possible lines and choose the one that fits the data best. Thankfully, you have an algorithm that does this for you (computers use it in their calculations). Formulas also exist for finding the slope and $y$-intercept of the best-fitting line by hand. (You can find those formulas in your intro stats text or in *Statistics For Dummies* [Wiley].)

To run a linear regression analysis in Minitab, go to Stat>Regression>Regression. Highlight the response *(y)* variable in the left-hand box, and click on Select. The variable shows up in the Response Variable box. Then highlight your explanatory *(x)* variable, and click on Select. This variable shows up in the Predictor Variable box. Click OK.

The equation of the line that best describes the relationship between average textbook weight and average student weight is: $y = 3.69 + 0.113x$, where $x$ is the average student weight for that grade, and $y$ is the average textbook weight. Figure 4-2 shows the Minitab output of this analysis.

```
The regression equation is
textbook wt = 3.69 + 0.113 student wt


Predictor        Coef  SE Coef     T      P
Constant        3.694    1.395  2.65  0.024
student wt    0.11337  0.01456  7.78  0.000


S = 1.51341     R-Sq = 85.8%     R-Sq(adj) = 84.4%
```

**Figure 4-2:** Simple linear regression analysis for the textbook weight example.

By writing $y = 3.69 + 0.113x$, you mean that this equation represents your estimated value of $y$, given the value of $x$ that you observe with your data. Statisticians write this equation by using a carrot (or *hat* as statisticians call it), like $\hat{y}$, so everyone can know it's an estimate, not the actual value of $y$. This *y-hat* is your estimate of the average value of $y$ over the long term, based on the observed values of $x$. However, in many intro stats texts, the hat is left off because statisticians have an unwritten understanding as to what $y$ represents. This issue comes up again in Chapters 5 through 8. (By the way, if you think *y-hat* is a funny term here, it's even funnier in Mexico, where statisticians call it *y-sombrero* — no kidding!)

## The y-intercept of the regression line

Selected parts of that Minitab output shown in Figure 4-2 are of importance to you at this point. First, you can see that under the column "Coef" you have the numerical values on the right side of the equation of the line — in other words, the slope and $y$-intercept. The number 3.69 represents the coefficient of "Constant," which is a fancy way of saying that's the $y$-intercept (because the $y$-intercept is just a constant, it never changes). The $y$-intercept is the point where the line crosses the $y$-axis, in other words, the value of $y$ when $x$ equals 0.

The $y$-intercept of a regression line may or may not have a practical meaning depending on the situation. To determine whether the $y$-intercept of a regression line has practical meaning, look at the following:

- Does the $y$-intercept fall within the actual values in the data set? If yes, then it has practical meaning.
- Does the $y$-intercept fall into negative territory where negative $y$-values aren't possible? For example if the $y$-values, are weights, they can't be negative. Then the $y$-intercept has no practical meaning. It is still correct though, because it just happens to be the place where the line, if extended to the $y$-axis, crosses the $y$-axis.
- Does the value $x = 0$ have practical meaning? For example, if $x$ is temperature at a football game in Green Bay, then $x = 0$ is a value that's relevant to examine. If $x = 0$ has practical meaning, then the $y$-intercept would also because it represents the value of $y$ when $x = 0$. If not, for example, when $x$ represents height of a toddler, then the $y$-intercept has no practical meaning.

In the textbook example, the $y$-intercept doesn't really have a practical meaning because students don't weigh zero pounds, so you don't really care what the estimated textbook weight is for that situation. But you do need to find a line that fits the data you do have (where average student weights go from

48.5 pounds to 142 pounds). That best-fitting line must include a *y*-intercept, and for this problem, that *y*-intercept happens to be 3.69.

### The slope of the regression line

The value 0.113 from Figure 4-2 indicates the coefficient (or number in front of) of the student-weight variable. This number is also known as the *slope.* It represents the change in *y* (textbook weight) due to a one-unit increase in *x* (student weight). As student weight increases by one pound, textbook weight increases by about 0.113 pounds, on average. To make this relationship more meaningful, you can multiply both quantities by ten to say that as student weight increases by 10 pounds, the textbook weight goes up by about 1.13 pounds on average.

TIP

Whenever you get a number for the slope, just take that number and put it over 1. Doing this can help you get started on a proper interpretation of slope. For example, a slope of 0.113 is rewritten as $\frac{0.113}{1}$. Using the idea that slope equals rise over run, or change in *y* over change in *x,* you can interpret the value of 0.113 in the following way: As *x* increases by one pound, *y* increases by 0.113 pounds.

### Making estimates by using the regression line

Now that you have a line that estimates *y* given *x,* you can use it to estimate the (average) value of *y* for a given value of *x.* The basic idea is to take a reasonable value of *x,* plug it in to the equation of the regression line, and see what the value of *y* gives you.

In the textbook-weight example, the best-fitting line (or model) is the line $y = 3.69 + 0.113x$. For an average student that weighs 60 pounds, for example, the estimated average textbook weight is $3.69 + 0.113 * 60 = 10.47$ pounds (those poor little kids!). If the average student weighs 100 pounds, the estimated average textbook weight is $3.69 + 0.113 * 100 = 14.99$, or nearly 15 pounds.

## Checking the Model's Fit (The Data, Not the Clothes!)

After you've established a relationship between *x* and *y* and have come up with an equation of a line that represents that relationship, you may think your job is done. (Many researchers erringly stop here, so I'm depending on you to break the cycle on this!) But the most-important job remains to be completed: checking to be sure that the conditions of the model are truly met and that the model fits well in more specific ways than the scatterplot and correlation measure. This section presents methods for defining and assessing the fit of a simple linear regression model.

## Defining the conditions

Two major conditions must be met before you apply a simple linear regression model to a data set:

- The *y*'s have to have a normal distribution for each value of *x*.
- The *y*'s have to have a constant amount of spread (standard deviation) for each value of *x*.

In the following sections, you look at these important conditions in depth.

### Normal y's for every x

For any value of *x*, the population of possible *y*-values must have a normal distribution. The mean of this distribution is the value for *y* that is on the best-fitting line for that *x*-value. That is, some of your data falls above the best-fitting line, some data falls below the best fitting line, and a few may actually land right on the line.

If the regression model is fitting well, the data values should be scattered around the best-fitting line in such a way that about 68 percent of the values lie within one standard deviation of the line, about 95 percent of the values should lie within two standard deviations of the line, and about 99.7 percent of the values should lie within three standard deviations of the line. This specification, as you may recall from your intro stats course, is called the 68-95-99.7 rule, and it applies to all bell-shaped data (for which the normal distribution applies).

You can see in Figure 4-3 how for each *x*-value, the *y*-values you may observe tend to be located near the best-fitting line in greater numbers, and as you move away from the line, you see fewer and fewer *y*-values, both above and below the line. More than that, they're scattered around the line in a way that reflects a bell-shaped curve, the normal distribution.

Why does this condition makes sense? The data you collect on *y* for any particular *x*-value varies from individual to individual (for example, not all students' textbooks weigh the same, even for students who weigh the exact same amount). But those values aren't allowed to vary any way they want to. To fit the conditions of a linear regression model, for each given value of *x*, the data should be scattered around the line according to a normal distribution. Most of the points should be close to the line, and as you get farther and farther from the line, you can expect fewer and fewer data points to occur. So condition number one is that the data have a normal distribution for each value of *x*.

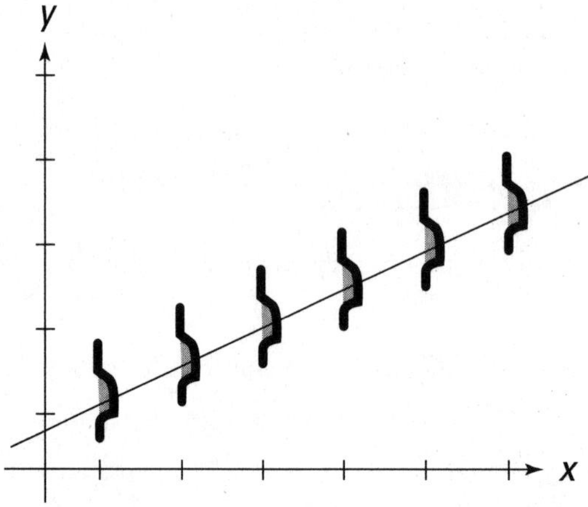


**Figure 4-3:** Conditions of a simple linear regression model.

### Same spread for every x

The second condition for being able to use the simple linear regression model is the following: As you move from left to right on the *x*-axis, the spread in the *y*-values around the line should be the same, no matter which value of *x* you're looking at. This requirement is called the *homoscedasticity condition.* (How they came up with that mouthful of a word just for describing the fact that the standard deviations stay the same across the *x*-values, I'll never know.) This condition ensures that the best-fitting line works well for all relevant values of *x*, not just in certain areas where the *y*-values lie close to each other.

You can see in Figure 4-3 that no matter what the value of *x* is, the spread in the *y*-values stays the same throughout. If the spread got bigger and bigger as *x* got larger and larger, for example, the line would lose its ability to fit well for those large values of *x*.

In the next sections, you can find out how to check the two conditions for simple linear regression, so keep reading.

## Finding and exploring the residuals

To check to see whether the *y*-values come from a normal distribution, you need to measure how far off your predictions were from the actual data that came in, and you need to check those errors and see how they stack up.

In the following sections, you center on finding a way to measure these errors that the model makes. You also explore the errors to identify particular problems that occurred in the process of trying to fit a straight line to the data. In other words, you can discover that looking at errors helps you assess the fit of the model and diagnose problems that caused a bad fit, if that was the case.

### Finding the residuals

A *residual* is the difference between the observed value of $y$ (from the best-fitting line) and the predicted value of $y$ (from the data set). Specifically, for any data point, you take its observed $y$-value (from the data) and subtract the expected $y$-value (from the line). If the residual is large, the line doesn't fit well in that spot. If the residual is small, the line fits well in that spot.

For example, suppose you have a point in your data set (2, 4) and the equation of the best-fitting line is $y = 2x + 1$. The expected value of $y$ in this case is $2 * 2 + 1 = 5$. The observed value of $y$ from the data set is 4. Taking the observed value minus the estimated value you get $4 - 5 = -1$. The residual for that particular data point (2, 4) is $-1$. If you observe a $y$-value of 6 and use the same straight line to estimate $y$, then the residual would be $6 - 5 = +1$.

In general, a positive residual means you underestimated $y$ at that point, and a negative residual means you overestimated $y$ at that point.

### Standardizing the residuals

To make interpreting the residuals easier, statisticians typically *standardize* them; that is, subtract the mean of the residuals (zero) and divide by the standard deviation of all the residuals. The residuals are a data set just like any other data set, so you can find their mean and standard deviation like you always do. Standardizing just means converting to a $Z$-score, so you see where it falls on the standard normal distribution.

### Making residual plots

You can plot the residuals on a graph called a *residual plot.* (If you've standardized the residuals, you call it a *standardized residual plot.*) Figure 4-4 shows Minitab output for a variety of standardized residual plots, all getting at the same idea: checking to be sure the conditions of the simple linear regression model are met.

### Checking normality

If the condition of normality is met, you can see on the residual plot lots of (standardized) residuals close to zero; as you move farther and farther away from zero, you can see fewer and fewer residuals. ***Note:*** A standardized residual at or beyond +3 or –3 is something you shouldn't expect to see. If this occurs, you can consider that point an outlier, which warrants further investigation. (For more on outliers, see the section "Scoping for outliers.")

The residuals should also occur at random — some above the line, some below the line. If a pattern occurs in the residuals, the line may not be fitting right.

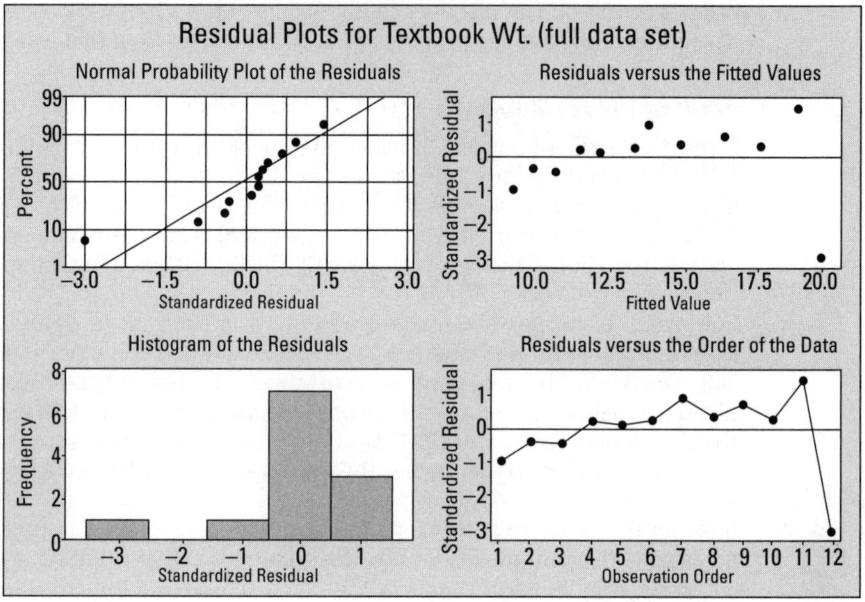


**Figure 4-4:** Standardized residual plots for textbook weight data.

The plots in Figure 4-4 seem to have an issue with the very last observation, the one for twelfth graders. In this observation, the average student weight (142) seemed to follow the pattern of increasing with each grade level, but the textbook weight (16.06) was less than for eleventh graders (20.79) and is the first point to break the pattern.

You can also see in the plot in the upper-right corner of Figure 4-4 that the very last data value has a residual that sticks out from the others and has a value of –3.0 (something that should be a very rare occurrence). So the value you expected for *y* based on your line was off by a factor of 3 standard deviations. And because this residual is negative, what you observed for *y* was much lower than you may have expected it to be using the regression line.

The other residuals seem to fall in line with a normal distribution, as you can see in the upper-right plot of Figure 4-4. The residuals concentrate around zero, with fewer appearing as you move farther away from zero. You can also see this pattern in the upper-left plot of Figure 4-4, which shows how close to normal the residuals are. The line in this graph represents the *equal-to-normal line.* If the residuals follow close to the line, then normality is okay. If not, you have problems (in a statistical sense, of course). You can see the residual with the highest magnitude is –3, and that number falls outside the line quite a bit.

The lower-left plot in Figure 4-4 makes a histogram of the standardized residuals, and you can see it doesn't look much like a bell-shaped distribution. It doesn't even look *symmetric* (the same on each side when you cut it down the middle). The problem again seems to be the residual of –3, which makes the histogram be skewed to the left.

The lower-right plot of Figure 4-4 plots the residuals in the order presented in the data set in Table 4-1. Because the data was ordered already, the lower-right residual plot looks like the upper-right residual plot in Figure 4-4, except the dots are connected. This lower-right residual plot makes the residual of –3 stand out even more.

### Checking the spread of the y's for each x

The graph in the upper-right corner of Figure 4-4 also addresses the homoscedasticity condition. If the condition is met, then the residuals for every *x*-value have about the same spread. If you cut a straight line down through each *x*-value, the residuals have about the same spread (standard deviation) each time, except for the last *x*-value, which again represents grade twelve. That means the condition of equal spread in the *y*-values is met for the backpack example.

TIP

If you look at only one residual plot, choose the one in the upper-right corner of Figure 4-4, the plot of the fitted values (the values of *y* on the line) versus the standardized residuals. Most problems with model fit pop up on that plot because a residual is defined as the difference between the observed value of *y* and the fitted value of *y*. In a perfect world, all the fitted values have no residual at all; a large residual (such as the one where the estimated weight is 20 pounds for twelfth graders; see Figure 4-4) is indicated by a point far off from zero. This graph also shows you deviations from the overall pattern of the line; for example, if large residuals are on the extremes of this graph (very low or very high fitted values), that shows the line isn't fitting in those areas.

## Using $r^2$ to measure model fit

One important way to assess how well the model fits is to measure the value of $r^2$, where $r$ is the correlation coefficient. Statisticians measure how well a model fits by looking at what percentage of the variability in *y* is explained by the model.

The *y*-values of the data you collect have a great deal of variability in and of themselves. You look for another variable *(x)* that helps you explain that variability in the *y*-values. After you put that *x* variable into the model, and you find it's highly correlated with *y*, you want to find out how well this model did at explaining why the values of *y* are different.

As it turns out, the value of $r^2$, gives you that measure of model fit. Because squaring a number between 0 and +1 makes the result get smaller (except for 0 and +1), how do you interpret $r^2$? A value of $r = +0.9$ or $-0.9$ is quite high;

note that when you square either one of them, you get 0.81, which you should also interpret as being high.

TIP

The following are some general guidelines for interpreting the value of $r^2$:

- If the model containing $x$ explains a lot of the variability in the $y$-values, then $r^2$ is high (in the 80 to 90 percent range is considered to be extremely high). Values like 0.70 are still considered fairly high. A high percentage of variability means that the line fits well because there is not much left to explain about the value of $y$ other than using $x$ and its relationship to $y$. So a larger value of $r^2$ is a good thing.
- If the model containing $x$ doesn't help much in explaining the difference in the $y$-values, then the value of $r^2$ is small (closer to zero; say between 0.00 and 0.30 roughly). The model, in this case, would not fit well. You need another variable to explain $y$ other than the one you already tried.
- Values of $r^2$ that fall in the middle (between, say, 0.30 and 0.70) mean that $x$ does help somewhat in explaining $y$, but it doesn't do the job well enough on its own. In this case, statisticians would try to add one or more variables to the model to help explain $y$ more fully as a group (read more about this in Chapter 5).

For the textbook weight example, the value of $r$ (the correlation coefficient) is 0.93. Squaring this result, you get $r^2 = 0.8649$. That number means approximately 86 percent of the variability you find in average textbook weights for all students ($y$-values) is explained by the average student weight ($x$-values). This percentage tells you that the model of using year in school to estimate backpack weight is a good bet.

In the case of simple linear regression, you have only one $x$ variable, but in Chapter 5, you can see models that contain more than one $x$ variable. In this situation, you use $r^2$ to help sort out the contributions each individual variable brings to the model.

## Scoping for outliers

Sometimes life isn't perfect (oh really?), and you may find a residual in your otherwise tidy data set that totally sticks out, which is called an *outlier.* That is, it has a standardized value at or beyond +3 or –3. It threatens to blow the conditions of your regression model and send you crying to your professor.

Before you panic, the best thing to do is to examine that outlier more closely. First, can you find an error in that data value? Did someone report her age as 642, for instance? (After all, mistakes do happen.) If you do find a certifiable error in your data set, you remove that data point (or fix it if possible) and analyze the data without it. However, if you can't explain away the problem by finding a mistake, you must think of another approach.

If you can't find a mistake that caused the outlier, you don't necessarily have to trash your model; after all, it's only one data point. What you do is analyze the data with that data point and analyze the data again without it. Then report and compare both analyses. This comparison can give you a sense of how influential that one data point is. It may lead other researchers to conduct more research to zoom in on the issue you brought to the surface.

In Figure 4-1, you can see the scatterplot of the full data set for the textbook weights example. Figure 4-5 shows the scatterplot for the data set minus the outlier. The scatterplot fits the data better without the outlier. The correlation increases to 0.993. The value of $r^2$ increases to 0.986. The equation for the regression line for this data set is $y = 1.78 + 0.139x$.

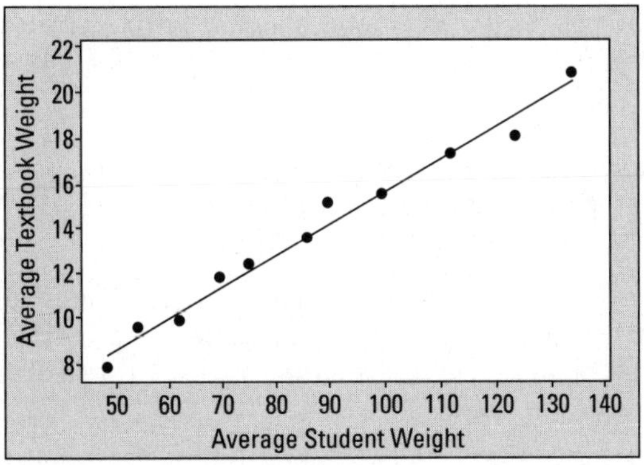


**Figure 4-5:** Scatterplot of textbook weight data (minus the outlier).

The slope of the regression line hasn't changed much by removing the outlier (compare it to Figure 4-2, where the slope is 0.113). However, the *y*-intercept has changed; it's now 1.78 without the outlier compared to 3.69 with the outlier. The slope of the lines are about the same, but the lines cross the *y*-axis in different places. It appears that the outlier (the last point in the data set) has quite an affect on the best-fitting line.

Figure 4-6 shows the residual plots for the regression line for the data set without the outlier. Each of these plots shows a much better fit of the data to the model compared to Figure 4-4. This result tells you that the data for grade twelve is influential in this data set, and that outlier needs to be noted and perhaps explored further. Do students peak out when they're juniors in high school? Or do they just decide when they're seniors that it isn't cool to carry books around? (A statistician's job isn't to wonder why, but to do and analyze.)

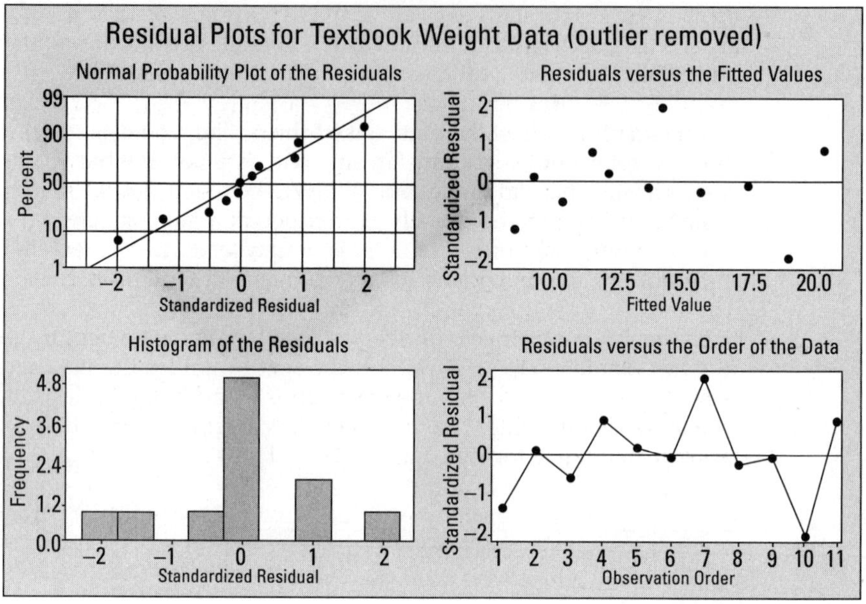


**Figure 4-6:** Residual plots for textbook weight data (minus the outlier).

# Making Correct Conclusions

The bottom line of any data analysis is to make the correct conclusions given your results. When you're working with a simple linear regression model, three major errors can be made. In this section, you see those errors and how to avoid them.

## Avoiding slipping into cause-and-effect mode

In a simple linear regression, you investigate whether *x* is related to *y*, and if you get a strong correlation and a scatterplot that shows a linear trend, then you find the best-fitting line and use it to estimate the value of *y* for reasonable values of *x*.

WARNING!

There is a fine line, however (no pun intended), that you don't want to cross with your interpretation of regression results. Be careful to not interpret slope in a cause-and-effect mode when you're using the regression line to estimate the value of *y* using *x*. Doing so can result in a leap of faith that can send you into the frying pan. Unless you have used a controlled experiment to get the data, you can only assume that the variables are correlated; you can't really give a stone-cold guarantee why they are related.

In the textbook weight example, you estimate the average weight of the students' textbooks by using the students' average weight, but that doesn't mean that increasing a particular child's weight causes his textbook weight to increase. For example, because of the strong positive correlation, you do know that students with lower weights are associated with lower total textbook weights, and students with higher weights tend to have higher textbook weights. But you can't take one particular third-grade student, increase his weight, and presto — suddenly his textbooks weigh more.

The variable that is underlying the relationship between a child's weight and the weight of his backpack is the grade level of the student; as grade level increases, so does the size of his books. Student grade level drives both student weight and textbook weight. In this situation, student grade level is what statisticians call a *confounding variable:* it's a variable that wasn't included in the study but is related to both the outcome and the response, and the variable confounds or confuses the issue of what is causing what to happen.

If the collected data was the result of a well-designed experiment that controls for possible confounding variables, you can establish a cause-and-effect relationship between $x$ and $y$ if they're strongly correlated. Otherwise, you can't.

## Extrapolation: The ultimate no-no

Plugging values of $x$ into the model that fall outside of the reasonable boundaries of $x$ is called *extrapolation.* And one of my colleagues sums up this idea very well, "Friends don't let friends extrapolate."

When you determine a best-fitting line for your data, you come up with an equation that allows you to plug in a value for $x$ and get a predicted value for $y$. In algebra, if you found the equation of a line and graphed it, the line would typically have an arrow on each end indicating it goes on forever in either direction. But that doesn't work for statistical problems ('cause statistics represents the *real* world). What I mean is that when you're dealing with real-world units like height, weight, IQ, GPA, house prices, and the weight of your statistics textbook, only certain numbers make sense.

So the first point is, don't plug in values for $x$ that don't make any sense. For example, if you're estimating the price of a house *(y)*, using its square footage *(x)*, you wouldn't think of plugging in a value of $x$ like 10 square feet or 100 square feet, because houses simply aren't that small. You also wouldn't think about plugging in values like 1,000,000 square feet for $x$ (unless your "house" is the Ohio State football stadium or the like). It wouldn't make sense. If you're estimating tomorrow's temperature using today's temperature, negative numbers for $x$ could possibly make sense, but if you're estimating the amount of precipitation tomorrow given the amount of precipitation today, negative numbers for $x$ (or $y$ for that matter) don't make sense.

Second, choose only reasonable values of $x$ for which you try to make estimates about $y$. That is, look at the values of $x$ for which your data was collected and stay within those bounds when making predictions. In the textbook weight example, the smallest average student weight is 48.5 pounds, and the largest average student weight is 142 pounds. Choosing student weights between 48.5 and 142 to plug in for $x$ in the equation is okay, but choosing values less than 48.5 or above 142 isn't a good idea. You can't guarantee that the same linear relationship (or any linear relationship for that matter) continues outside the given boundaries.

Think about it: If the relationship you found actually continued for any value of $x$, no matter how large, then a 250-pound linebacker from OSU would have to carry $3.69 + 0.113 * 250 = 31.94$ pounds of books around in his backpack. Of course this would be easy for him, but what about the rest of us?

## Knowing the limitations of a simple linear regression model

A simple linear regression model is just what it says it is: simple. I don't mean easy to work with, necessarily, but simple in the uncluttered sense. The model tries to estimate the value of $y$ by only using one variable, $x$. However, the number of real-world situations that can be explained by using a simple, one-variable linear regression is small. Oftentimes one variable just can't do all the predicting.

If one variable alone doesn't result in a model that fits, add more variables. Oftentimes it takes many variables to make a good estimate for $y$. In the case of stock market prices, they're still looking for that ultimate prediction model.

As another example, health insurance companies try to estimate how long you will live by asking you a series of questions (each of which represents a variable in the regression model). You can't find one single variable that estimates how long you'll live; you must consider many factors: your health, your weight, whether or not you smoke, genetic factors, how much exercise you do each week, and the list goes on and on and on.

The point is, regression models don't always use just one variable, $x$, to estimate $y$. Some models use two, three, or even more variables to estimate $y$. Those models aren't called simple linear regression models; they're called *multiple linear regression models,* because of their employment of multiple variables to make an estimate. (You can explore multiple linear regression models in Chapter 5.)

# Chapter 5

# When Two Variables Are Better than One: Multiple Regression

### In This Chapter

- Getting the basic ideas behind a multiple regression model
- Finding, interpreting, and testing coefficients
- Checking model fit

The idea of regression is to build a model that estimates or predicts one quantitative variable *(y)* by using at least one other quantitative variable *(x)*. Simple linear regression uses exactly one *x* variable to estimate the *y* variable. (See Chapter 4 for all the information you need on simple linear regression.) Multiple linear regression, on the other hand, uses more than one *x* variable to estimate the value of *y*.

In this chapter, you see how multiple regression works and how to apply it to build a model for *y*. You see all the steps necessary for the process, including determining which *x* variables to include, estimating their contributions to the model, finding the best model, using the model for estimating *y*, and assessing the fit of the model. It may seem like a mountain of information, but you won't regress on the topic of regression if you take this chapter one step at a time.

## The Multiple Regression Model

Before being able to jump right into using the multiple regression model, its good to get a feel for what it's all about. In this section, you see the usefulness of multiple regression as well as the basic elements of the multiple regression model. Some of the ideas are just an extension of the simple linear regression model (Chapter 4). Some of the concepts are a little more complex, as you may guess because the model is more complex. But the concepts and the results should make intuitive sense, which is always good news.

## Discovering the uses of multiple regression

One situation in which multiple regression is useful is when the $y$ variable is hard to track down; that is, its value can't be measured straight up, and you need more than one other piece of information to help get a handle on what its value will be. For example, you may want to estimate the price of gold today. It would be hard to imagine being able to do that with only one other variable. You may base it on recent gold prices, the price of other commodities on the market that move with or against gold, and a host of other possible economic conditions associated with the price of gold.

Another case for using multiple regression is when you want to figure out what factors play a role in determining the value of $y$. For example, what information is important to real estate agents in setting a price for a house going on the market?

## Looking at the general form of the multiple regression model

The general idea of simple linear regression is to fit the best straight line through that data that you possibly can and use that line to make estimates for $y$ based on certain $x$-values. The equation of the best-fitting line in simple linear regression is $y = b_0 + b_1x_1$, where $b_0$ is the $y$-intercept and $b_1$ is the slope. (The equation also has the form $y = a + bx$; see Chapter 4.)

In the multiple regression setting, you have more than one $x$ variable that is related to $y$. Call these $x$ variables $x_1, x_2, \ldots x_k$. In the most basic multiple regression model, you use some or all of these $x$ variables to estimate $y$ where each $x$ variable is taken to the first power. This process is called finding the best-fitting linear function for the data. This linear function looks like the following: $y = b_0 + b_1x_1 + b_2x_2 + \ldots + b_kx_k$, and you can call it the *multiple (linear) regression model.* You use this model to make estimates about $y$ based on given values of the $x$ variables.

TECHNICAL STUFF

A *linear* function is an equation whose $x$ terms are taken to the first power only. For example $y = 2x_1 + 3x_2 + 24x_3$ is a linear equation using three $x$ variables. If any of the $x$ terms are squared, the function would be a *quadratic* one; if an $x$ term is taken to the third power, the function would be a *cubic* function, and so on. In this chapter, I consider only linear functions.

## Stepping through the analysis

Your job in conducting a multiple regression analysis is to do the following (the computer can help you do steps three through six):

1. **Come up with a list of possible $x$ variables that may be helpful in estimating $y$.**
2. **Collect data on the $y$ variable and your $x$ variables from step one.**
3. **Check the relationships between each $x$ variable and $y$ (using scatterplots and correlations) and use the results to eliminate those $x$ variables that aren't strongly related to $y$.**
4. **Look at possible relationships between the $x$ variables themselves to make sure you aren't being redundant (in statistical terms, you're trying to avoid the problem of multicolinearity).**

   If two $x$ variables relate to $y$ the same way, you don't need both in the model.
5. **Use those $x$ variables (from step four) in a multiple regression analysis to find the best-fitting model for your data.**
6. **Use the best-fitting model (step five) to predict $y$ for given $x$-values by plugging those $x$-values into the model.**

I outline each of these steps in the sections to follow.

# Looking at X's and Y's

The first step of a multiple regression analysis comes way before the number crunching on the computer; it occurs even before the data is collected. Step one is where you sit down and think about what variables may be useful in predicting your response variable $y$. This step will likely take more time than any other step, except maybe the data-collection process. Deciding which $x$ variables may be candidates for consideration in your model is a deal-breaking step, because you can't go back and collect more data after the analysis is over.

REMEMBER

Always check to be sure that your response variable, $y$, and at least one of the $x$ variables are quantitative. For example, if $y$ isn't quantitative but at least one $x$ is, a logistic regression model may be in order (see Chapter 8).

Suppose you're in the marketing department for a major national company that sells plasma TVs. You want to sell as many TVs as you can, so you want to figure out which factors play a role in plasma TV sales. In talking with your advertising people and remembering what you learned in those college classes on business, you know that one powerful way to get sales is through advertising. You think of the types of advertising that may be related to sales of plasma TVs and your team comes up with two ideas:

- **TV ads:** Of course, how better to sell a TV than through a TV ad?
- **Newspaper sales:** Hit 'em on Sunday when they're watching the game through squinty eyes that are missing all the good plays and the terrible calls the referees are making.

By coming up with a list of possible *x* variables to predict *y,* you have just completed step one of a multiple regression analysis, according to the list in the previous section. Note that all three variables I use in the TV example are quantitative (the TV ad and newspaper sales variables and the TV sales response variable), which means you can go ahead and think about a multiple regression model by using the two types of ads to predict TV sales.

# Collecting the data

Step two in the multiple regression analysis process is to collect the data for your *x* and *y* variables. To do this, make sure that for each individual in the data set, you collect all the data for that individual at the same time (including the *y*-value and all *x*-values) and keep the data all together for each individual, preserving any relationships that may exist between the variables. You must then enter the data into a table format by using Minitab or any other software package (each column represents a variable and each row represents all the data from a single individual) to get a glimpse of the data and to organize it for later analyses.

To continue with the TV sales example from the preceding section, say that you start thinking about all the reams of data you have available to you regarding the plasma TV industry. You remember you've worked with the advertising department before to do a media blitz by using, among other things, TV and newspaper ads. So you have data on these variables from a variety of store locations. You take a sample of 22 store locations in different parts of the country and put together the data on how much money was spent on each type of advertising, along with the plasma TV sales for that location. You can see the data in Table 5-1.

**Table 5-1 Advertising Dollars and Sales of Plasma TVs**

| *Location* | *Sales ($ mil)* | *TV Ads ($1,000)* | *Newspaper Ads ($1,000)* |
|---|---|---|---|
| 1 | 9.73 | 0 | 20 |
| 2 | 11.19 | 0 | 20 |
| 3 | 8.75 | 5 | 5 |
| 4 | 6.25 | 5 | 5 |
| 5 | 9.10 | 10 | 10 |
| 6 | 9.71 | 10 | 10 |
| 7 | 9.31 | 15 | 15 |
| 8 | 11.77 | 15 | 15 |
| 9 | 8.82 | 20 | 5 |
| 10 | 9.82 | 20 | 5 |
| 11 | 16.28 | 25 | 25 |
| 12 | 15.77 | 25 | 25 |
| 13 | 10.44 | 30 | 0 |
| 14 | 9.14 | 30 | 0 |
| 15 | 13.29 | 35 | 5 |
| 16 | 13.30 | 35 | 5 |
| 17 | 14.05 | 40 | 10 |
| 18 | 14.36 | 40 | 10 |
| 19 | 15.21 | 45 | 15 |
| 20 | 17.41 | 45 | 15 |
| 21 | 18.66 | 50 | 20 |
| 22 | 17.17 | 50 | 20 |

The question is, can the amount of money spent on these two forms of advertising do a good job of estimating sales (in other words, are the ads worth the money)? And if so, do you need to include spending for both types of ads to estimate sales, or is one of them enough? Looking at the numbers in Table 5-1, you can see that higher sales may be related at least to higher amounts spent on TV advertising; the situation with newspaper advertising may not be so clear. So will the final multiple regression model contain both $x$ variables or only one? In the following sections, you can find out.

# Pinpointing Possible Relationships

The third step in doing a multiple regression analysis (see the list in the "Stepping through the analysis" section) is to find out which (if any) of your possible *x* variables are actually related to *y*. If an *x* variable has no relationship with *y*, including it in the model is pointless. Data analysts use a combination of scatterplots and correlations to examine relationships between pairs of variables (as you can see in Chapter 4). While these two techniques can be viewed under the heading of looking for relationships, I walk you through each one separately in the following sections to discuss their nuances.

## Making scatterplots

You make scatterplots in multiple linear regression to get a handle on whether your possible *x* variables are even related to the *y* variable you're studying. To investigate these possible relationships, you make one scatterplot of each *x* variable with the response variable *y*. If you have *k* different *x* variables being considered for the final model, you make *k* different scatterplots.

To make a scatterplot in Minitab, enter your data in columns, where each column represents a variable and each row represents all the data from one individual. Go to Graph>Scatterplots>Simple. Select your *y* variable on the left-hand side and click Select. That variable appears in the *y*-variable box on the right-hand side. Then select your *x* variable on the left-hand side and click Select. That variable appears in the *x*-variable box on the right-hand side. Click OK.

Scatterplots of TV ad spending versus TV sales and newspaper spending versus TV sales are shown in Figure 5-1.

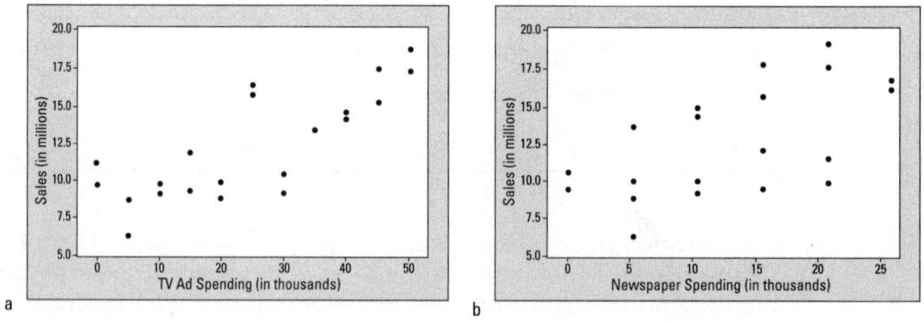


**Figure 5-1:** Scatterplots of TV and newspaper ad spending versus plasma TV sales.

You can see from Figure 5-1a that TV spending does appear to have a fairly strong linear relationship with sales. This observation gives evidence that TV ad spending may be useful in estimating plasma TV sales. Figure 5-1b shows a linear relationship between newspaper ad spending and sales, but the relationship isn't as strong as the one between TV ads and sales. However it may be somewhat helpful in estimating sales.

## Correlations: Examining the bond

The second portion of step three involves calculating and examining the correlations between the *x* variables and the *y* variable. (Of course, if a scatterplot of an *x* variable and the *y* variable fails to come up with a pattern, then you drop that *x* variable altogether and don't proceed to find the correlation.)

Whenever you employ scatterplots to explore possible linear relationships, correlations are typically not far behind. The *correlation coefficient* is a number that measures the strength and direction of the linear relationship between two variables, *x* and *y*. (See Chapter 4 for all the information you need on correlation.) This process involves two parts:

- Finding and interpreting the correlations
- Testing the correlations to see which ones are statistically significant (thereby determining which *x* variables are significantly related to *y*)

I explain these two steps in the following sections.

### Finding and interpreting correlations

You can calculate a set of all possible correlations between all pairs of variables in Minitab. This set of all possible correlations between all pairs of variables in a given set is called a *correlation matrix*. You can see the correlation matrix output for the TV data from Table 5-1 in Figure 5-2. You can see the correlations between the *y* variable (sales) and each *x* variable (TV = TV ads; and Newspaper = newspaper ads). You also get the correlation between TV ads and newspaper ads.

**Figure 5-2:** Correlation values and *p*-values for the TV sales example.

```
Correlations: Sales, TV, Newspaper

                Sales   TV
TV              0.791
                0.000


Newspaper       0.594 0.058
                0.004 0.799
```

Minitab can find a correlation matrix between any pairs of variables in the model, including the *y* variable and all the *x* variables as well. To calculate a correlation matrix for a group of variables in Minitab, first enter your data in columns (one for each variable). Then go to Stat>Basic Statistics>Descriptive Statistics>Correlation. Highlight the variables from the left-hand side for which you want correlations, and click on Select. Typically you also want to test those correlations, so check the Display *p*-values box as well. (I discuss how to interpret those *p*-values later in this section.)

To interpret the values of the correlation matrix from the computer output, intersect the row and column variables you want to find the correlation for, and the top number in that intersection is the correlation of those two variables. (I discuss the bottom number later in this section.) For example, the correlation between TV ads and TV sales is 0.791, because it intersects the TV row with the Sales column in the correlation matrix in Figure 5-2. This result indicates a fairly strong positive linear relationship between these two variables. (That is, as dollars spent on TV ads increase, so do plasma TV sales.) You can also see that the correlation between newspaper ads and plasma TV sales is 0.594, showing a moderately strong positive linear relationship. This correlation isn't as strong as that of the TV ads, but it's still worth examining further. These results together indicate that TV and newspaper ads are each somewhat related to TV sales.

### Testing correlations for significance

Many times in statistics a rule-of-thumb approach to interpreting a correlation coefficient is sufficient. However, you're in the big leagues now, so you need a more precise tool for determining whether or not a correlation coefficient is large enough to be statistically significant — that's the real test of any statistic. Not that the relationship is fairly strong or moderately strong in the sample, but whether or not the relationship can be generalized to the population.

Now that phrase *statistically significant* should ring a bell in your memory. It's your old friend the hypothesis test calling to you (see Chapter 3 for a brush-up on hypothesis testing). Just like a hypothesis test for the mean of a population or the difference in the means of two populations, you also have a test for the correlation between two variables within a population.

The null hypothesis to test a correlation is Ho: $\rho = 0$ versus Ha: $\rho \neq 0$. If you can't reject Ho based on your data, you can't conclude that the correlation between *x* and *y* differs from zero, indicating you don't have evidence that the two variables are related and *x* shouldn't be in the multiple regression model. However, if you can reject Ho, you conclude that the correlation isn't equal to zero, based on your data, so the variables are related. More than that, their relationship is deemed to be statistically significant; that is, the relationship would occur very rarely in your sample just by chance.

The letter $\rho$ is the Greek version of $r$ and represents the true correlation of $x$ and $y$ in the entire population; $r$ is the correlation coefficient of the sample.

Any statistical software package can calculate a hypothesis test of a correlation for you. The actual formulas used in that process are beyond the scope of this book. However the interpretation is the same as for any test: If the *p*-value is smaller than your prespecified value of $\alpha$ (typically 0.05), reject Ho and conclude $x$ and $y$ are related. Otherwise you can't reject Ho, and you conclude you don't have enough evidence that the variables are related.

In Minitab, you can conduct a hypothesis test for a correlation by clicking on Stat>Basic Statistics>Correlation, and checking the Display *p*-values box. Choose the variables you want to find correlations for, and click Select. You'll get output that is in the form of a little table that shows the correlations between the variables for each pair with the respective *p*-values under each one. You can see the correlation output for the ads and sales example in Figure 5-2.

Looking at Figure 5-2, the correlation of 0.791 between TV ads and sales has a *p*-value of 0.000, which means it's actually less than 0.001. That's a highly significant result, much less than 0.05 (your predetermined $\alpha$ level). So TV ad spending is strongly related to sales. The correlation between newspaper ad spending and sales was 0.594, which is also found to be statistically significant with a *p*-value of 0.004.

# Checking for Multicolinearity

You have one more very important step to complete in the relationship-exploration process before going on to using the multiple regression model. That is, you need to complete step four: looking at the relationship between the $x$ variables themselves and checking for redundancy. Failure to do so can lead to problems during the model-fitting process.

*Multicolinearity* is a term you use if two $x$ variables are highly correlated. Not only is it redundant to include both related variables in the multiple regression model, but it's also problematic. The bottom line is this: If two $x$ variables are significantly correlated, only include one of them in the regression model, not both. If you include both, the computer won't know what numbers to give as coefficients for each of the two variables, because they share their contribution to determining the value of $y$. Multicolinearity can really mess up the model-fitting process and give answers that are inconsistent and oftentimes not repeatable in subsequent studies.

To head off the problem of multicolinearity, along with the correlations you examine regarding each $x$ variable and the response variable $y$, also find the correlations between all pairs of $x$ variables. If two $x$ variables are highly correlated, don't leave them both in the model, or multicolinearity will result. To see the correlations between all the $x$ variables, have Minitab calculate a correlation matrix of all the variables (see the section "Finding and interpreting correlations"). You can ignore the correlations between the $y$ variable and the $x$ variables and only choose the correlations between the $x$ variables shown in the correlation matrix. Find those correlations by intersecting the rows and columns of the $x$ variables for which you want correlations.

If two $x$ variables $x_1$ and $x_2$ are strongly correlated (that is their correlation is beyond +0.7 or –0.7), then one of them would do just about as good a job of estimating $y$ as the other, so you don't need to include them both in the model.

Now if $x_1$ and $x_2$ aren't strongly correlated, then both of them working together would do a better job of estimating sales than either variable alone. For the ad spending example, you have to examine the correlation between the two $x$ variables, TV ad spending and newspaper ad spending, to be sure no multicolinearity is present. The correlation between these two variables (as you can see in Figure 5-2) is only 0.058. You don't even need a hypothesis test to tell you whether or not these two variables are related; they're clearly not. However, if you want to know, the $p$-value for the correlation between the spending for the two ad types is 0.799 (see Figure 5-2), which is much, much larger than 0.05 ever thought of being and therefore not statistically significant.

The large $p$-value for the correlation between spending for the two ad types confirms your thoughts that both variables together may be helpful in estimating $y$ because each makes its own contribution. It also tells you that keeping them both in the model will not create any multicolinearity problems. (This completes step four of the multiple regression analysis, as listed in the "Stepping through the analysis" section.)

# Finding the Best-Fitting Model

After you have a group of $x$ variables that are all related to $y$ and not related to each other (see previous sections), you're ready to perform step five of the multiple regression analysis (as listed in the "Stepping through the analysis" section). That is, you're ready to find the best-fitting model that fits the data.

In the multiple regression model with two $x$ variables, you have the general equation $y = b_0 + b_1x_1 + b_2x_2$, and you already know which $x$ variables to include in the model (by doing step four); the task now is to figure out which coefficients (numbers) to put in for $b_0$, $b_1$, and $b_2$, so you can use the resulting equation to estimate $y$. This specific model is the *best-fitting multiple linear regression model.* In this section, you see how to get, interpret, and test those coefficients in order to complete step five in the multiple regression analysis.

Finding the best-fitting linear equation is like finding the best-fitting line in simple linear regression, except that you're not finding a line. When you have two $x$ variables in multiple regression, for example, you're estimating a best-fitting plane for the data.

## Getting the multiple regression coefficients

In the simple linear regression model, you have the straight line $y = b_0 + b_1x$; the coefficient of $x$ is the slope, and it represents the change in $y$ per unit change in $x$. In a multiple linear regression model, the coefficients $b_1$, $b_2$, and so on quantify in a similar matter the sole contribution that each corresponding $x$ variable ($x_1$, $x_2$) makes in predicting $y$. The coefficient $b_0$ indicates the amount by which to adjust all of these values in order to provide a final fit to the data (like the $y$-intercept does in simple linear regression).

Computer software does all the nitty-gritty work for you to find the proper coefficients ($b_0$, $b_1$, and so on) that fit the data best. The coefficients that Minitab settles on to create the best-fitting model are the ones that as a group minimize the sum of the squared residuals (sort of like the variance in the data around the selected model). The equations for finding these coefficients by hand are too unwieldy to include in this book; a computer can do all the work for you. The results appear in the regression output in Minitab. You can find the multiple regression coefficients ($b_0$, $b_1$, $b_2$, . . . , $b_k$) on the computer output under the column labeled *Coef.*

To run a multiple regression analysis in Minitab, click on Stat>Regression>Regression. Then choose the response variable $(y)$ and click on Select. Then choose your predictor variables ($x$ variables), and click Select. Click on OK, and the computer will carry out the analysis.

For the plasma TV sales example from the previous sections, Figure 5-3 shows the multiple regression coefficients in the Coef column for the multiple regression model. The first coefficient (5.257) in Figure 5-3 is just the constant term (or $b_0$ term) in the model and isn't affiliated with any $x$ variable. This constant just sort of goes along for the ride in the analysis — the number that you tack on the end to make the numbers work out right. The second coefficient in the Coef column of Figure 5-3 is 0.162; this value is the coefficient of the $x_1$ (TV ads) term, also known as $b_1$. The third coefficient in the Coef column of Figure 5-3 is 0.249, which is the value for $b_2$ in the multiple regression model and is the coefficient that goes with $x_2$ (newspaper ad amount).

**Figure 5-3:** Regression output for the ads and plasma TV sales example.

```
The regression equation is
Sales = 5.267 + 0.162 TV ads + 0.249 Newsp ads

Predictor      Coef   SE Coef      T      P
Constant     5.2574    0.4984  10.55  0.000
TV ads      0.16211   0.01319  12.29  0.000
Newsp ads   0.24887   0.02792   8.91  0.000

S = 0.976613    R-Sq = 92.8%    R-Sq(adj) = 92.0%
```

Putting these coefficients into the multiple regression equation, you see the regression equation is Sales = 5.267 + 0.162 (TV ads) + 0.249 (Newspaper ads).

So you have your coefficients (no sweat, right?), but where do you go from here? What does it all mean? Keep reading.

## Interpreting the coefficients

In simple linear regression (Chapter 4), the coefficients represented the slope and *y*-intercept of the best-fitting line and were straightforward to interpret. The slope in particular represents the change in *y* due to a one-unit increase in *x*, because you can write any slope as a number over one (and slope is rise over run).

In the multiple regression model, the interpretation's a little more complicated. Due to all the mathematical underpinnings of the model and how it's finalized (believe me you don't want to go there unless you want a PhD in statistics), the coefficients have a different meaning.

REMEMBER

The coefficient of an *x* variable in a multiple regression model is the amount by which *y* changes if that *x* variable increases by one and the values of *all* other *x* variables in the model *don't change.* So basically, you're looking at the marginal contribution of each *x* variable when you hold the other variables in the model constant.

In the ads and sales regression analysis (see Figure 5-3), the coefficient of $x_1$ (TV ad spending) equals 0.16211. So $y$ (plasma TV sales) increases by 0.16211 million dollars when TV ad spending increases by \$1,000 and spending on newspaper ads doesn't change. (Note that keeping more digits after the decimal point reduces rounding error when in units of millions.)

You can more easily interpret the number 0.16211 million dollars by converting it to a dollar amount without the decimal point: \$0.16211 million is equal to \$162,110. (To get this value, I just multiplied 0.16211 by 1,000,000.) So plasma TV sales increases by \$162,110 for each \$1,000 increase in TV ad spending and newspaper ad spending remains the same.

Similarly, the coefficient of $x_2$ (newspaper ad spending) equals 0.24887. So plasma TV sales increases by 0.24887 million dollars (or \$248,870) when newspaper ad spending increases by \$1,000 and TV ad spending remains the same.

Don't forget the units of each variable in a multiple regression analysis. This mistake is one of the most common in intermediate statistics. If you forgot about units in the ads and sales example, you would think that sales increased by 0.24887 dollars with a dollar in newspaper ad spending!

Knowing the multiple regression coefficients ($b_1$ and $b_2$, in this case) and their interpretation, you can now answer the original question: Is the money spent on TV or newspaper ads worth it? The answer is a resounding *Yes!* Not only that, but you can also say how much you expect sales to increase per \$1,000 you spend on TV or newspaper advertising. Note that this conclusion assumes the model fits the data well. You have some evidence of that through the scatterplots and correlation tests, but more checking needs to be done before you can run to your manager and tell her the good news. (See the section "Testing the coefficients" to figure out what to do next.)

## Testing the coefficients

Another step in determining whether you have the right $x$ variables in your multiple regression model is to do a formal hypothesis test to make sure the coefficients are not equal to zero. Note that if the coefficient of an $x$ variable is zero, then when you put that coefficient into the model, you get zero times that $x$ variable (which equals zero). This result is essentially saying that if an $x$ variable's coefficient is equal to zero, you don't need that $x$ variable in the model.

The computer performs all the necessary hypothesis tests for the regression coefficients automatically with any regression analysis. Along with the regression coefficients you can find on the computer output, you see the test statistics and $p$-values for a test of each of those coefficients in the same row for each coefficient. Each one is testing Ho: Coefficient = 0 versus Ha: Coefficient $\neq$ 0.

The general format for finding a test statistic in most any situation is to take the statistic (in this case, the coefficient), subtract the value in Ho (zero), and divide by the standard error of that statistic (for this example, the standard error of the coefficient). (For more info on the general format of hypothesis tests, see Chapter 3.)

To test a regression coefficient, the test statistic (using the labels from Figure 5-3) is (Coef – 0)/SE Coef. In non-computer language, that means you take the coefficient, subtract zero, and divided by the standard error (SE) of the coefficient. The standard error of a coefficient here is a measure of how much the coefficient is expected to vary when you take a new sample. (See Chapter 3 for more on standard error.)

The test statistic has a *t*-distribution with $n - k - 1$ degrees of freedom, where $n$ equals the sample size and $k$ is the number of predictors ($x$ variables) in the model. This number of degrees of freedom works for any coefficient in the model (except you don't bother with a test for the constant, because it has no $x$ variable associated with it).

The test statistic for testing each coefficient is listed in the column marked *T* (because it has a *t*-distribution) on the Minitab output. You compare the value of the test statistic to the *t*-distribution with $n - k - 1$ degrees of freedom (using Table A-1 in the Appendix) and come up with your *p*-value. If the *p*-value is less than your prespecified $\alpha$ (usually 0.05), then you reject Ho and conclude that the coefficient of that $x$ variable isn't zero and that variable makes a significant contribution toward estimating $y$ (given the other variables are also included in the model). If the *p*-value is larger than 0.05, you can't reject Ho, so that $x$ variable makes no significant contribution toward estimating $y$ (when the other variables are included in the model).

In the case of the ads and plasma TV sales example, Figure 5-3 shows that the coefficient for the TV ads is 0.1621 (the second number in column two). The standard error is listed as being 0.0132 (the second number in column three). To find the test statistic for TV ads, take 0.1621 minus zero and divide by the standard error, 0.0132. You get a value of $t = 12.29$, which is the second number in column four). Comparing this value of $t$ to a *t*-distribution with $n - k - 1 = 22 - 2 - 1 = 19$ degrees of freedom (Table A-1 in the Appendix), you see the value of $t$ is way off the scale. That means the *p*-value is smaller than can be measured on Table A-1. Minitab lists the *p*-value in column five of Figure 5-3 as 0.000 (meaning it's less than 0.001). This result leads you to conclude that the coefficient for TV ads is statistically significant, and TV ads should be included in the model for predicting TV sales.

The newspaper ads coefficient is also significant with a *p*-value of 0.000 by the same reasoning; these results can be found by looking across the newspaper ads row of Figure 5-3. From this you should include both the TV ads variable and the newspaper ads variable in the model for estimating TV sales.

# Predicting Y by Using the X Variables

By now, you should have your multiple regression model. You're finally ready to complete step six of the multiple regression analysis: to predict the value of $y$ given a set of values for the $x$ variables. To make this prediction, you take those $x$ values for which you want to predict $y$, plug them into the multiple regression model, and simplify.

In the ads and plasma TV sales example (see analysis from Figure 5-3), the best-fitting model is $y = 5.26 + 0.162x_1 + 0.249x_2$. In the context of the problem, the model is Sales = 5.26 + 0.162 TV ad spending ($x_1$) + 0.249 newspaper ad spending ($x_2$).

Remember that the units for plasma TV sales is in millions of dollars and the units for ad spending for both TV and newspaper ads is in the thousands of dollars. That is, \$20,000 spent on TV ads means $x_1 = 20$ in the model. Similarly, \$10,000 spent on newspaper ads means $x_2 = 10$ in the model. Forgetting the units can lead to serious miscalculations.

Suppose you want to estimate plasma TV sales if you spend \$20,000 on TV ads and \$10,000 on newspaper ads. Plug $x_1 = 20$ and $x_2 = 10$ into the multiple regression model, and you get $y = 5.26 + 0.162(20) + 0.249(10) = 10.99$. In other words, if you spend \$20,000 on TV advertising and \$10,000 in newspaper advertising, you estimate that sales will be \$10.99 million dollars.

This estimate at least makes some sense in terms of the data shown in Table 5-1. At location ten, they spent \$20,000 on TV ads and \$5,000 on newspaper ads (short of what you had) and got sales of \$9.82 million. Location eleven spent a little more on TV ads and a lot more on newspaper ads than what you had, and got sales of \$16.28 million. Your spending amounts fall between the amounts of locations ten and eleven, and your estimated sales fall in between theirs also.

Be careful to put in only values for the $x$ variables that fall in the range of where the data lies. In other words, Table 5-1 shows data for TV ad spending between \$0 and \$50,000; newspaper ad spending goes from \$0 to \$25,000. It would not be appropriate, say, to try to estimate sales for spending amounts of \$75,000 for TV ads and \$50,000 for newspaper ads, respectively. The reason is that the regression model you came up with only fits the data that you collected; you have no way of knowing whether that same relationship continues outside that area. This no-no of estimating $y$ for values of the $x$ variables outside their range is called *extrapolation.* As one of my colleagues says, "Friends don't let friends extrapolate."

# Checking the Fit of the Model

Before you run to your boss in triumph saying you've slam-dunked the question of how to estimate plasma TV sales, you first have to make sure all your *i*'s are dotted and all your *t*'s are crossed, as you do with any other statistical procedure. In this case, you have to check the conditions of the multiple regression model. These conditions mainly focus on the *residuals* (the difference between the estimated values for *y* and the observed values of *y* from your data). If the model is close to the actual data you collected, you can feel somewhat confident that if you collected more data, it would fall in line with the model as well, and your predictions shouldn't be too bad.

In this section, you see what the conditions are for multiple regression, and specific techniques statisticians use to check each of those conditions. The main character in all of this condition checking is the residual.

## Noting the conditions

The conditions for multiple regression concentrate on the error terms, or residuals. The residuals are the amount that's left over after the model has been fit. They represent the difference between the actual value of *y* observed in the data set and the estimated value of *y* based on the model. The conditions of the multiple regression model are the following (note that all need to be met in order to give the go-ahead for a multiple regression model):

- The residuals have a normal distribution with mean zero.
- The residuals have the same variance for each fitted (predicted) value of *y*.
- The residuals are independent (don't affect each other).

## Plotting a plan to check the conditions

It may sound like you have a ton of things to check here and there, but luckily, Minitab gives you all the info you need to know in a series of four graphs, all presented at one time. These plots are called the *residual plots,* and they graph the residuals against the values of a normal distribution to see whether the normality condition fits.

You can get the set of residual plots in two flavors:

- **Regular residuals:** The regular residual plots (the vanilla-flavored ones) show you exactly what the residuals are for each value of *y*. Figure 5-4 shows the plots of the regular residuals for the TV sales example. Use these plots if you want to mainly look for patterns in the data.
- **Standardized residuals:** The standardized residual plots (the strawberry-flavored kind) take each residual and convert it to a *Z*-score by subtracting the mean and dividing by the standard deviation of all the residuals. Figure 5-5 shows the plots of the standardized residuals for the TV sales example. Use these plots if you want to not only look for patterns in the data, but you want to assess the standardized values of the residuals in terms of values on a *Z*-distribution to check for outliers. (Most statisticians use standardized residual plots.)

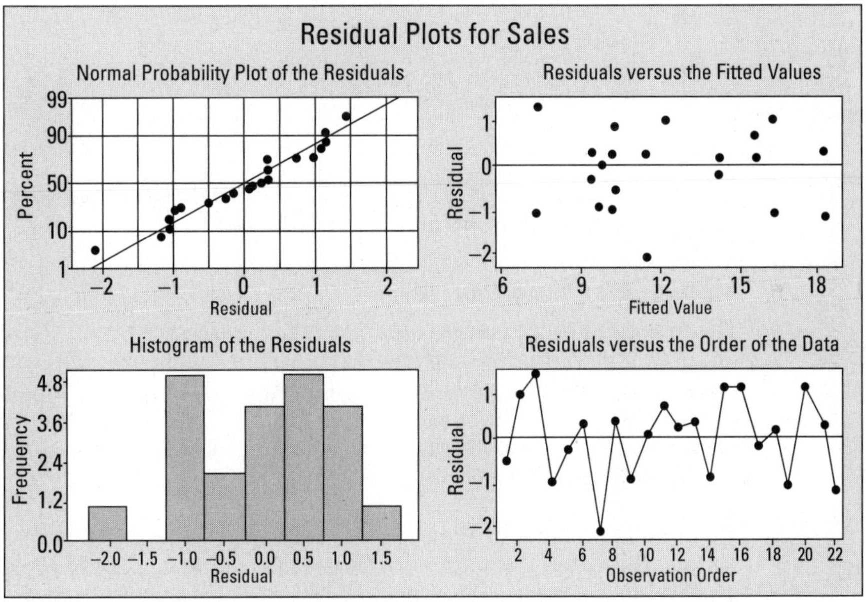


**Figure 5-4:** Residual plots for the ads and plasma TV sales example.

To make residual plots in Minitab, go to Stat>Regression>Regression. Select your response *(y)* variable and your predictor variables *(x)* variables. Click on Graphs, and choose either Regular or Standardized for the residuals, depending on which one you want. Then click on Four-in-one, which indicates you want to get all four residual plots shown in Figure 5-4 (using regular residuals) and Figure 5-5 (using standardized residuals).

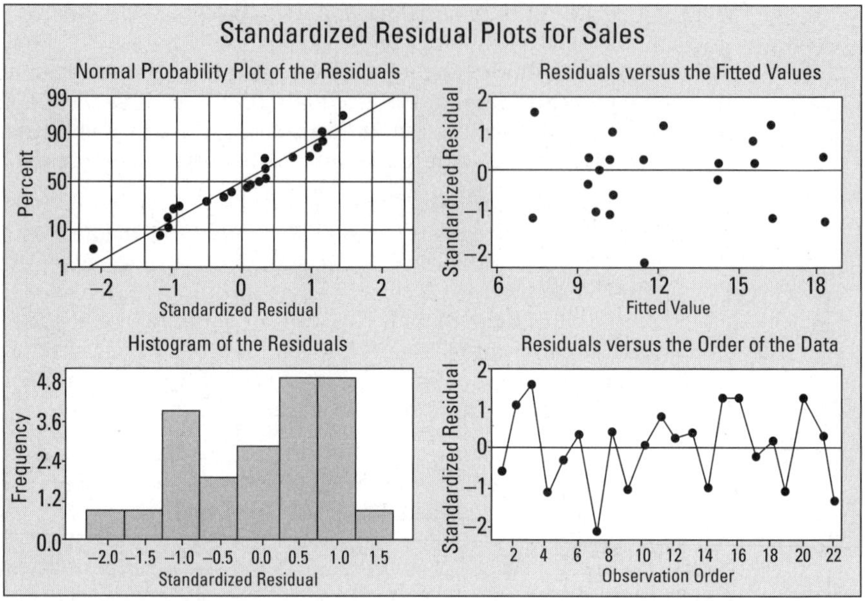


**Figure 5-5:** Standardized residual plots for the ads and plasma TV example.

In the following sections, you see how to check the residuals to see whether these three conditions are met in your data set.

### *Meeting the first condition*

The first condition to meet (outlined in the previous section "Noting the conditions") is that the residuals must have a normal distribution with mean zero. The upper-left plot of Figure 5-4 shows how well the residuals match a normal distribution. If the residuals fall in a straight line, that means the normality condition is met. By the looks of this plot, I'd say that condition is met for the ad and sales example.

The upper-right plot of Figure 5-4 shows what the residuals look like for the various estimated *y* values. Look at the horizontal line going across that plot; it's at zero as a marker. The residuals should average out to be at that line (zero). This Residuals versus Fitted Values plot checks the mean-of-zero condition and holds for the ads and sales example looking at Figure 5-4.

REMEMBER

You need the regular residual plots to see whether the mean of the residuals equals zero (via the plot on the upper right of Figure 5-4). If you look at the standardized residuals, they will always have mean zero due to the fact that they have been standardized to have a mean of zero. If the mean of zero condition isn't met for the regular residuals, that means that many of the estimated values are off in the same direction by a certain amount, which would not be good.

As an alternative check for normality apart from using the regular residuals, you can look at the standardized residuals plot (Figure 5-5) and check out the upper-right plot. It shows how the residuals are distributed across the various estimated (fitted) values of *y*. Standardized residuals are supposed to follow a standard normal distribution. That is, they should have mean zero and standard deviation one. So when you look at the standardized residuals, they should be centered around zero in a way that has no predictable pattern, with the same amount of variability around the horizontal line that crosses at zero as you move from left to right.

You should also find looking at the upper-right plot of Figure 5-5 that most (95 percent) of the standardized residuals fall within two standard deviations of the mean, which in this case is –2 to +2 (via the 68-95-99.7 Rule — remember that from intro stats?). You should see more residuals hovering around zero (where the middle lump would be on a standard normal distribution), and you should have fewer and fewer of the residuals as you go away from zero. The upper-right plot in Figure 5-5 confirms a normal distribution for the ads and sales example on all the counts I just mentioned.

The lower-left plot of Figures 5-4 and 5-5 show histograms of the regular and standardized residuals, respectively. These histograms should reflect a normal distribution; that is, the shape of the histograms should be approximately symmetric and look like a bell-shaped curve. Note that if the data set is small (as is the case here with only 22 observations), the histogram may not be as close to normal as you would like; in that case, consider it part of the body of evidence that all four residual plots show you. The histograms shown in the lower-left plots of Figure 5-4 and 5-5 aren't terribly normal looking; however, because you can't see any glaring problems with the upper-right plots, don't be worried.

### Satisfying the second condition

To look at the variance issue (condition two from a previous section), you can look again at the upper-right plot of Figure 5-4 (or Figure 5-5). You shouldn't see any change in the amount of spread (variability) in the residuals around that horizontal line as you move from left to right. Looking at Figure 5-4, the upper-right graph, you can see no reason to say that condition number two (the residuals have the same variance for each combination of the *x* variables) hasn't been met.

One particular problem that raises a red flag is if the residuals fan out, or increase in spread, as you move from left to right on the upper-right plot. This fanning out means that the variability increases more and more for higher and higher predicted values of *y*, so the condition of equal variability around the fitted line isn't met, and the regression model would not fit well in that case.

### *Checking the third condition*

The third condition is that the residuals are independent (in other words, they don't affect each other). Looking at the lower-right plot on either Figure 5-4 or 5-5, you can see the residuals plotted by *observation number,* which is the order in which the data came in the sample. If you see a pattern (if you were to connect the dots so to speak, you get a straight line, or a curve, or any kind of predictable up or down trend), you have trouble. You can see no patterns in the lower-right plots, so the independence condition is met for the ads and plasma TV sales example.

If the data must be collected over time, such as stock prices over a ten-year period, the independence condition may be a big problem because the data from the previous time period may be related to the data from the next time period. This kind of data requires time series analysis and is beyond the scope of this book.

# Chapter 6

# One Step Forward and Two Steps Back: Regression Model Selection

### *In This Chapter*

- Evaluating different methods for choosing a multiple regression model
- Understanding how forward selection and backward selection works
- Using the best subsets methods to find a good model

Suppose you're trying to estimate some quantitative variable, *y,* and you have many *x* variables available at your disposal. You have so many variables related to *y,* in fact, that you feel like I do in my job every day — overwhelmed with opportunity. Where do you go? What do you do? Never fear, this chapter is for you.

In this chapter, you see three different procedures statisticians use to find a best possible model — forward selection, backward selection, and best subsets selection. Each procedure can lead you to a different final model, and you can't find one single procedure that everyone agrees is *the one* to use. Each selection method has positives and negatives associated with it, as you can see in this chapter. No matter what method you choose, each method has the same goal: to get the best possible model for *y* by using a set of *x* variables. Yet the road that each procedure takes to get there is a bit different, so read on!

REMEMBER

Note that the term *best* has many connotations here. You can't find one end-all-be-all model that everyone comes up with in the end. That is to say that each data analyst can come up with a different model, and each model still does a good job of predicting *y.*

# Getting a Kick out of Estimating Punt Distance

Before you jump into a model selection procedure to predict *y* by using a set of *x* variables, you have to do some legwork. The variable of interest is *y,* and that's a given. But where do the *x* variables come from? How do you choose which ones to investigate as being possible candidates for predicting *y?* And how do those possible *x* variables interact with each other toward making that prediction? All of these questions must be answered before any model selection procedure can be used. However, this part is the most challenging and the most fun; a computer can't think up *x* variables for you!

Suppose you're at a football game and the opposing team has to punt the ball. You see the punter line up and get ready to kick the ball, and a question comes to you. "Gee, I wonder how far this punt will go? I wonder what factors influence the distance of a punt? Can I use those factors in a multiple regression model to try to estimate punt distance? Hmm, I think I'll consult my *Intermediate Statistics For Dummies* book on this and analyze some data during half-time. . . ." Well, maybe that's pushing it, but it's still an interesting question for football players, golfers, soccer players, and even baseball players. Everyone's looking for more distance and a way to get it.

In the following sections, you can see how to identify and assess different *x* variables in terms of their potential contribution to predicting *y.*

## Brainstorming variables and collecting data

Starting with a blank slate and trying to think of a set of *x* variables that may be related to *y* may sound like a daunting task, but in reality, this task is probably not as bad as you think. Most researchers who are interested in predicting *y* in the first place have some ideas about which variables may be related to it. After you come up with a set of logical possibilities for *x,* you collect data on those variables, as well as *y,* to see what their actual relationship with *y* may be.

The Virginia Polytechnic Institute did a study to try to estimate the distance of a punt in football (something Ohio State fans aren't familiar with). Possible variables they thought may be related to the distance of a punt included the following: hang time (time in the air, in seconds), right leg strength (measured in pounds of force), left leg strength (in pounds of force), right leg flexibility (in degrees), left leg flexibility (in degrees), and overall leg strength (in pounds). The data collected on a sample of 13 punts (by right-footed punters) is shown in Table 6-1. (Distance is measured in feet.)

**Table 6-1 Data Collected for Punt Distance Study**

| *Distance* | *Hang* | *R Strength* | *L Strength* | *R Flexibility* | *L Flexibility* | *O Strength* |
|---|---|---|---|---|---|---|
| 162.50 | 4.75 | 170 | 170 | 106 | 106 | 240.57 |
| 144.00 | 4.07 | 140 | 130 | 92 | 93 | 195.49 |
| 147.50 | 4.04 | 180 | 170 | 93 | 78 | 152.99 |
| 163.50 | 4.18 | 160 | 160 | 103 | 93 | 197.09 |
| 192.00 | 4.35 | 170 | 150 | 104 | 93 | 266.56 |
| 171.75 | 4.16 | 150 | 150 | 101 | 87 | 260.56 |
| 162.00 | 4.43 | 170 | 180 | 108 | 106 | 219.25 |
| 104.93 | 3.20 | 110 | 110 | 86 | 92 | 132.68 |
| 105.67 | 3.02 | 120 | 110 | 90 | 86 | 130.24 |
| 117.59 | 3.64 | 130 | 120 | 85 | 80 | 205.88 |
| 140.25 | 3.68 | 120 | 140 | 89 | 83 | 153.92 |
| 150.17 | 3.60 | 140 | 130 | 92 | 94 | 154.64 |
| 165.17 | 3.85 | 160 | 150 | 95 | 95 | 240.57 |

Other variables you may think of that are related to punt distance may include the direction and speed of the wind at the time of the punt, the angle at which the ball was snapped, the average distance of punts made in the past by this punter, whether the game is at home or away in a hostile environment, and so on. However, these researchers seem to have enough information on their hands to build a model to estimate punt distance. For the sake of simplicity, you can assume the kicker is right-footed, which isn't always the case, but it represents the overwhelming majority of kickers.

Looking just at this raw data set in Table 6-1, you can't figure out which variables, if any, are related to distance of the punt or how those variables may be related to punt distance. You need more analyses to get a handle on this.

## Examining scatterplots and correlations

After you've identified a set of possible *x* variables, the next step is to find out which of these variables are highly related to *y* in order to start trimming down the set of possible candidates for the final model. In the punt distance example, the goal is to see which of the six variables in Table 6-1 are strongly related to punt distance. The two ways to look at these relationships are the following:

- **Scatterplots:** A graphical technique
- **Correlation:** A one-number measure of the linear relationship between two variables

Both of these elements are important, and I discuss each of them in the following sections.

### Seeing relationships through scatterplots

To begin examining the relationships between the *x* variables and *y,* you use a series of scatterplots. Figure 6-1 shows all the scatterplots, not only of each *x* variable with *y,* but each *x* variable with itself. The scatterplots are in the form of a *matrix,* which is a table made of rows and columns. For example, the first scatterplot in row two of Figure 6-1 looks at the variables of distance (which appears in column one) and hang time (which appears in row two). This scatterplot shows a possible positive (uphill) linear relationship between distance and hang time.

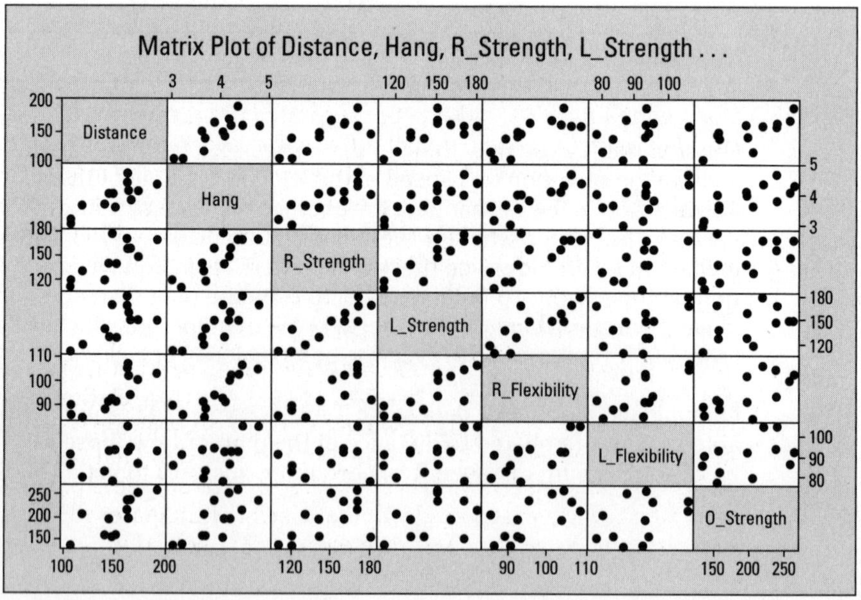


**Figure 6-1:** A matrix of all scatterplots between pairs of variables in the punting distance example.

Note that Figure 6-1 is essentially a symmetric matrix across the diagonal line. That is, the scatterplot for distance and hang time is the same as the scatterplot for hang time and distance; the *x* and *y* axes are just switched. The essential relationship shows up either way. So you only have to look at all the scatterplots below the diagonal (where the variable names appear) or all the scatterplots above the diagonal. You need not examine both.

To get a matrix of all scatterplots between a set of variables in Minitab, go to Graph>Matrix Plot> and choose Matrix of Plots>Simple. Highlight all the variables in the left-hand box for which you want scatterplots by clicking on them; click Select, and then click OK. You will see the matrix of scatterplots with a format similar to Figure 6-1.

Looking across row one of Figure 6-1, you can see that all the variables seem to have a positive linear relationship with punt distance except left leg flexibility. Perhaps the reason left leg flexibility isn't much related to punt distance is because the left foot is planted into the ground when the kick is made — for a right-footed kicker, the left leg doesn't have to be nearly as flexible as the right leg, which does the kicking. So it doesn't appear that left leg flexibility contributes a great deal to the estimation of punt distance on its own.

You can also see in Figure 6-1 that the scatterplots showing relationships between pairs of *x* variables are to the right of column one and below row one. (Remember you need to look on only the bottom part of the matrix or the top part of the matrix to see the relevant scatterplots.) It appears that hang time is somewhat related to each of the other variables (except left leg flexibility, which doesn't contribute to estimating *y*). So hang time could possibly be the most important single variable in estimating the distance of a punt.

You also need to look at the scatterplots showing the relationships between each pair of *x* variables. It's important to be mindful that if two *x* variables are strongly related to each other, then including them both in the model is not a good idea. First, adding the second of those two variables adds virtually nothing toward helping predict *y*. But more important than that, if two *x* variables are highly correlated and both are included in the model, the computer gets confused and doesn't know how much of the model to attribute to which *x* variable. This problem is called *multicolinearity.* (See Chapter 5 for more on how you can spot multicolinearity and avoid it.)

### Finding connections by using correlations

Scatterplots can give you some general ideas as to whether two variables are related in a linear way. However, pinpointing that relationship requires a numerical value to tell you how strongly the variables are related (in a linear fashion) as well as the direction of that relationship. That numerical value is the *correlation* (also known as *Pearson's correlation;* see Chapter 4). So the next step toward trimming down the possible candidates for *x* variables is to calculate the correlation between each *x* variable and *y*.

To get a set of all the correlations between any set of variables in your model by using Minitab, go to Stat>Basic Statistics>Correlation. Then highlight all the variables you want correlations for and click Select. (To include the *p*-values for each correlation, click the Display *p*-values box.) Then click OK. You can see a listing of all the variables' names across the top row and down the first column. Intersect the row depicting the first variable with the column depicting the second variable, and you can find the correlation for that pair.

Table 6-2 shows the correlations you can calculate between $y$ = punt distance and each of the $x$ variables. These results confirm what the scatterplots were telling you. Distance seems to be related to all the variables except left leg flexibility, because that's the only variable that didn't have a statistically significant correlation with distance using the $\alpha$ level 0.05. (For more info on the test for correlation, see Chapter 5.)

**Table 6-2 Correlations between Distance of a Punt and Other Variables**

| *X Variable* | *Correlation with Punt Distance* | *P-value* |
|---|---|---|
| Hang time | 0.819 | 0.001* |
| Right leg strength | 0.791 | 0.001* |
| Left leg strength | 0.744 | 0.004* |
| Right leg flexibility | 0.806 | 0.001* |
| Left leg flexibility | 0.408 | 0.167 |
| Overall leg strength | 0.796 | 0.001* |

** statistically significant at level $\alpha = 0.05$*

If you take a look at Figure 6-1, you can see that hang time is related to other variables such as right foot and left foot strength, right leg flexibility, and so on. This is where things start to get sticky. You have hang time related to distance, and lots of other variables related to hang time. While hang time is clearly the most related to distance, the final multiple regression model may not include hang time. Here's one possible scenario: You find a combination of other $x$ variables that can do a good job estimating $y$ together. And all of those other variables are strongly related to hang time. This result might mean that in the end you don't need to include hang time in the model. Strange things happen when you have many different $x$ variables to choose from.

After you narrow down the set of possible $x$ variables for inclusion in the model to predict punt distance, the next step is to put those variables through a selection procedure of some sort, which trims down the list to a set of essential variables for predicting $y$. The next sections show various techniques for going through this model selection process.

# Using the Forward Model Selection Procedure

The first of the three model selection procedures I present in this chapter is called *forward selection.* This process gives a systematic way of selecting a good model to predict *y.* It starts out with no variables at all, and then adds one variable, then another one, and then another one — each time including the variable that contributes the highest amount toward estimating *y,* given the other variables that are already in the model.

This section shows you how the forward selection procedure works for selecting a final regression model, and what the philosophy is for doing so. It also shows you how to assess the fit of the final model by using some new criterion.

## Adding variables — one at a time

The forward selection procedure starts with a model that contains no *x* variables and then adds *x* variables one at a time until the final model has been reached.

Here's how the forward selection procedure works in general, but before the hair begins to stand up on the back of your neck, note that Minitab or any other statistical software takes care of all the heavy lifting used for this and all the other model selection procedures:

1. **Choose a prespecified value of $\alpha$ for determining when to add a variable to the model.**

   This $\alpha$ is called the *entry level* for a variable. Typically you want to choose the value $\alpha = 0.05$ or 0.10 as the entry level. The higher the $\alpha$ level, the easier it is to add a variable to the model.

2. **Start with the model containing no variables: $y = b_0$.**

   You are left with just the constant $b_0$ term.

3. **Go through each possible *x* variable that could be included in the model and test each one's coefficient to see whether it's statistically significant by using a *t*-test.**

   If the variable is statistically significant, it has a significant contribution to determining *y,* given that the rest of the variables in the model are fixed. Any variable that isn't statistically significant is out of the running to be added to the model at this point. (See Chapter 5 on conducting *t*-tests for regression coefficients.)

4. **Examine the *p*-values from each of the *t*-tests in step three (listed on the Minitab output) and choose the smallest one.**

   The variable associated with that *p*-value is the best candidate to be added to the model, because that variable is the most statistically significant of all the possible *x* variables at this point.

5. **If the *p*-value for the *x* variable found in step four is smaller than the prespecified $\alpha$, add that *x* variable to the model.**

   After the first round, you have the model $y = b_0 + b_i x_i$ where $x_i$ refers to the first *x* variable you added to the model.

6. **Repeat steps three through five, using the new model from step five, and keep adding variables one at a time as long as the smallest *p*-value of each round is less than the prespecified $\alpha = 0.05$.**

   If the smallest *p*-value is larger than the prespecified $\alpha$, don't add any more variables to the model and stop the forward selection process. Your final model contains all of the *x* variables that were added during each phase of the forward selection process.

To find a best multiple linear regression model by using the forward selection procedure in Minitab, go to Stat>Regression>Stepwise. Highlight which variable is the response *(y)* variable and click Select. This variable will show up in the Response box. Then highlight which variables are the predictor *(x)* variables and click Select. These variables will show up in the Predictor box. Click on Methods, and click on Forward Selection. In the Alpha to Enter box, put in your prespecified value of $\alpha$ you want to require to allow an *x* variable to be included in the model. Typically statisticians would set this value at between 0.05 and 0.10. (I use 0.05.) This prespecified $\alpha$ level is called the *entry level* for the forward selection procedure. The higher the entry level, the easier it is for a variable to be entered, but the greater chance that the variable was just significant by random chance. (In the *F*-value box, the default is 4.0, which should be fine. The *F*-value is beyond the scope of this book in this context, although you do work with it when you do analysis of variance — see Chapter 10. ) Click OK and you get the output from the forward selection procedure.

You use a prespecified $\alpha$ level as the entry criteria for adding a variable because it represents the chance of making a Type I error and inadvertently putting in a variable based on your sample when it shouldn't be included. (See Chapter 3 for more on Type I errors.) You choose a small $\alpha$ level because you don't want to make it too easy to add a variable, because it increases the chance of adding something that isn't truly meaningful. (You have to put a lid on it somehow!)

## How well does the model fit?

The details regarding the formulas used behind the model selection procedures in this chapter are beyond the scope of this book. However, knowing what the procedure is doing and how to interpret the results are what's most important. To assess the fit of any multiple regression model, you can use the following three techniques: $R^2$, $R^2$ adjusted, and Mallows's C-p. You can find all three on the bottom line of the Minitab output when you do any sort of model selection procedure.

I describe these techniques in the following:

- **$R^2$:** $R^2$ is the percentage of the variability in the *y* values that's explained by the model. It falls between 0 percent and 100 percent (0 and 1.0). Values closer to 0 mean the model doesn't do a good job of explaining *y*. Values closer to 1.0 mean the model does an excellent job. Typically, I say that you can consider $R^2$ values higher than 0.70 to be good.
- **$R^2$ adjusted:** $R^2$ adjusted is the value of $R^2$, adjusted down for a higher number of variables in the model (which makes it much more useful than the regular value of $R^2$). A high value of $R^2$ adjusted means the model you have is fitting the data very well. I typically find a value of 0.70 to be considered high for $R^2$ adjusted.
- **Mallow's C-p:** Mallow's C-p is another measure of how well a model fits. It basically looks at how much error is left unexplained by a model with *k* predictor *(x)* variables compared to the average error left over from the full model (with all the *x* variables) and adjusts it for the number of variables in the model. The smaller Mallow's C-p is, the better. Because when it comes to the amount of error in your model, less is more.

Always use $R^2$ adjusted rather than the regular $R^2$ to assess the fit of a multiple regression model. With every addition of a new variable into a multiple regression model, the value of $R^2$ stays the same or increases; it will never go down. That's because a new variable will either help explain some of the variability in the *y*'s (thereby increasing $R^2$ by definition), or it will do nothing (leaving $R^2$ exactly where it was before). So theoretically, you could just keep adding more and more variables into the model just for the sake of getting a larger value of $R^2$. Here's why the $R^2$ adjusted is important: It keeps you from adding more and more variables by taking into account how many values are in the model. This way, the value of $R^2$ adjusted can actually decrease if the added value of the additional variable is outweighed by the number of variables in the model. This gives you an idea of how much or how little added value you get from a bigger model (bigger isn't always better).

The goal of any model selection procedure is to have the smallest number of *x* variables in the model as possible, with a high enough value of $R^2$ adjusted and a small enough Mallow's C-p to feel good about it.

## Applying forward selection to punt distances

To get a better feel for the forward selection procedure, you can apply it to the punt distance example. The researchers turn their data over to your capable hands for model selection. Using Minitab, you decide to apply the forward selection procedure to the punt distance data shown in Table 6-1, using an entry level of $\alpha = 0.05$. You can now examine your results, shown in Figure 6-2.

In this section, you see the step-by-step process Minitab used to come up with your results; you also see how to interpret those results in a way your client researchers will appreciate and understand (which is the goal of all things data analytical). You also get a heads up on how your choice of entry level can impact your results.

**Stepwise Regression: Distance versus Hang, R_Strength . . .**

```
Forward Selection. Alpha-to-Enter: 0.05
Response is Distance on 6 predictors, with N = 13

Step               1
Constant      -22.33

Hang            43.5
T-Value         4.73
P-Value        0.001

S               15.6
R-Sq           67.05
R-Sq(adj)      64.06
Mallows C-p      1.7
```

**Figure 6-2:** Forward selection results for the punt distance data with entry level 0.05.

### Breaking down the results

You can see in Figure 6-2 that the procedure you asked Minitab to use is forward selection (line one) and that you set the $\alpha$ level for entering a new variable to be 0.05. In line two, you can see the response *(y)* variable is distance, and you have six predictor *(x)* variables to start with, all based on a sample of $N = 13$ observations.

In the next part of the output, you see that at Step 1 the model has the constant listed as –22.33. You can also see it includes hang time as the first variable in the model. In the section "Exploring scatterplots and correlations," you can see that hang time is one of the more prominent variables, so you may not be surprised that it shows up in the model selection process right away.

The *p*-value of hang time is 0.001, indicating that the variable is significant (less than $\alpha = 0.05$). However, no Step 2 is in this output. That means after hang time was included, no other variables made a significant enough contribution beyond hang time. The other variables' *p*-values were all greater than 0.05.

The forward selection procedure's modus operandi is that you have to be in the in-crowd in order to be added to the model. The model is like an A-list in a way.

The final model for the punt distance data using the forward selection procedure with $\alpha = 0.05$ is $y = -22.33 + 43.50x$ where $y$ = punt distance and $x$ = punt hang time. Note that this is a simple linear regression model (Chapter 4 style), because it has only one $x$ variable in it.

You can now use this final model to predict punt distance by using hang time. Say the hang time is three seconds. That means the punt is expected to go $y = -22.33 + 43.50 * 3 = 108.17$ feet, or 36.06 yards. (Hang times for punts can range anywhere from 0 seconds if the punt is blocked to around 5.00 seconds (see Table 6-1), so don't put numbers into this equation like 8 seconds. That would make for an *unbelievable* punt distance — seriously!).

You can find the coefficient of an $x$ variable by looking at the value in the output directly across from the name of the variable. Under that value is the *t*-value of this coefficient, and its *p*-value follows.

### Looking at the fit of the final model

The value of $R^2$ adjusted for this model as shown in Figure 6-2 is 64.06 percent, which may not seem all that great. However, you're dealing with a simple linear regression model, and the value of $R$ in this case is the correlation coefficient between hang time and distance. This value of $R$ (denoted by small $r$ in its own simple regression context) is the square root of 0.6406, which is 0.80. This correlation is somewhat strong, actually, so the model fits fairly well. Mallow concurs, with a relatively small value of 1.7, as you can see on the last line of Figure 6-2.

### A cautionary word about entry level

So you can have an example where you see more than one variable added to a model via forward selection, I conducted a forward selection procedure on the punt distance data. I bumped the entry level of $\alpha$ up to 0.25. (Don't try this at home; it's much too high of an entry level for practical use. I reran the analysis, and I've included the results in Figure 6-3.

**Stepwise Regression: Distance versus Hang, R_Strength . . .**

Forward selection. Alpha-to-Enter: 0.25
Response is Distance on 6 predictors, with N = 13

| Step | 1 | 2 | 3 |
|---|---|---|---|
| Constant | -22.326 | -1.300 | 1.672 |
| Hang | 43.5 | 26.9 | 8.9 |
| T-Value | 4.73 | 2.07 | 0.50 |
| P-Value | 0.001 | 0.065 | 0.630 |
| O_Strength | | 0.22 | 0.24 |
| T-Value | | 1.69 | 1.86 |
| P-Value | | 0.122 | 0.096 |
| R_Strength | | | 0.44 |
| T-Value | | | 1.41 |
| P-Value | | | 0.191 |
| S | 15.6 | 14.4 | 13.7 |
| R-Sq | 67.05 | 74.38 | 79.03 |
| R-Sq(adj) | 64.06 | 69.26 | 72.04 |
| Mallows C-p | 1.7 | 1.3 | 1.8 |

**Figure 6-3:** Forward selection results for the punt data, using entry level 0.25.

Looking at Figure 6-3, you see the coefficient of the variables in the final model, located in the Step 3 column. The final model, using forward selection with this way-too-large entry level of $\alpha = 0.25$, is $y = 1.67 + 8.9x_1 + 0.24x_2 + 0.44x_3$ where $y$ = punt distance, $x_1$ = punt hang time, $x_2$ = overall leg strength, and $x_3$ = right leg strength. With this three-variable model, the $R^2$ adjusted is 72.04 percent (this number is found in Figure 6-3 in the third column, second value up from the bottom). This value of $R^2$ adjusted is a fairly small increase over the one-variable model you found by doing the forward selection procedure, using the more reasonable entry level of 0.05 (see Figure 6-2).

# Shifting into Reverse: The Backward Model Selection Procedure

The *backward selection procedure* for selecting a best multiple linear regression model works in a similar way as the forward selection procedure from the previous section. The big difference is that instead of starting with no *x* variables and adding $x$ variables one by one until you stop, you start with all the $x$ variables in the model and remove $x$ variables one by one until you stop. You may think that the forward selection procedure and the backward selection procedure would give you the same final model, but in many cases they don't, which you can discover in the sections that follow.

## Eliminating variables one by one

The backward selection procedure starts out with the full multiple regression model containing all of the $x$ variables (of which there are $k$ of them.) The starting model is $y = b_0 + b_1x_1 + \ldots + b_kx_k$. The object is to whittle down the model so it includes the fewest number of variables needed to still fit well. (Statisticians, as mysterious, mystical, and complicated as they may seem, actually like their models to be as simple as possible!)

The computer does all the work for all model selection procedures, but you have to set the criteria for when to allow a variable to be removed. You're also left standing with the output that needs to be interpreted. Don't worry though. It's all a step-by-step process that you take one at a time. (Hopefully those steps are forward and not backward, right? Right.)

In general, here's how the backward selection procedure works (note that Minitab does all the work for you on this procedure; all you have to do is interpret the results and understand the process by which those results were attained):

1. **Choose a prespecified value of $\alpha$ for determining when to remove a variable from the model.**

   In the backward selection procedure, you call $\alpha$ the *removal level.* Typically you want to choose the removal level $\alpha = 0.10$. The higher the $\alpha$ level, the easier it is to remove a variable from the model. Statisticians warn against using a removal level higher than the traditional value of 0.10 for fear of dropping variables out of the model too quickly, removing important contributions that may be made by those variables. However, if $\alpha$ is too small, the model could wind up being overly complex.

2. **Start with the model containing all of the $x$ variables: $y = b_o + b_1x_1 + b_2x_2 + \ldots + b_kx_k$, where $k$ is the total number of $x$ variables.**

   Remember that this model is called the *full model.*

3. **Conduct a $t$-test on the coefficient of each $x$ variable to see whether it's statistically significant (see Chapter 5 for conducting $t$-tests on coefficients of a multiple regression model), and note the $p$-value of each $t$-test.**

If the $x$ variable is statistically significant (its $p$-value is less than the preselected $\alpha$ level), it makes a significant contribution to determining $y$, given that the rest of the variables in the model are fixed. In that case, that $x$ variable remains a possible candidate for inclusion in the model at this point. If the $x$ variable isn't statistically significant, then it is considered for removal at this particular point.

4. **Find the variable with the largest *p*-value on the Minitab output.**

   This variable is the one that has the least contribution toward $y$ given the rest of the variables in the model.

5. **If the *p*-value for the variable found in step four is larger than the removal level, then remove the variable from the model.**

6. **Repeat steps three through five on the new model, removing one variable at a time; after the largest *p*-value from step four falls below the removal level, stop the backward selection process and don't remove that variable or more variables.**

   You now have your final model, which will include some subset of $x$ variables from the full model in step two.

To find a best multiple linear regression model by using the backward selection procedure in Minitab, go to Stat>Regression>Stepwise. Highlight the variable that is the response *(y)* variable, and click Select. Then highlight the variables that are the predictor *(x)* variables, and click Select. Click on Methods, and choose Backward Selection. Choose the $\alpha$ to remove (the removal level for a variable chosen by you). The *F*-value for removal has a default at 4.0, which should be fine. Click OK, and you get the output for the backward selection procedure similar to Figure 6-4.

## Assessing model fit

The fit of the models at each stage of the backward selection procedure are the same as those for the forward selection procedure in the previous section. The computer output shows you the value of $R^2$, the value of $R^2$ adjusted, and Mallow's C-p. (See an earlier section "How well does the model fit?" for more information on each of these measures.)

## Kicking variables out to estimate punt distance

This section applies the backward selection procedure to the punt distance data so you can see how the process works and how to interpret the results at each step. Note that each type of model selection procedure can produce a different final model, which is normal. After all, if all the techniques led you to the same result, why bother having more than one technique?

Using the punt distance data presented in Table 6-1, imagine that you analyzed the data by using the backward selection procedure with level of removal $\alpha = 0.10$. I show your results in Figure 6-4. Each stage in the model selection process is represented by a column in the results.

### Examining the x variables: The Step 1 column

The Step 1 column of Figure 6-4 shows all the $x$ variables in the model. Looking at the *p*-values in that first column, you can see that the largest one turns out to be 0.953. This *p*-value is associated with the left leg strength variable. (Check out the next section on the Step 2 column to find out what happens to this variable.)

Stepwise Regression: Distance versus Hang, R_Strength . . .

Backward elimination. Alpha-to-Remove: 0.1
Response is Distance on 6 predictors, with N = 13

| Step | 1 | 2 | 3 | 4 | 5 |
|---|---|---|---|---|---|
| Constant | -31.26 | -33.29 | -33.30 | -35.25 | 12.77 |
| Hang | 3 | 4 | | | |
| T-Value | 0.10 | 0.16 | | | |
| P-Value | 0.927 | 0.874 | | | |
| R_Strength | 0.28 | 0.29 | 0.33 | 0.39 | 0.56 |
| T-Value | 0.56 | 0.78 | 1.08 | 1.46 | 2.64 |
| P-Value | 0.596 | 0.461 | 0.310 | 0.178 | 0.025 |
| L_Strength | 0.04 | | | | |
| T-Value | 0.06 | | | | |
| P-Value | 0.953 | | | | |
| R_Flexibility | 1.24 | 1.28 | 1.34 | 0.86 | |
| T-Value | 0.79 | 0.96 | 1.10 | 0.99 | |
| P-Value | 0.457 | 0.371 | 0.303 | 0.346 | |
| L_Flexibility | -0.41 | -0.42 | -0.41 | | |
| T-Value | -0.50 | -0.57 | -0.59 | | |
| P-Value | 0.634 | 0.588 | 0.574 | | |
| O_Strength | 0.21 | 0.21 | 0.22 | 0.22 | 0.27 |
| T-Value | 1.21 | 1.50 | 1.87 | 2.00 | 2.71 |
| P-Value | 0.271 | 0.177 | 0.098 | 0.077 | 0.022 |
| S | 15.8 | 14.6 | 13.7 | 13.2 | 13.2 |
| R-Sq | 81.47 | 81.45 | 81.38 | 80.58 | 78.45 |
| R-Sq(adj) | 62.93 | 68.21 | 72.07 | 74.11 | 74.14 |
| Mallows C-p | 7.0 | 5.0 | 3.0 | 1.3 | -0.0 |

**Figure 6-4:** Backward selection procedure for estimating punt distance.

### Removing one variable: The Step 2 column

Notice in the Step 2 column of Figure 6-4 that the left leg strength variable no longer appears as a result (and it stays that way), because it has the highest *p*-value at Step 1 and that *p*-value is larger than the entry level of 0.10. This is the work of the backward selection procedure. It operates in the only-the-strong-survive mode when it comes to variable elimination.

In looking at the *p*-values for this new model in the Step 2 column, you see the variable with the highest *p*-value is hang time (0.874). This result doesn't make sense at first because in Table 6-2 you saw hang time had the strongest relationship with punt distance.

However, remember what the *p*-value represents here — the significance of the variable in its contribution to *y*, given all the other variables already in the model. Because so many of the other variables in the model were shown to be correlated with hang time (see Figure 6-1), it makes sense that hang time could possibly be eliminated somewhere near the beginning of this procedure.

### Working down to the final model: The Step 3 column and beyond

The Step 3 column of Figure 6-4 shows the model without left leg strength or hang time. The next variable to be removed is left leg flexibility, which has a *p*-value = 0.574. Looking at the Step 4 column of Figure 6-4, the next variable to be removed is right leg flexibility, which has a *p*-value of 0.346.

After right leg flexibility is removed from the model, you can see the result in Step 5 of Figure 6-4. All the remaining variables in the model have *p*-values smaller than the level for removal, which is 0.10. This means you stop the backward selection procedure and keep the model you've got. The final model for the punt distance data using the backward selection procedure with removal level 0.10 is $y = 12.77 + 0.56x_1 + 0.27x_2$, where $x_1$ = right leg strength and $x_2$ = overall leg strength. The final value of $R^2$ adjusted is 74.14 percent, which isn't all that bad. (I've seen higher values of $R^2$, but I've also seen a lot worse.) Mallow cheers this model on with a C-p value of 0, which has been rounded off a bit.

TIP

Always remember to use the $R^2$ adjusted rather than $R^2$ to assess the fit of your model at each step of any selection procedure, and here's why: In the punt distance example, the values of $R^2$ and $R^2$ adjusted appear on the second and third lines from the bottom of the Minitab output in Figure 6-4. You can see that with each step, the values of $R^2$ decrease because fewer variables are in the model to contribute something to predicting *y*. However, the values of $R^2$ adjusted increase because the adjustment needed for the number of variables in the model goes down. Each variable left in the model is providing more bang for the buck in terms of helping predict *y*.

# Using the Best Subsets Procedure

The *best subsets procedure* presents yet another way to find a best multiple regression model. It basically examines the fit of every single possible model that could be formulated from your *x* variables. You then use those model-fitting results to make a decision about which model is the best one to use.

In this section, you see how the best subsets procedure works for model selection in a step-by-step manner. Then you see how to take all the information given to you and wade through it to make your way to the answer — the best-fitting model based on a subset of the available *x* variables. Finally, you see how this procedure is applied to find a model to predict punt distance.

## Forming all models and choosing the best one

The best subsets procedure has fewer steps than the forward or backward selection model because the computer formulates and analyzes all possible models in a single step. In this section, you see how to get the results and then use them to come up with a best multiple regression model for predicting *y*.

Here are the steps for conducting the best subsets model selection procedure to select a multiple regression model (note that Minitab does all the work for you to crunch the numbers):

1. **Conduct the best subsets procedure in Minitab, using all possible subsets of the *x* variables being considered for inclusion in the final model (see the nearby Computer Output icon).**

   The output contains a listing of all models that contain one *x* variable, all models that contain two *x* variables, all models that contain three *x* variables, and so on, all the way up to the full model (containing all the *x* variables). Each model is presented in one row of the output.

2. **Choose the best of all the models shown in the best subsets Minitab output by finding the model with the largest value of $R^2$ adjusted and the smallest value of Mallow's C-p; if two competing models are about equal, choose the model with the fewer number of variables.**

   Mallow's C-p is a measure of the amount of error in the predicted values compared to the overall amount of variability in the data. If the model fits well, the amount of error in the predicted values is small compared to the overall variability in the data, and Mallow's C-p will be small. So look for a model that has a small value of Mallow's C-p compared to its competitors. $R^2$ adjusted measures how much of the variability in the *y*-values can be explained by the model, adjusted for the number of variables included. ($R^2$ adjusted ranges from 0 to 100 percent; see the section "How well does the model fit?" earlier in this chapter.) If the model fits well, $R^2$ adjusted is high. So you also want to look for the smallest possible model that has a high value of $R^2$ adjusted, and a small value of Mallow's C-p compared to its competitors. And if it comes down to two similar models, you always want to make your final model as easy to interpret as possible by selecting the model with the fewer variables.

To carry out the best subsets selection procedure in Minitab, go to Stat>Regression>Best Subsets. Highlight the response variable *(y),* and click Select. Highlight all the predictor *(x)* variables, and click Select. Click on OK.

## *Applying best subsets to the punt distance example*

Say that you analyzed the punt distance data by using the best subsets model selection procedure. Your results are shown in Figure 6-5. This section follows Minitab's footsteps in getting these results, and provides you with a guide for interpreting the results.

### *Pouring over the output*

Assuming that you already used Minitab to carry out the best subsets selection procedure on the punt distance data, you can now analyze the output from Figure 6-5. Each variable shows up as a column on the right side of the output. Each row represents the results from a model containing the number of variables shown in column one. The *X*'s at the end of each row tell you which variables were included in that model. The number of variables in the model starts at one and increases to six because six *x* variables are available in the data set.

The models with the same number of variables are ordered by their values of $R^2$ adjusted and Mallows C-p, from best to worst. The top-two models (for each number of variables) are included in the computer output.

For example, rows one and two of Figure 6-5 (both marked 1 in the Vars column) show the top-two models containing one *x* variable; rows three and four show the top two models containing two *x* variables (and so on). Finally the last row of Figure 6-5 shows the results of the full model containing all six variables. (Only one model contains all six variables, so you don't have a second-best model in this case.)

**Best Subsets Regression: Distance versus Hang, R_Strength . . .**

Response is Distance

| Vars | R-Sq | R-Sq(adj) | Mallows C-p | S | Hang | R_Strength | L_Strength | R_Flexibility | L_Flexibility | O_Strength |
|---|---|---|---|---|---|---|---|---|---|---|
| 1 | 67.1 | 64.1 | 1.7 | 15.570 | X | | | | | |
| 1 | 65.0 | 61.8 | 2.3 | 16.043 | | | | X | | |
| 2 | 78.5 | 74.1 | -0.0 | 13.206 | | X | | | | X |
| 2 | 78.2 | 73.8 | 0.1 | 13.294 | | | X | | | X |
| 3 | 80.6 | 74.1 | 1.3 | 13.214 | | X | | X | | X |
| 3 | 79.5 | 72.7 | 1.6 | 13.581 | | X | X | | | X |
| 4 | 81.4 | 72.1 | 3.0 | 13.724 | | X | | X | X | X |
| 4 | 80.7 | 72.0 | 3.3 | 13.977 | | X | X | X | | X |
| 5 | 81.5 | 68.2 | 5.0 | 14.643 | X | X | | X | X | X |
| 5 | 81.4 | 68.2 | 5.0 | 14.650 | | X | X | X | X | X |
| 6 | 81.5 | 62.9 | 7.0 | 15.812 | X | X | X | X | X | X |

**Figure 6-5:** Best subsets procedure results for the punt distance example.

Looking at the first two rows of Figure 6-5, the top one-variable model is the one including hang time only. The second-best one-variable model includes only right foot flexibility. The right foot flexibility model has a lower value of $R^2$ and a higher Mallow's C-p than the hang time model, which is why it's the second best.

Row three shows that the best two-variable model for estimating punt distance is the model containing right leg strength and overall leg strength. The best three-variable model is in row five. It shows that the best three-variable model includes right foot strength, right foot flexibility, and overall leg strength. The best four-variable model is found in row seven, and includes right foot strength, right and left foot flexibility, and overall foot strength. The best five-variable model is found in row nine and includes every variable except left foot strength. The only six-variable model is listed in the last row.

### Choosing the best model by using $R^2$ adjusted and Mallow's C-p

Now among the best one-variable, two-variable, three-variable, four-variable, and five-variable models, which one should you choose for your final multiple regression model? That is, which model is the best of the best? With all these results, it would be easy to have a major freak out over which one to pick, but never fear — Mallow's is here (along with his friendly sidekick, the $R^2$ adjusted).

Looking at Figure 6-5, you see that as the number of variables in the model increase, $R^2$ adjusted peaks out and then drops way off. That's because $R^2$ adjusted takes into account the number of variables in the model and reduces $R^2$ accordingly. You can see that $R^2$ adjusted peaks out at a level of 74.1 percent for two models. The corresponding models are the top two-variable model (right leg strength and overall leg strength) and the best three-variable model (right foot strength, right foot flexibility, and overall leg strength).

Now look at Mallow's C-p for these two models. Notice that Mallow's C-p is 0 for the two-variable model and 1.3 for the three-variable model. Both values are small compared to others in Figure 6-5, but because Mallow's C-p is smaller for the two-variable model and because it has one less variable in it, you should choose the two-variable model (right leg strength and overall leg strength) as the final model, using the best subsets procedure.

# Comparing Model Selection Procedures

Upon examining the results of the previous sections, the first concern you may have is why you don't get the same results with all three model selection procedures. (I suppose one could argue that if you got the same results all the time, you would have no need for three different procedures, right? But that's beside the point.) All attempts at humor aside, I address this issue, as well as compare how the procedures (from the previous sections) stack up against one another here in this section.

## Why don't all the procedures get the same results?

The forward and backward selection procedures' overall goals and general process are similar. In both the forward and backward selection procedures, you're trying to fit a good model to the data. In both procedures, you evaluate each new model based on how it compares to the previous model that you examined (which has only a one-variable difference). But because the forward selection model starts at one end of the number of $x$ variables spectrum and the backward selection model starts at the other end, the two procedures build their final models differently, one variable at a time. Therefore these two models might meet in the middle and give the same model, but it is certainly not the norm.

In the punt distance example, you can see that in Figure 6-2 (forward selection) the computer includes hang time first because it makes the biggest contribution toward estimating *y*. But in Figure 6-4 (backward selection), all the variables are in the model from the get-go, and after the weakest variable (on all counts) was eliminated (left foot flexibility), the remaining variables were the ones strongly related to hang time (see Figure 6-1). That made hang time a redundant variable, so it was removed.

The best subsets model takes a totally different approach from forward and backward selection. It just looks at all possible models you could have and chooses the best ones at each level (one, two, three variables, and so on). This model selection procedure has no building process that goes on where subsequent models depend on what was selected in previous steps. That means the best subsets procedure can easily give different results than either of the other two procedures simply because it has many more possible models to choose from.

## How do the procedures stack up against each other?

So the big question is which model selection procedure is the *best* one? You can't find a straight answer to that. The debate over this issue goes on and on among the various research groups that analyze their data by using model selection procedures. All three procedures, for example, are available in Minitab, so they are considered viable procedures. However, many statisticians do prefer one model selection procedure over the others, which I reveal to you later in this section along with the positives and negatives of each procedure.

### Looking at the positives

What is nice about each of these procedures is that they have some order to them and they make sense. You don't take a haphazard approach with any of the procedures, and any two people choosing the same procedure for building the best model with the same data set would get the same answer, which is reassuring. All three procedures also usually provide results that are reasonable and final models that have interpretative value, and each has its own plus side. The forward selection keeps the models as simple as possible; backward selection helps you not miss any important variables; and the best subsets model examines every possible model and makes straight comparisons between them.

Because all three model selection procedures are available in Minitab, the temptation may be to just run all three procedures, see what you get, and choose the one you like the best. This approach wouldn't be a good idea and is called *data fishing* or *data snooping,* which can lead to conclusions that others can't confirm (for more on these no-no's, flip to Chapter 1).

### Examining the downsides

The forward and backward selection procedures are somewhat limiting in the way they build their models. After hang time, for example, is eliminated in the backward selection procedure (in Figure 6-4), it never appears again in any later models. After hang time is added in the forward selection procedure, it stays in every model from then on. The best subsets procedure (in Figure 6-5), on the other hand, examines all possible models including those containing hang time and those that don't.

### Standing out above the rest: The best subsets procedure

Because of its versatility and the comprehensive way it looks at all possible models, the best subsets model is generally the model of choice by statisticians. With six possible variables having two possibilities for each one (being included or not being included in the model), you have $2 * 2 * 2 * 2 * 2 * 2 = 64$ possible models to look at in the best subsets procedure. Notice that this set of all possible (64) models includes all the models shown in the step-by-step process of forward and backward selection.

# Chapter 7

# When Data Throws You a Curve: Using Nonlinear Regression

### *In This Chapter*

- Determining when a straight-line regression model isn't enough
- Fitting a polynomial to your data set
- Exploring exponential models to fit your data

In introductory statistics, you concentrate on the *simple linear regression model,* where you look for one quantitative variable, *x,* that you can use to make a good estimate of another quantitative variable, *y.* The examples you look at fell right in line with this kind of model, such as using height to estimate weight or using GPA to estimate exam score. (For information on simple linear regression models, see Chapter 4.)

Nonlinear regression comes into play in situations where you have graphed your data on a *scatterplot* (a two dimensional graph showing the *x* variable on the *x*-axis and the *y* variable on the *y*-axis), and you see a pattern emerging that doesn't look like a straight line, but instead looks like some type of curve. Examples of data that follow a curve include population sizes over time, demand for a product as a function of supply, or the length of time that a battery lasts. When a data set follows a curved pattern, the time has come to move away from the linear regression models (Chapters 4 and 5) and move on to a nonlinear regression model.

In this chapter, you see how to make your way around the curved road of data that leads to nonlinear regression models. The good news is that you can use many of the same techniques you use for regular regression and that Minitab, in the end, does the analysis for you.

# Starting Out with Scatterplots

As with any type of data analysis, before you plunge in and select a model that you think fits the data, or that is supposed to fit the data, you have to step back and take a look at the data and see whether any patterns emerge. To do this, look at a scatterplot of the data, and see whether or not you can draw a smooth curve through the data and find that most of the points follow along that curve.

Suppose you're interested in modeling how quickly a rumor spreads. One person knows a secret, tells another person, and now two know the secret; each of them tells a person, and now four know the secret; some of those people may pass it on, and so it goes on down the line. Pretty soon, a large number of people know the secret (which is a secret no longer). To collect your data, you count the number of people who know a secret by tracking who tells who over a six-day period. You can see a scatterplot of the data in Figure 7-1.

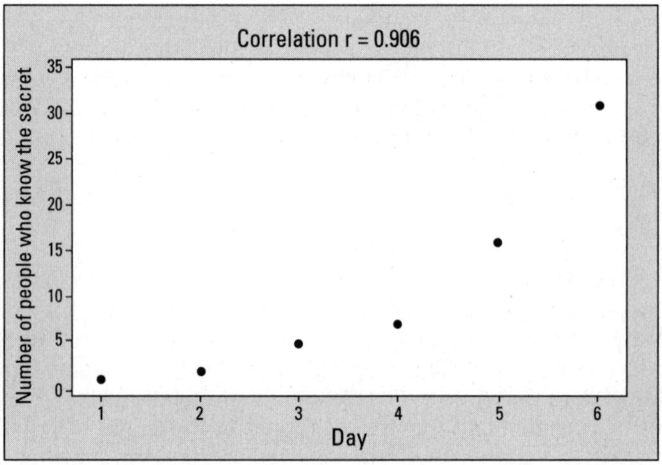


**Figure 7-1:** A scatterplot showing the spread of a secret over a six-day period.

In this situation, the explanatory variable, *x,* is day, and the response variable, *y,* is the number of people who know the secret. Looking at Figure 7-1, you can see a pattern between the values of *x* and *y.* But this pattern isn't linear. It curves upwards. If you tried to fit a line to this data set anyway, how well would it fit?

To figure this out, you can look at the correlation coefficient between *x* and *y,* which is found on Figure 7-1 to be 0.906 (see Chapter 4 for more on correlation). You can interpret this correlation as a strong, positive (uphill) linear relationship between $x$ and $y$. However in this case, the correlation is misleading, because the scatterplot appears to be curved. As with any regression analysis, taking into account both the scatterplot and the correlation when making a decision about how well the model being considered would fit the data is very important. The contradiction in this example between the scatterplot and the correlation is a red flag telling you that a straight-line model isn't the best idea.

The correlation coefficient measures only the strength and direction of the *linear* relationship between $x$ and $y$ (see Chapter 4). However, you may run into situations (like the one shown in Figure 7-1) where a correlation can be strong, yet the scatterplot shows a curve would fit better. Don't rely solely on either the scatterplot or the correlation coefficient alone to make your decision about whether to go ahead and fit a straight line to your data.

The bottom line here is that fitting a line to data that appears to have a curved pattern isn't the way to go. What you need to do in this situation is explore models that have curved patterns themselves. In the following sections, you see two major types of nonlinear (or curved) models that are used to model curved data: polynomials (beyond a straight line) and exponential models (that start out small and quickly increase, or the other way around). Because the pattern of the data in Figure 7-1 starts low and bends upward, the correct model to fit this data is an exponential regression model. (This model would also be appropriate for data that starts out high and bends down low.)

# Handling Curves in the Road with Polynomials

One major family of nonlinear models is the *polynomial* family. You use these models when a polynomial function (beyond a straight line) best describes the curve in the data. (For example, the data may follow the shape of a parabola, which is a second-degree polynomial.) You typically use polynomial models when the data follow a pattern of curves going up and down a certain number of times. For example, suppose a doctor examines the occurrence of heart problems in patients as it relates to their blood pressure. She finds that patients with very low or very high blood pressure had a higher occurrence of problems, while patients whose blood pressure fell in the middle, constituting the normal range, had fewer problems. This pattern of data has a U-shape, and a parabola would fit this data well.

In this section, you see what a polynomial regression model is, how you can search for a good-fitting polynomial for your data, and how you can assess polynomial models.

## Bringing back polynomials

You may recall from algebra that a *polynomial* is a sum of $x$ terms raised to a variety of powers, and each $x$ is preceded by a constant called the *coefficient* of that term. For example, the model $y = 2x + 3x^2 + 6x^3$ is a polynomial. The general form for a polynomial regression model is $y = \beta_0 + \beta_1 x^1 + \beta_2 x^2 + \beta_3 x^3 + \ldots + \beta_k x^k$. Here, $k$ represents the total number of terms in the model.

An example of a polynomial regression model is $y = 2x + 3x^2$. This model is called a *second-degree* (or *quadratic*) polynomial, because the largest exponent is a 2. A second-degree polynomial forms a parabola shape — either an upside-down or right-side-up bowl; it changes direction one time (see Figure 7-2a). A *third-degree* polynomial typically (those having 3 as the highest power of $x$) has a sideways *S*-shape, changing directions two times (see Figure 7-2b). *Fourth-degree* polynomials (those involving $x^4$) typically change directions in curvature three times to look like the letter *W* or the letter *M,* depending on whether they're upside down or right-side up (see Figure 7-2c). In general, if the largest exponent on the polynomial is $n$, the number of curve changes in the graph is typically $n - 1$. (For more information on graphs of polynomials, see your algebra textbook or *Algebra For Dummies* by Mary Jane Sterling [Wiley].)

TECHNICAL STUFF

The nonlinear models in this chapter involve only one explanatory variable, $x$. You can include more explanatory variables in a nonlinear regression, raising each separate variable to a power. These models are beyond the scope of this book; I give you information on basic multiple regression models in Chapter 5.

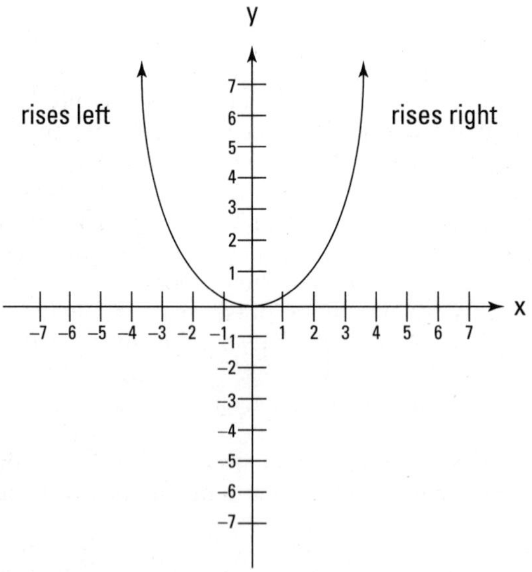


a.

**Figure 7-2:** Examples of second-, third-, and fourth-degree polynomials.

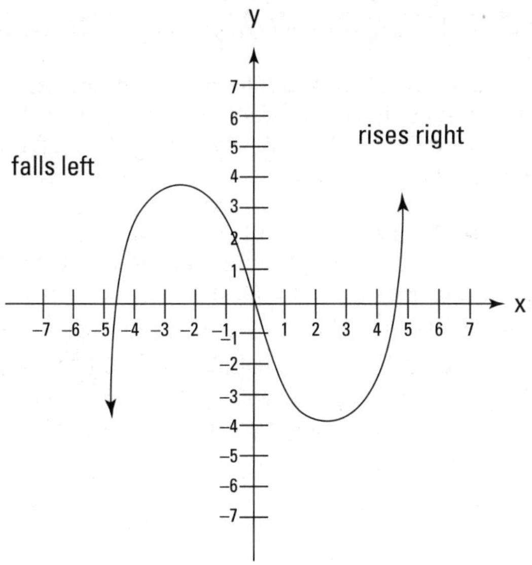
y
x
falls left
rises right
-7 -6 -5 -4 -3 -2 -1 1 2 3 4 5 6 7
7 6 5 4 3 2 1 -1 -2 -3 -4 -5 -6 -7

b.

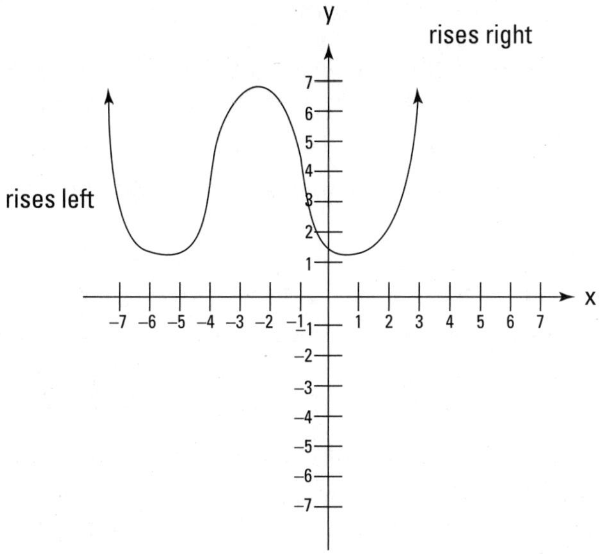
y
x
rises left
rises right
-7 -6 -5 -4 -3 -2 -1 1 2 3 4 5 6 7
7 6 5 4 3 2 1 -1 -2 -3 -4 -5 -6 -7

c.

## Searching for the best polynomial model

When fitting a polynomial regression model to your data, the most important idea is to always start with the simplest model possible and work your way up as you need to. Don't plunge in with a high-order polynomial regression model right off the bat. Here are a couple reasons why:

- **High-order polynomials are hard to interpret, and their models are complex.** For example, with a straight line you can interpret the values of the *y*-intercept and slope easily, but interpreting a tenth-degree polynomial is hard (putting it mildly).
- **High-order polynomials also tend to cause overfitting.** If you're fitting the model as close as you can to every single point in a data set, your model may not hold for a new data set; your estimates for *y* could be way off.

To fit a polynomial to a dataset in Minitab, go to Stat>Regression>Fitted Line Plot> and click on the type of regression model you want: linear, quadratic, or cubic. (It doesn't go beyond a second-degree polynomial; however, these options should cover 90 percent of the cases.) Click on the *y* variable from the left-hand box and click Select; this variable will appear in the Response (*y*) box. Click on the *x* variable from the left-hand box and click Select; it will appear in the Predictor (*x*) box. Click OK.

Following are the steps for fitting a polynomial model to your data (statistical software can jump in and fit the models for you after you tell it which ones to fit):

1. **Try to fit a first-degree polynomial (straight line) to the data first: $y = b_0 + b_1x$.**

   This model is for a straight line. If it doesn't fit (using both the correlation coefficient, *r*, and the scatterplot), move to step two.

2. **Try to fit a second-degree polynomial (parabola): $y = b_0 + b_1x + b_2x^2$.**

   If the data fits the model well, stop here (see the section on assessing model fit). If the model still doesn't fit well, go to step three.

3. **Try to fit a third-degree polynomial: $y = b_0 + b_1x + b_2x^2 + b_3x^3$.**

   If the data fits the model well (check out the section on assessing model fit), don't go on to the next polynomial. If the model still doesn't fit well, go to step four.

4. **Continue trying to fit higher-order polynomials until you find one that fits or until the order of the polynomial (largest exponent) is simply getting too large to find a reliable pattern.**

How large is too large? Typically, if you can't fit the data by the time the degree of the polynomial reaches three, then perhaps a different type of model would work better. Or you may determine that you observe too much scatter and haphazard behavior in the data to try to fit any model.

Minitab can do each of these steps for you up to degree two (step two); from there, you need a more sophisticated statistical software program, such as SAS or SPSS. However, most of the models you need to fit go up to the second-degree polynomials. In the next section, you use a second-degree polynomial to predict a student's quiz score based on his or her study time.

## Using a second-degree polynomial to pass the quiz

The first step in fitting a polynomial model is to graph the data in a scatterplot and see whether the data fall into a particular pattern. Many different types of polynomials exist to fit data that has a curved type of pattern. One of the most common patterns found in curved data is the quadratic pattern, or second-degree polynomial, which goes up and comes back down, or goes down and comes back up, as the $x$ values move from left to right (see Figure 7-2a). The second-degree (quadratic) polynomial is the simplest and most commonly used polynomial beyond the straight line, so it deserves special consideration.

This section is dedicated to looking at a second-degree polynomial. You can see the exploratory process of graphing data and looking at the graph's shape by using the data involving quiz scores and study time. (After you master the basic ideas based on second-degree polynomials, you can apply them to polynomials with higher powers.)

Suppose 20 students take a statistics quiz. You record the quiz scores (which have a maximum score of ten) and the number of hours students reported studying for the quiz. (You can see the results in Figure 7-3.)

Looking at Figure 7-3, it appears that three camps of students are in this class. Camp One, on the left end of the $x$-axis, understands the stuff (as reflected in their higher scores) but didn't have to study hardly at all (because their study time on the $x$-axis is low). Camp Three also did very well on the quiz (as indicated by their high quiz scores), but had to study a great deal to get that grade (as seen on the far-right end of the $x$-axis). The students in the middle, Camp Two, didn't seem to fare well. All in all, from the scatterplot in Figure 7-3, it does appear that study time may explain quiz scores on some level, and explains it in a way indicative of a second-degree polynomial. So a quadratic regression model may fit this data.

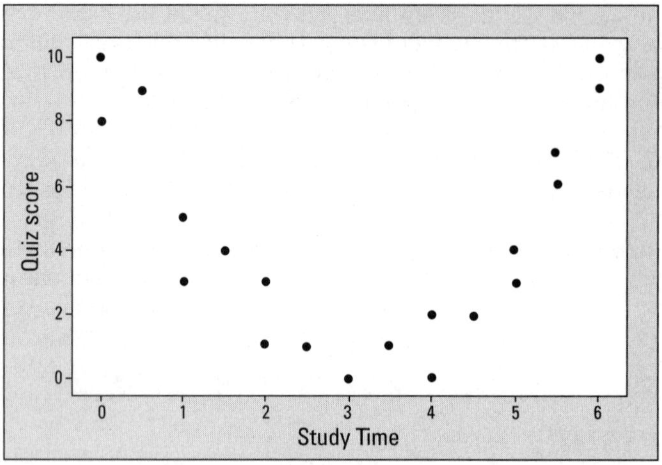


**Figure 7-3:** Scatterplot showing study time and quiz scores.

Suppose a data analyst (not you!) doesn't know about polynomial regression and just tries to fit a straight line to the quiz-score data. In Figure 7-4, you can see the data and the straight line that he tried to fit it in. The correlation as shown in the figure is –0.033, which is basically zero. This correlation means that no linear relationship lies between $x$ and $y$. (It doesn't mean that no relationship is present at all, just not a linear relationship — see Chapter 4 for more on linear relationships.) So trying to fit a straight line here was indeed a bad idea.

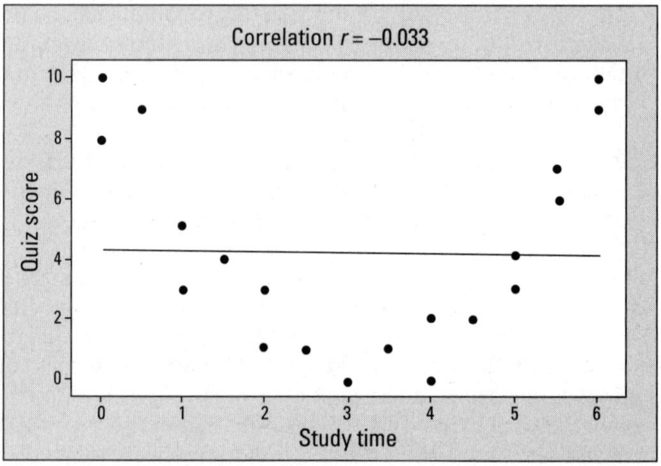


**Figure 7-4:** Trying to fit a straight line to quadratic data.

After you know that a quadratic polynomial seems to be a good fit for the data, the next challenge is finding the equation for that particular parabola that fits the data, among all the possible parabolas out there. Remember from algebra that the general equation of a parabola is $y = ax^2 + bx + c$. Now you have to find the values of *a, b,* and *c* that create the best-fitting parabola to the data (just like you find the *a* and the *b* that create the best-fitting line to data in a linear regression model). That is the object of the regression model.

Say that you fit a quadratic regression model to the quiz-score data by using Minitab (see the Minitab output in Figure 7-5 and the instructions for using Minitab to fit this model in the previous section). On the top line of the output, you can see that the equation of the best-fitting parabola is quiz score = 9.82 – 6.15 * study time + 1.00 * study time squared. (Note that *y* is quiz score and *x* is study time in this example because you're using study time to predict quiz score.)

**Figure 7-5:** Minitab output for fitting a parabola to the quiz-score data.

**Polynomial Regression Analysis: Quiz Score versus Study Time**

```
The regression equation is
Quiz score = 9.823-6.149 study time + 1.003 study time**2

S = 1.04825    R-Sq = 91.7%   R-Sq(adj) = 90.7%
```

The scatterplot of the quiz-score data and the parabola that was fit to the data via the regression model is shown in Figure 7-6. From algebra, you may remember that a positive coefficient on the quadratic term (here $a = 1.00$) means the bowl is right-side-up, which you can see is the case here.

Looking at Figure 7-6, it appears that the quadratic model fits this data pretty well, because the data fall closely to the curve that Minitab found. However, data analysts can't live by scatterplots alone. In the next section, you figure out how to assess the fit of a polynomial model in more detail.

## Assessing the fit of a polynomial model

You have made a scatterplot of your data, and you saw a curved pattern. You used polynomial regression to fit a model to the data; the model appears to fit well because the points follow closely to the curve Minitab found. But don't stop there. To make sure your results can be generalized to the population from which your data was taken, you need to do a little more checking beyond just the graph to make sure your model fits well.

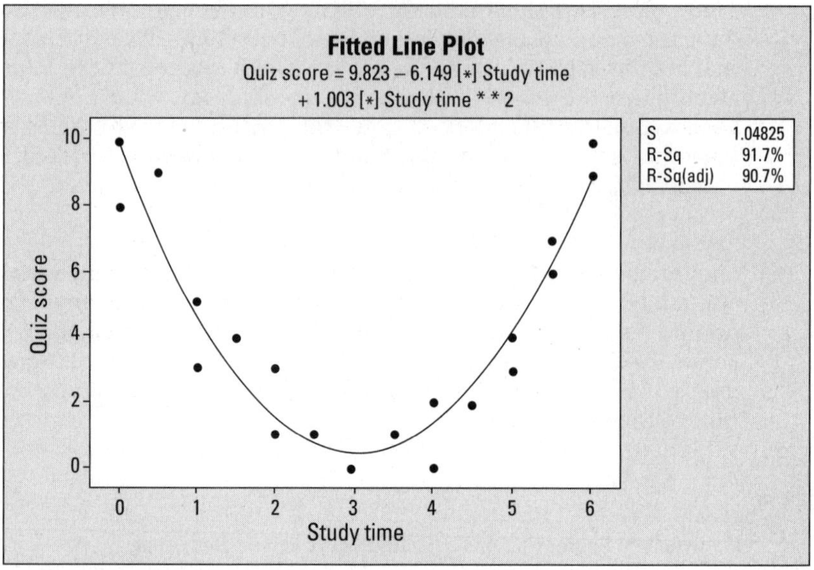


**Figure 7-6:** The parabola appears to fit the quiz-score data nicely.

To assess the fit of any model beyond the usual suspect, a scatterplot of the data, you look at two additional items. Those items are the value of $R^2$ adjusted and the residual plots, which you typically check in that order after assessing the scatterplot.

All three assessments must agree before you can conclude that the model fits. If the three assessments don't agree, you'll likely have to use a different model to fit the data besides a polynomial model, or you'll have to change the units of the data to help a polynomial model fit better. However, the latter fix is outside the scope of intermediate statistics, and you probably will not encounter that situation.

In the following sections, you take a deeper look at the value of $R^2$ adjusted and the residual plots and figure out how you can use them to assess your model's fit. (You can find more info on the scatterplot in the section "Starting out with Scatterplots" earlier in this chapter.)

### Examining $R^2$ and $R^2$ adjusted

Finding $R^2$, the coefficient of determination (see Chapter 5 for full details), is like the day of reckoning for any model. You can find $R^2$ on your regression output, listed as "R-Sq" right under the portion of the output where the coefficients of the variables are shown (see Figure 7-5).

Figure 7-5 shows the Minitab output for the quiz-score data example; the value of $R^2$ in this case is 91.7 percent. The value of $R^2$ tells you what percentage of the variation in the *y*-values the model can explain. To interpret this percentage, the closer a value of $R^2$ is to 100 percent, the better. You can consider values of $R^2$ over 80 percent good. Values under 60 percent aren't good. Those in between I'd consider to be so-so; they could be better. (This assessment is just my rule of thumb; opinions may vary a bit from one statistician to another.)

However, you can find such a thing in statistics as too many variables spoiling the pot. Right beside $R^2$ on the computer output from any regression analysis is the value of $R^2$ adjusted, which adjusts the value of $R^2$ down a notch for each variable (and each power of each variable) entered into the model. That way, you can't just throw in a ton of variables into a model whose tiny increments all add up to an acceptable $R^2$ value, without taking a hit for throwing everything in the model but the kitchen sink.

REMEMBER

To be on the safe side, you can always use $R^2$ adjusted to assess the fit of your model, rather than $R^2$. But you should always use $R^2$ adjusted if you have more than one *x* variable in your model (or more than one power of an *x* variable). The values of $R^2$ and $R^2$ adjusted will be close if you have only a couple of different variables (or powers) in the model, but as the number of variables (or powers) increases, so does the gap between $R^2$ and $R^2$ adjusted. In that case, $R^2$ adjusted is the most fair and consistent coefficient to use to examine model fit.

In the quiz-score example (analysis shown in Figure 7-5), the value of $R^2$ adjusted is 90.7 percent, still a very high value, meaning the quadratic model fits this data very well. (See Chapter 6 for more on $R^2$ and $R^2$ adjusted.)

### Checking the residuals

You've looked at the scatterplot of your data and the value of $R^2$ is high. What's next? Now you want to examine how well the model fits each individual point in the model, to make sure you can't find any spots where the model is way off or places where you missed another underlying pattern in the data.

A *residual* is the amount of error, or leftover, that occurs when you fit a model to a data set. For each observed *y*-value in the data set, you also have a predicted value from the model, typically called *y-hat.* The residual is the difference between value of *y* and *y*-hat. Each *y*-value in the data set has a residual; you examine all the residuals together as a group, looking for patterns or unusually high values (indicating a big difference between the observed *y* and the predicted *y* at that point; see Chapter 4 for the full info on residuals and their plots).

In order for the model to fit well, the residuals need to meet two conditions:

- **The residuals are independent.** The independence of residuals means that as you plot the residuals you don't see any pattern; they don't affect each other and should be random.
- **The residuals have a normal distribution centered at zero, and the standardized residuals follow suit.** Having a normal distribution with mean zero means that most of the residuals should be centered around zero, with fewer of them occurring the farther from zero you get. You should observe about as many residuals above the zero line as below it. If the residuals are standardized, their standard deviation is one; you should expect about 95 percent of them to lie between –2 and +2, following the 68-95-99.7 Rule (see your intro stats text).

The way to determine whether or not these two conditions are met for the residuals is by using a series of four graphs called *residual plots.* (The residuals are the distances between the predicted values in the model and the observed values of the data themselves.) Most statisticians prefer to standardize the residuals (convert them to *Z*-scores by subtracting their mean and dividing by their standard deviation) before looking at them, because then you can compare them with values on a *Z*-distribution. Hence, you can ask Minitab to give you a series of four standardized residual plots with which to check the conditions.

Figure 7-7 shows the standardized residual plots for the quadratic model, using the quiz-score data from previous sections. The upper-left plot shows that the standardized residuals follow one-to-one with a normal distribution. The upper-right plot shows that most of the standardized residuals fall between –2 and +2 (see Chapter 4 for more on standardized residuals). The lower-left plot shows that the residuals bear some resemblence to a normal distribution, and the lower-right plot demonstrates how the residuals have no pattern. They appear to occur at random. All of these plots together suggest that the conditions on the residual are met to apply the selected quadratic regression model.

## Making predictions

After you've found the model that fits well, you can now use that model to make predictions for *y* given *x* by simply plugging in the desired *x*-value, and out comes your predicted value for *y*. (Make sure any values you plug in for *x* occur within the range of where data was collected; if not, you can't guarantee the model holds.)

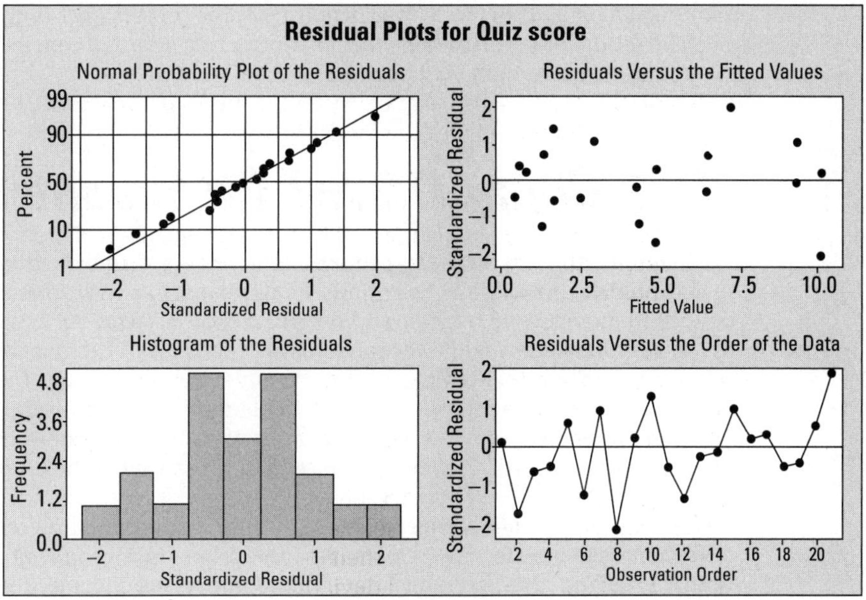


**Figure 7-7:** Standardized residual plots for the quiz-score data, using the quadratic model.

Returning to the quiz-score data from previous sections, can you use study time to predict quiz score by using a quadratic regression model? By looking at the scatterplot and the value of $R^2$ adjusted (see Figures 7-5 and 7-6, respectively), you can see the quadratic regression model appears to fit the data well (isn't it nice when you find something that fits?). By looking at the residual plots (Figure 7-7), the conditions seem to be met to fit this model; you can find no major patterns in the residuals, they appear to center at one, and most of them stay within the normal boundaries of standardized residuals: –2 and +2.

With all this evidence together, study time does appear to have a quadratic relationship with quiz score in this case. You can now use the model to make estimates of quiz score given study time. For example, because the model (shown in Figure 7-5) is $y = 9.82 - 6.15x + 1.00x^2$, if your study time is 5.5 hours, then your estimated quiz score is $9.82 - 6.15 * 5.5 + 1.00 * 5.5^2 = 9.82 - 33.83 + 30.25 = 6.25$. That value corresponds to what you see on the graph in Figure 7-3 if you look at the place where $x = 5.5$; the $y$-values are in the vicinity of 6 to 7.

WARNING!

As with any regression model, you can't estimate the value of $y$ for $x$-values outside the range of where data was collected. This error is called *extrapolation.* You can't be sure that the model you fit to your data actually continues ad infinitum for any old value of $x$. In the quiz-score example (see Figure 7-3), it doesn't make sense to estimate quiz scores for study times higher than six hours when using this model because the scores on this quiz don't go above ten. The model likely levels off after six hours to a score of ten, indicating that studying more than six hours is overkill.

# Going Up? Going Down? Go Exponential!

Exponential models work well in situations where a $y$ variable either increases or decreases exponentially over time. That means, the $y$ variable starts out slow, then increases at a faster and faster rate, or it starts out high and decreases at a faster and faster rate. Many processes in the real world behave like an exponential model: for example, population size over time, average household incomes over time, the length of time a product lasts, or the level of patience one has as the number of statistics homework problems goes up (of course, using this book should cut that time in half, no?).

In this section, you familiarize yourself with the exponential regression model, and see how to use it to fit data that either rises or falls at an exponential rate. You also discover how to build and assess exponential regression models to make accurate predictions for a response variable $y$, using an explanatory variable $x$.

## Recollecting exponential models

Exponential models have the form $y = \alpha\beta^x$. These models involve a constant, $\beta$, raised to higher and higher powers of $x$ multiplied by a constant, $\alpha$. The constant $\beta$ represents the amount of curvature in the model. The constant $\alpha$ is a multiplier in front of the model that shows where the model crosses the $y$-axis (because when $x = 0$, $y = \alpha * 1$).

REMEMBER

An exponential model generally looks like the upper part of a hyperbola (remember those from advanced algebra?). A *hyperbola* is a curve that crosses the $y$-axis at a point and curves downward toward zero or starts at some point and curves upward to infinity (see Figures 7-8a and 7-8b for examples). If $\beta$ is greater than one in an exponential model, the graph curves upward toward infinity. If $\beta$ is less than one, the graph curves downward toward zero. All exponential models stay above the $x$-axis.

For example, the model $y = 1 * 3^x$ is an exponential model. Here, say you made $\alpha = 1$, indicating that the model crosses the $y$-axis at 1 (because plugging $x = 0$ into the equation gives you 1). You set the value of $\beta$ equal to three, indicating that you want a bit of curvature to this model. The *y*-values curve upward quickly from the point (0, 1). For example, when $x = 1$, you get $1 * 3^1 = 3$; for $x = 2$, you get $1 * 3^2 = 9$; for $x = 3$, you get $1 * 3^3 = 27$, and so on. Figure 7-8a shows a graph of this model. Notice the huge scale needed on the *y*-axis when *x* is only 10.

Now suppose you let $\alpha = 1$ and $\beta = 0.5$. These values give you the model $y = 1 * 0.5^x$. This model takes 0.5 (a fraction between 0 and 1) to higher and higher powers, which makes the *y*-values smaller and smaller, never reaching zero but always getting closer. (For example, 0.5 to the second power is 0.25, which is less than 0.50, and 0.50 to the tenth power is 0.00098.) Figure 7-8b shows a graph of this model.

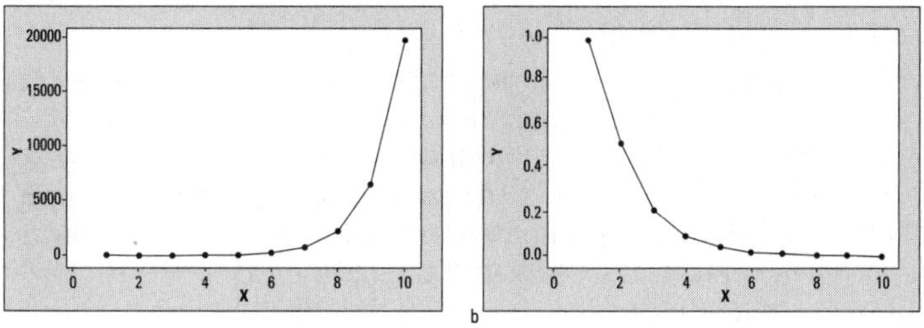


a b

**Figure 7-8:** The exponential regression model for different values of β.

# *Searching for the best exponential model*

Finding the best-fitting exponential model requires a bit of a twist compared to finding the best-fitting line by using simple linear regression (Chapter 4). Because fitting a straight-line model is much easier than fitting an exponential model directly from data, you transform the data into something for which a line fits. Then you fit a straight-line model to that transformed data. Finally you undo the transformation, getting you back to an exponential model. The transformation used is *logarithms* (because they are the inverse of exponentials). But before you start sweating, don't worry; these math gymnastics aren't something you do by hand — the computer does most of the grit work for you.

The exponential model looks like this (if you're using base 10): $y = 10^{b_0 + b_1 x}$. Follow these steps for fitting an exponential model to your data and using it to make predictions:

1. **Make a scatterplot of the data and see whether the data appears to have a curved pattern that resembles an exponential curve.**

If the data follows an exponential curve, proceed on to the next step; otherwise, consider alternative models (such as multiple regression in Chapter 5).

To see how to make a scatterplot in Minitab, check out Chapter 4. For more details on what shape to look for, see the section "Recollecting exponential models."

2. **Use Minitab to fit a line to the log(*y*) data.**

   In Minitab, you go to the regression model (curve fit). Under Options, select Logten of *y*. Then select Using scale of logten to give you the proper units for the graph.

   Understanding the basic idea of what Minitab does during this step is important; being able to calculate it by hand isn't. You can see what Minitab does during this step in the following:

   - Minitab applies the log (base 10) to the *y*-values. For example, if *y* is equal to 100, $\log_{10}100$ equals 2 (because 10 to the second power equals 100). Note that if the *y*-values fell close to an exponential model before, the log(*y*) values will fall close to a straight-line model. This phenomenon occurs because the logarithm is the inverse of the exponential function, so they basically cancel each other out, and you're left with a straight line.
   - Minitab fits a straight line to the log(*y*) values by using simple linear regression (from Chapter 4). The equation of the best-fitting straight line for the log(*y*) data is $\log(y) = b_0 + b_1x$. Then Minitab passes this model on to you in its output; you take it from here.

3. **Transform the model back to an exponential model by starting with the straight-line model, $\log(y) = b_0 + b_1x$, that was fit to the $\log_{10}(y)$ data and then applying ten to the power of the left side of equation and ten to the power of the right side.**

   By the definition of logarithm, you get *y* on the left side of the model and ten to the power of $b_0 + b_1x$ on the right side. The resulting exponential model for *y* is $y = 10^{b_0 + b_1 x}$.

4. **Use the exponential model found in step three to make predictions for *y* (your original variable) by plugging your desired value of *x* into the model.**

   Only plug in values for *x* that are in the range of where the data are located.

5. **Assess the fit of the model by looking at the scatterplot of the log(*y*) data, checking out the value of $R^2$ adjusted for the straight-line model for log(*y*), and checking the residual plots for the log(*y*) data.**

The techniques and criteria you use to do this are the same as those I discuss in a previous section "Assessing the fit of a polynomial model."

The math magic from step three works courtesy of the definition of logarithm, which says $\log_b(a) = y \Leftrightarrow b^y = a$. Suppose you have the equation $\log_{10}y = 2 + 3x$. Now if you take ten to the power of each side, you get $10^{\log_{10}(y)} = 10^{2+3x}$. By the definition of logarithm, the tens cancel out on the left side and you get $y = 10^{2+3x}$. This model is exponential because $x$ is in the exponent. You can take step two up another notch to include the general form of the straight line model $y = b_0 + b_1x$. Using the definition of logarithm on this line, you get $\log_{10}(y) = b_0 + b_1x \Leftrightarrow 10^{b_0 + b_1 x} = y$.

If these steps seem dubious to you, stick with me. By looking at the example in the next section, you can see each step firsthand and that will help a great deal. In the end, actually finding predictions by using an exponential model is a lot easier to do than it is to explain.

## Spreading secrets at an exponential rate

Often, the best way to figure something out is to see it in action. By using the secret-spreading quiz example from Figure 7-1, you can work through the series of steps from the preceding section to find the best-fitting exponential model and use it to make predictions.

### Checking the scatterplot

Your goal in step one is to make a scatterplot of the secret-spreading data and determine whether the data resembles the curved function of an exponential model. Figure 7-1 shows the data for the spread of a number of people knowing the secret, as a function of the number of days. You can see that the number of people starts out small, but then as more and more people tell more and more people, the number grows quickly until the secret isn't a secret anymore. This is a good situation for an exponential model, due to the amount of upward curvature in this graph.

### Letting Minitab do its thing to log(y)

In step two, you let Minitab find the best-fitting line to the log($y$) data (see the section "Searching for the best exponential model" to find out how to do this in Minitab). The output for the analysis of the secret-spreading data is in Figure 7-9. You can see in Figure 7-9 that the best-fitting line is $\log(y) = -0.19 + 0.28 * x$, where $y$ is the number of people knowing the secret and $x$ is the number of days.

**Figure 7-9:** Minitab fits a line to the log(*y*) for the secret-spreading data.

**Regression Analysis: Day versus Number**

```
The regression equation is
logten (number) = -0.1883 + 0.2805 day

S = 0.157335   R-Sq = 93.3%   R-Sq(adj) = 91.6%
```

### *Going exponential*

After you have your Minitab output, you're ready for step three. You transform the model $\log(y) = -0.19 + 0.28 * x$ into a model for *y*. Do this by taking 10 to the power of the left-hand side and 10 to the power of the right-hand side. Transforming the log(*y*) equation for the secret-spreading data, you get $y = 10^{-0.19+0.28x}$.

### *Making predictions*

By using the exponential model from step three, you can move on to step four: Make predictions for appropriate values of *x* (within the range of where data was collected). Continuing to use the secret-spreading data, suppose you want to estimate the number of people knowing the secret on day five (see Figure 7-1). Just plug $x = 5$ into the exponential model to get $y = 10^{-0.19+0.28*5} = 10^{1.21} = 16.22$. Looking at Figure 7-1, you can see that this estimation falls right in line with the graph.

### *Assessing the fit of your exponential model*

Now that you've found the best-fitting exponential model, you have the worst behind you. You have arrived at step five and are ready to further assess the model fit (beyond the scatterplot of the original data) to make sure no major problems arise.

In general, to assess the fit of an exponential model, you do three things, in the following order:

1. **Check the scatterplot of the log(*y*) data to see how well it resembles a straight line.**
2. **Examine the value of $R^2$ adjusted for the model of the best-fitting line for log(*y*), done by Minitab.**
3. **Look at the residual plots from the fit of a line to the log(*y*) data.**

If you look at the section "Assessing the fit of a polynomial model," you can figure out how to apply these assessment strategies to the straight-line fit of log(*y*).

You assess the fit of the log(*y*) for the secret spreading first through the scatterplot shown in Figure 7-10. The scatterplot shows that the model appears to fit the data well, because the points are scattered in a tight pattern around a straight line.

The value of $R^2$ adjusted for this model is found in Figure 7-10 to be 91.6 percent. This value also indicates a good fit because it is very close to 100 percent. Therefore, 91.6 percent of the variation in the number of people knowing the secret is explained by how many days it has been since the secret spreading started. (Makes sense.)

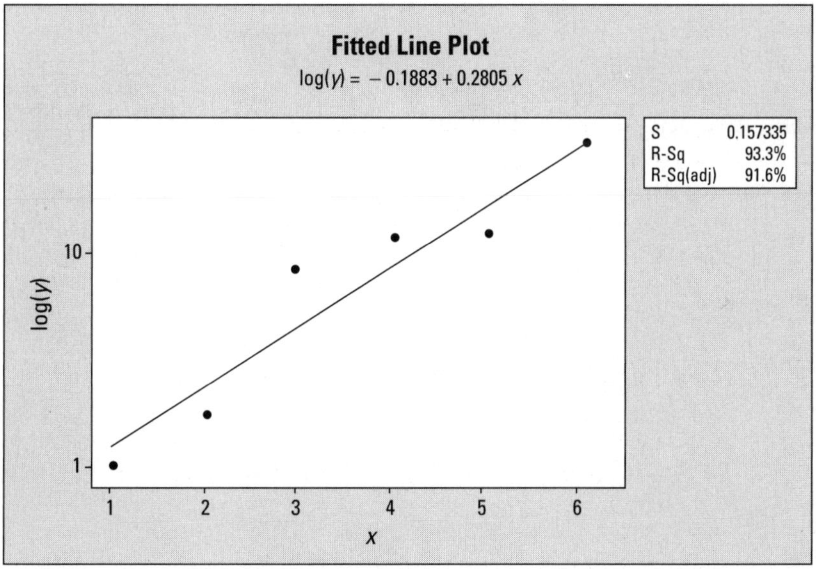


**Figure 7-10:** A scatterplot showing the fit of a straight line to log(*y*) data.

The residual plots from this analysis (see Figure 7-11) show no major departures from the conditions that the errors are independent and have a normal distribution. Note that the histogram in the lower-left corner doesn't look all that bell-shaped, but you don't have a lot of data in this example, and the rest of the residual plots seem okay. So, you have little cause to really worry.

All in all, it appears that the secret's out on the secret-spreading data, now that you have an exponential model that explains how it happens.

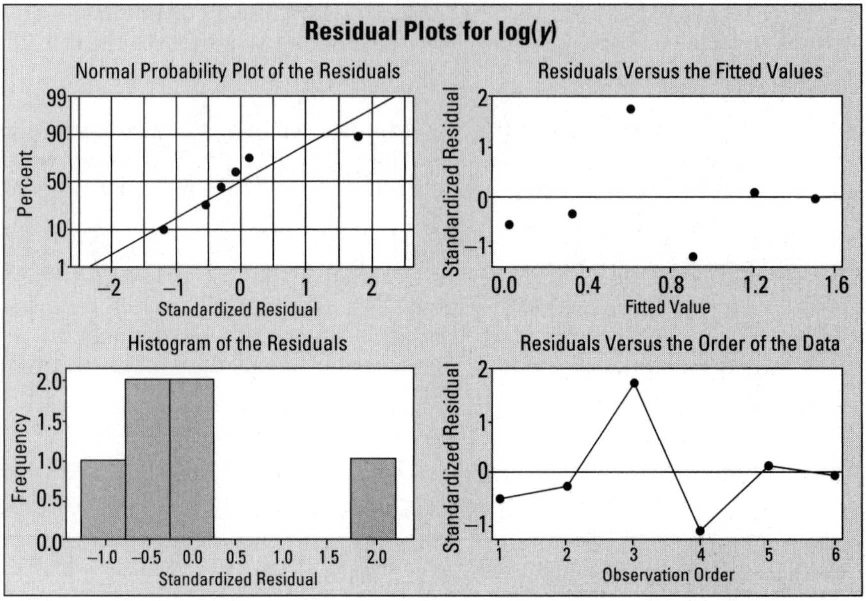


**Figure 7-11:** A residual plot showing the fit of a straight line to log(*y*) data.

# Chapter 8

# Yes, No, Maybe So: Making Predictions by Using Logistic Regression

### In This Chapter

- Knowing when logistic regression is appropriate
- Building logistic regression models for yes or no data
- Checking model conditions and making the right conclusions

Everyone (even yours truly) tries to make predictions about whether or not a certain event is going to happen. For example, what's the chance it's going to rain this weekend? What is our team's chances of winning our next game? What is the chance that I'll have complications during this surgery? These predictions are often based on *probability,* the long-term percentage of time an event is expected to happen. In the end, you want to estimate $p$, the probability of an event occurring. In this chapter, you see how to build and test models for $p$ based on a set of explanatory $(x)$ variables. This technique is called *logistic regression.*

## Setting Up the Logistic Regression Model

Yes or no data that comes from a random sample has a binomial distribution with probability of success (the event occurring) equal to $p$. In the binomial problems you saw in intro stats, you had a sample of size $n$ trials, you had yes or no data, and you had a probability of success on each trial, denoted by $p$. In your intro stat course, for any binomial problem the value of $p$ was somehow given to be a certain value, but in intermediate stats, you operate under the much more realistic scenario that it's not. In fact, because $p$ isn't known, your job is to estimate what it is and use a model to do that.

To estimate *p*, the chance of an event occurring, you need data that comes in the form of yes or no, indicating whether or not the event occurred for each individual in the data set. Now because yes or no data don't have a normal distribution, a condition needed for other types of regression, you need a new type of regression model to do this job — *logistic regression.* Keep reading this section to find out more about this model.

## Defining a logistic regression model

A logistic regression model ultimately gives you an estimate for *p*, the probability that a particular outcome will occur in a yes or no situation (for example, the chance that it will rain versus not). The estimate is based on information from one or more explanatory variables; you can call them $x_1$, $x_2$, $x_3$, . . . $x_k$. (For example, $x_1$ = humidity, $x_2$ = barometric pressure, $x_3$ = cloud cover, . . . and $x_k$ = wind speed.) ***Note:*** In this chapter, I present only the case where you use one explanatory variable. You can extend the ideas in exactly the same way as you can extend the simple linear regression model (Chapter 4) to a multiple regression model (Chapter 5).

## Using an S-curve to estimate probabilities

In a simple linear regression model, the general form of a straight line is $y = \beta_0 + \beta_1 x$. In the case of estimating *p*, the linear regression model is the straight line $p = \beta_0 + \beta_1 x$. However, it doesn't make sense to use a straight line to estimate the probability of an event occurring based on another variable, due to the following reasons:

- **The estimated values of *p* can never be outside of [0, 1], which goes against the idea of a straight line (a straight line continues on in both directions).**
- **It doesn't make sense to force the values of *p* to increase in a linear way based on *x*.** For example, an event may occur very frequently with a range of large values of *x* and very frequently with a range of small values of *x*, with very little chance of the event happening in an area in between. This type of model would have a U-shape, rather than a straight-line shape.

To come up with a more appropriate model for *p*, statisticians created a new function of *p* whose graph is called an S-curve. The *S-curve* is a function that involves *p*, but it also involves *e* (the natural logarithm) as well as a ratio of two functions. The values of the S-curve always fit between 0 and 1 and allows the probability, *p*, to change from low to high or high to low, according to a curve that is shaped like an *S*. The general form of the logistic regression model based on an S-curve is $p = \frac{e^{\beta_0 + \beta_1 x}}{1 + e^{\beta_0 + \beta_1 x}}$.

## Interpreting the coefficients of the logistic regression model

REMEMBER

The sign on the parameter $\beta_1$ tells you the direction of the S-curve. If $\beta_1$ is positive, the S-curve goes from low to high (see Figure 8-1a); if $\beta_1$ is negative, the S-curve goes from high to low (Figure 8-1b).

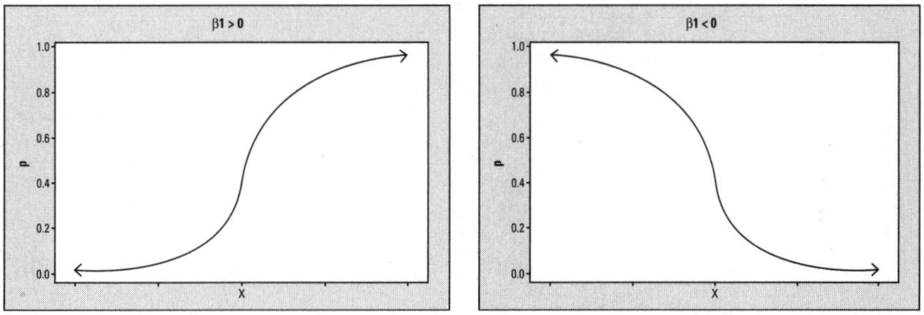


**Figure 8-1:** Two basic types of S-curves.

The magnitude of $\beta_1$ (indicated by its absolute value) tells you how much curvature is in the model. High values indicate a steep curvature and low values indicate slow curvature. The parameter $\beta_0$ just shifts the S-curve to the proper location to fit your data. It shows you the cutoff point where x-values change from high to low probability and vice versa.

## Estimating the chance a movie will be a hit by using logistic regression

Often, the best way to figure something out is to see it in action. In this section, I give you an example of a situation where you can use a logistic regression model to estimate a probability. (I expand on this example later in this chapter; for now, I'm just setting up a scenario for logistic regression.)

Suppose movie marketers want to estimate the chance that someone will enjoy a certain family movie, and you believe age may have something to do with it. Translating this research question into $x$'s and $y$'s, the response variable $(y)$ is whether or not a person will enjoy the movie, and the explanatory variable $(x)$ is the person's age. You want to estimate $p$, the chance of someone enjoying the movie. You collect data on a random sample of 40 people, shown in Table 8-1. Based on your data, it appears that younger people enjoyed the movie more than older people, and that at a certain age, the trend switches from liking the movie to disliking it; so, you can build a logistic regression model to estimate $p$.

**Table 8-1 Movie Enjoyment (Yes or No Data) Based on Age**

| *Age* | *Enjoyed the Movie* | *Total Number Sampled* |
|---|---|---|
| 10 | 3 | 3 |
| 15 | 4 | 4 |
| 16 | 3 | 3 |
| 18 | 2 | 3 |
| 20 | 2 | 3 |
| 25 | 2 | 4 |
| 30 | 2 | 4 |
| 35 | 1 | 5 |
| 40 | 1 | 6 |
| 45 | 0 | 3 |
| 50 | 0 | 2 |

# General Steps for Logistic Regression

The basic idea of any model-fitting process is to look at all possible models you can have under the general format and find the one that fits your data best. The general form of the best-fitting logistic regression model is $\hat{p} = \frac{e^{b_0 + b_1 x}}{1 + e^{b_0 + b_1 x}}$, where $\hat{p}$ is the estimate of $p$, $b_0$ is the estimate of $\beta_0$, and $b_1$ is the estimate of $\beta_1$ (from the previous section). The only values you have a choice about to form your particular model are the values of $b_0$ and $b_1$. These values are the ones you're trying to estimate through the logistic regression analysis.

To find the best-fitting logistic regression model for your data, complete the following steps:

1. **Run a logistic regression analysis on the data you collected (see the section "Running the analysis in Minitab" for these instructions.)**
2. **Find the coefficients of constant and *x*, where *x* is the name of your explanatory variable.**

   These coefficients are $b_0$ and $b_1$, the estimates of $\beta_0$ and $\beta_1$ in the logistic regression model.

3. **Plug the coefficients from step one into the logistic regression model:**

$$\hat{p} = \frac{e^{b_0 + b_1 x}}{1 + e^{b_0 + b_1 x}}$$

   This equation is your best-fitting logistic regression model for the data. Its graph is an S-curve (for more on the S-curve, see the section "Using an S-curve to estimate probabilities" earlier in this chapter).

In the sections that follow, you see how to ask Minitab to do the above steps for you. You also see how to interpret the resulting computer output, find the equation of the best-fitting logistic regression model, and use that model to make predictions (being ever mindful that all conditions are met).

## Running the analysis in Minitab

Using Minitab, here's how to perform a logistic regression (other statistical software packages are similar):

1. **Input your data in the spreadsheet as a table that lists each value of the *x* variable in column one, the number of yeses for that value of *x* in column two and the total number of trials at that *x*-value in column three.**

   These last two columns represent the outcome of the response variable *y*. (For an example of how to enter your data, see Table 8-1 based on the movie-age data.)

2. **Go to Stat>Regression>Binary Logistic Regression.**
3. **Beside the Success option, select your variable name from column two, and beside Trial, select your variable name for column three.**
4. **Under Model, select your variable name from column one, because that's the column containing the explanatory (*x*) variable in your model.**
5. **Click OK, and you get your logistic regression output.**

When you fit a logistic regression model to your data, the computer output is composed of two major portions:

- **The model-building portion:** In this part of the output, you can find the coefficients $b_0$ and $b_1$ (I describe coefficients in the section "Finding the coefficients and making the model").
- **The model-fitting portion:** You can see the results of a Chi-square goodness-of-fit test (see Chapter 15) as well as the percentage of concordant and discordant pairs in this section of the output. (A *concordant pair* means the predicted outcome from the model matches the observed outcome from the data. A *discordant pair* is one that doesn't match.)

In the case of the movie and age data, the model-building part of the Minitab output is shown in Figure 8-2. The model-fitting part of the Minitab output from the logistic regression analysis is in Figure 8-4. In the following sections, you see how to use this output to build the best-fitting logistic regression model for your data and to check the model's fit.

**Figure 8-2:** The model-building part of the movie and age data's logistic regression output.

Logistic Regression Table

| Predictor | Coef | SE Coef | Z | P | Odds Ratio | 95% CI Lower | 95% CI Upper |
|---|---|---|---|---|---|---|---|
| Constant | 4.86539 | 1.43434 | 3.39 | 0.001 | | | |
| Age | -0.175745 | 0.0499620 | -3.52 | 0.000 | 0.84 | 0.76 | 0.93 |

## Finding the coefficients and making the model

After you have Minitab run a logistic regression analysis on your data, you can find the coefficients $b_0$ and $b_1$ and put them together to form the best-fitting logistic regression model for your data.

Figure 8-2 shows part of the Minitab output for the movie enjoyment and age data. I call this portion of the output the *model-building* part of the output. (I discuss the remaining output in the section "Checking the fit of the model.") The first column of numbers is labeled *Coef,* which stands for the coefficients in the model. The first coefficient, $b_0$, is labeled *Constant.* The second coefficient is in the row labeled by your explanatory variable, $x$. (In the movie and age data, the explanatory variable is age. This age coefficient represents the value of $b_1$ in the model.)

According to the Minitab output in Figure 8-2, the value of $b_0$ is 4.87 and the value of $b_1$ is –0.18. After you've determined the coefficients $b_0$ and $b_1$ from the Minitab output to find the best-fitting S-curve for your data you put these two coefficients into the general logistic regression model: $\hat{p} = \frac{e^{b_0 + b_1 x}}{1 + e^{b_0 + b_1 x}}$. For the movie and age data, you get $\hat{p} = \frac{e^{4.87 - 0.18x}}{1 + e^{4.87 - 0.18x}}$, which is the best-fitting logistic regression model for this data set.

The graph of the best-fitting logistic regression model for the movie and age data is shown in Figure 8-3. Notice it has an S-shaped curve to it. Note that the graph's a downward-sloping S-curve, because higher probabilities of liking the movie are affiliated with lower ages and lower probabilities are affiliated with higher ages. The movie marketers now have the answer to their question. This movie has a higher chance of being well liked by kids (and the younger, the better) and a lower chance of being well liked by adults (and the older they are, the lower the chance of liking the movie).

The point where the probability changes from high to low is between ages 25 and 30. That means that the tide of probability of liking the movie appears to turn from higher to lower in that age range. Using calculus terms, this point is called the *saddle point* of the S-curve, which is the point where the graph changes from concave up to concave down, or vice versa.

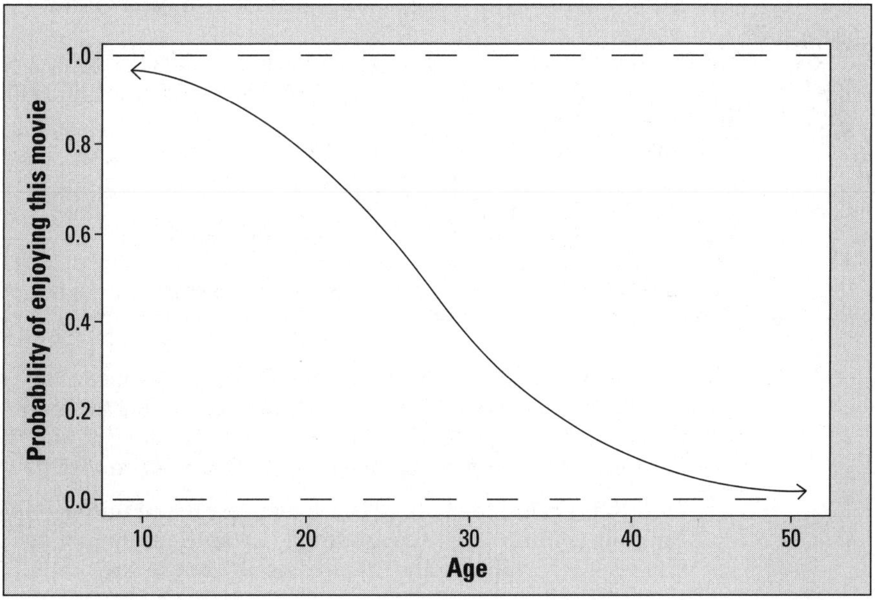


**Figure 8-3:** The best-fitting S-curve for the movie and age data.

# Estimating p

You've determined the best-fitting logistic regression model for your data, obtained the values of $b_0$ and $b_1$ from the logistic regression analysis, and know the precise S-curve that fits your data best (check out the previous sections). You're now ready to estimate $p$ and make predictions about the probability that the event of interest will happen, given the value of the explanatory variable $x$.

To estimate $p$ for a particular value of $x$, plug that value of $x$ into your equation (the best-fitting logistic regression model) and simplify it by using your algebra skills. The number you get is the estimated chance of the event occurring for that value of $x$, and it should be a number between 0 and 1, being a probability and all.

Continuing with the movie and age example from the preceding sections, suppose you want to predict whether a child of age 15 would enjoy the movie. To estimate $p$, plug 15 in for $x$ in the logistic regression model $\hat{p} = \frac{e^{b_0 + b_1 x}}{1 + e^{b_0 + b_1 x}}$ to get $\hat{p} = \frac{e^{4.87 - 0.18 * 15}}{1 + e^{4.87 - 0.18 * 15}} = \frac{e^{2.17}}{1 + e^{2.17}} = \frac{8.76}{9.76} = 0.90$. That answer means you've found a 90 percent chance that a 15-year-old child will like the movie. You can see in Figure 8-3 that when $x$ is 15, $p$ is approximately 0.90. On the other hand, if the person is 50 years old, the chance he will like this movie is $\hat{p} = \frac{e^{4.87 - 0.18 * 50}}{1 + e^{4.87 - 0.18 * 50}}$, or 0.02 (shown in Figure 8-3 for $x = 50$), which is only a 2 percent chance.

REMEMBER

The results you get from a logistic regression analysis, as with any other data analysis, are all subject to the model fitting appropriately. The following section deals with that.

## Checking the fit of the model

To determine whether or not your logistic regression model fits, follow these steps:

1. **Locate the *p*-value of the goodness-of-fit test (found in the Goodness-of-Fit portion of the computer output; see Figure 8-4 for an example); if the *p*-value is larger than 0.05, conclude that your model fits, and if the *p*-value is less than 0.05, conclude that your model doesn't fit.**

2. **Find the *p*-value for the $b_1$ coefficient (it's listed under *P* in the row for your column one [explanatory] variable); if the *p*-value is less than 0.05, the $x$ variable is statistically significant in the model, so it should be included.**

   If the *p*-value is greater than or equal to 0.05, the $x$ variable isn't statistically significant and shouldn't be included in the model.

3. **Look later in the output at the percentage of concordant pairs to determine how well the model fits; the higher the percentage, the better the model fits.**

   That percentage pertains to the number of times that the data and the model actually agree with each other.

REMEMBER

The conclusion in step one based on the *p*-value may seem backwards to you, but here's what's happening: Chi-square goodness-of-fit tests measure the overall difference between what you expect to see via your model versus what you actually observe in your data. (Chapter 15 gives you the lowdown on Chi-square tests.) The null hypothesis (Ho) for this test says you have a difference of zero between what you observed and what you expected from the model; that is, your model fits. The alternative hypothesis, denoted Ha, says that the model doesn't fit. If you get a small *p*-value (under 0.05), reject Ho and conclude the model doesn't fit. If you get a larger *p*-value (above 0.05), you can stay with your model.

WARNING!

Failure to reject Ho here (having a large *p*-value) only means that you can't say your model doesn't fit the population from which the sample came. It doesn't necessarily mean the model fits with 100 percent certainty. Your data could be unrepresentative of the population just by chance.

```
Goodness-of-Fit Test
Method              Chi-Square   DF        P
Pearson                2.83474    9    0.970
Deviance               3.63590    9    0.934
Hosmer-Lemeshow        2.75232    6    0.839

Measures of Association:
 (Between the Response Variable and Predicted Probabilities)
Pairs          Number    Percent     Summary Measures
Concordant        349       87.3     Somers' D               0.80
Discordant         30        7.5     Goodman-Kruskal Gamma   0.84
Ties               21        5.3     Kendall's Tau-a         0.41
Total             400      100.0
```

**Figure 8-4:** The model-fitting part of the movie and age data's logistic regression output.

Using Figure 8-4 to complete the first step of checking the model's fit, you can see many different goodness-of-fit tests. The particulars of each of these tests are beyond the scope of this book; however, in this case (as with most cases), each test has only slightly different numerical results and the same conclusions. All the *p*-values in Column 4 of Figure 8-4 are over 0.80, which is much higher than the 0.05 you need to reject the model. After looking at the *p*-values, the model appears to fit this data.

For step two, you look at the significance of the *x* variable age. In Figure 8-2, you can see the constant for age, –0.18, and farther along in its row, you can see that the *Z*-value is –3.52; this *Z*-value is the test statistic for testing Ho: $\beta_1 = 0$ versus Ha: $\beta_1 \neq 0$. The *p*-value is listed as 0.000, which means it's smaller than 0.001 (a highly significant number). So you know that the coefficient in front of *x,* also known as $\beta_1$, is statistically significant (not equal to zero), and you should include *x* (age) in the model.

To complete step three of the fit-checking process, look at the percentage of concordant pairs reported in Figure 8-4. This value shows the percentage of times the data actually agreed with the model (87.3). To get this result make predictions as to whether the event should have occurred for each individual based on the model and compare those results to what actually happened. Now the logistic regression model is for $p$, the probability of the event occurring, so if $p$ is estimated to be > 0.50 for some value of $x$, your best guess is that the event will occur (versus not occurring). If the estimated value of $p$ is < 0.50 for a particular $x$-value, your best guess is that it won't occur.

For the movie and age data, the percentage of concordant pairs (that is, the percentage of times the model made the right decision in predicting what would happen) is 87.3 percent, which is quite high. The percentage of concordant pairs was obtained by taking the number of concordant pairs and dividing by the total number of pairs. I'd start getting excited if the percentage of concordant pairs got over 75 percent; the higher, the better.

Figure 8-5 shows the logistic regression model for the movie and age data, with the actual values of the observed data added as circles. Much of the time, the model made the right decision; probabilities above 0.50 are associated with more circles at the value of 1, and probabilities below 0.50 are associated with more circles at the value of 0. It's the outcomes that have $p$ near 0.50 that are hard to predict, because the results can go either way.

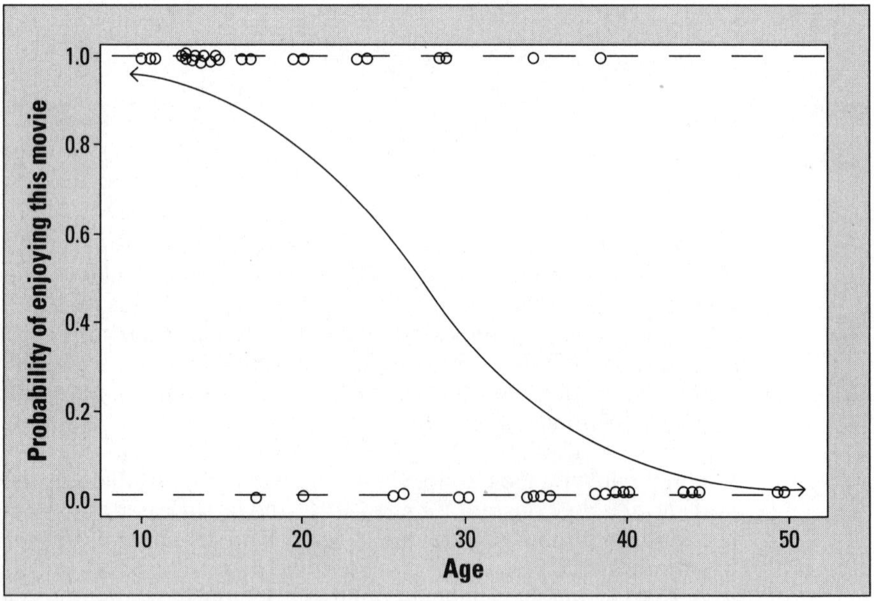


**Figure 8-5:** Actual observed values (0 and 1) compared to the model.

All of this evidence helps confirm that your model fits your data well. You can go ahead and make estimates predictions based on this model for the next individual that comes up, whose outcome you don't know. (See the section "Estimating $p$" earlier in this chapter.)

# Part III
# Comparing Many Means with ANOVA

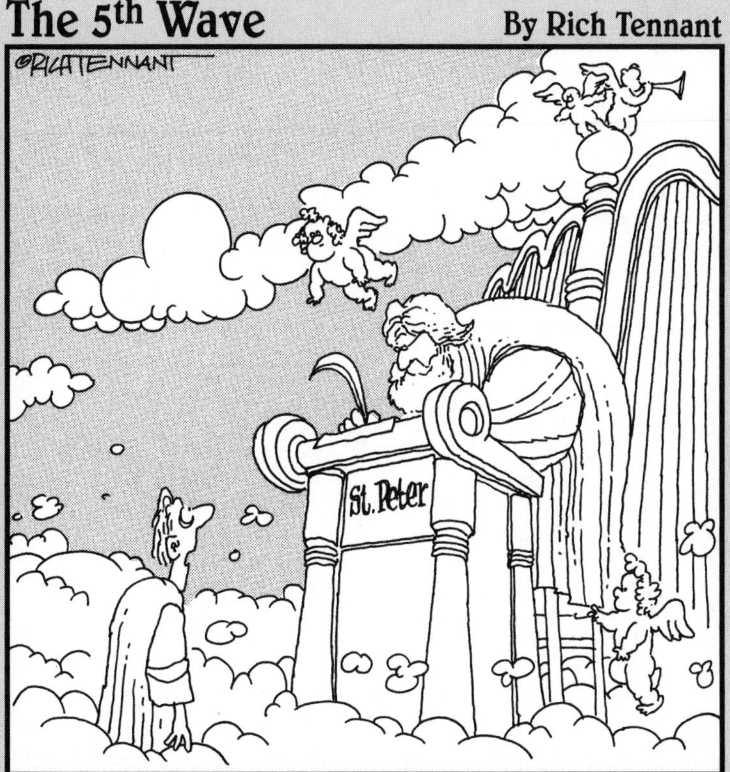


"What do you mean I don't fit your desired sample population at this time?"

## In this part . . .

You get all the nuts and bolts you need to understand one-way and two-way analyses of variance (also known as ANOVA), which compare the means of several populations at one time, based on one or two different characteristics. You see how to read and understand ANOVA tables and computer output and go behind the scenes to understand the big ideas behind the formulas used in ANOVA. (Don't sweat it, I always present formulas only on a need-to-know basis.)

# Chapter 9

# Going One-Way with Analysis of Variance

### In This Chapter

- Extending the $t$-test for comparing two means by using ANOVA
- Discovering and utilizing the ANOVA process
- Carrying out an $F$-test
- Navigating the ANOVA table

One of the most commonly used statistical techniques at the intermediate level is *analysis of variance* (affectionately known as ANOVA). Because the name has the word variance in it, you may think that this technique has something to do with *variance* — and you would be right. Analysis of variance is all about examining the amount of variability in a $y$ (response) variable and trying to understand where that variability is coming from.

One way that you can use ANOVA is to compare several populations regarding some quantitative variable, $y$. The populations you want to compare constitute different groups (denoted by an $x$ variable), such as political affiliations, age groups, or different brands of a product. ANOVA is also particularly suitable for situations involving an experiment where you apply certain treatments $(x)$ to subjects, and you measure a response $(y)$.

In this chapter, you start with the $t$-test for two population means, the precursor to ANOVA. Then you move on to the basic concepts of ANOVA: sums of squares, the $F$-test, and the ANOVA table. You apply these basics to the one-factor or one-way ANOVA, where you compare the responses based only on one treatment variable. (In Chapter 11, you can see them applied to a two-way ANOVA, which has two treatment variables.)

# Comparing Two Means with a t-Test

The *two sample t-test* is designed to test to see whether two population means are different. The conditions for the two sample *t*-test are the following:

- The two populations are independent (in other words, their outcomes don't affect each other).
- The response variable *(y)* is a quantitative variable (meaning that its values represent counts or measurements).
- The *y*-values for each population have a normal distribution (however, their means may be different; that is what the *t*-test determines).
- The variances of the two normal distributions are equal.

For large sample sizes when you know the variances, you use a *Z*-test for the two population means. However, a *t*-test allows you to test two population means when the variances are unknown or the sample sizes are small. This occurs quite often in situations where an experiment is performed and the number of subjects is limited.

Although you have seen *t*-tests before in your intro stats class, it may be good to review the main ideas. The *t*-test tests the hypotheses Ho: $\mu_1 = \mu_2$ versus Ha: $\mu_1$ is $\leq$, $\geq$, or $\neq \mu_2$, where the situation dictates which of these hypotheses you use. (Just a note that with ANOVA, you extend this idea to *k* different means from *k* different populations, and the only version of Ha of interest is $\neq$.)

To conduct the two sample *t*-test, you collect two data sets from the two populations, using two independent samples. To form the test statistic (the *t*-statistic), you subtract the two sample means and divide by the standard error (a combination of the two standard deviations from the two samples and their sample sizes). You compare the *t*-statistic to the *t*-distribution with $n_1 + n_2 - 2$ degrees of freedom and find the *p*-value.

If the *p*-value is less than the prespecified $\alpha$ level, say 0.05, you have enough evidence to say the population means are different. (For information on hypothesis tests, see Chapter 3.)

For example, suppose you're at a watermelon seed spitting contest where contestants each put watermelon seeds in their mouths and spit them as far as they can. Results are measured in inches and are treated with the reverence of the shot-put results at the Olympics. You want to compare the watermelon seed spitting distances of female and male adults. Your data set includes ten people from each group.

You can see the results of the *t*-test in Figure 9-1. The mean spitting distance for females was 47.8 inches; the mean for males was 56.5 inches. The *t*-statistic for the difference in the two means (females – males) is $t = -2.23$, which has a *p*-value of 0.039 (see last line of Figure 9-1 output). At a level of $\alpha = 0.05$, this difference is significant (because $0.039 < 0.05$). You conclude that males and females differ with respect to their mean watermelon seed spitting distance. And you can say males are likely spitting farther because their sample mean was higher.

**Figure 9-1:** A *t*-test comparing mean watermelon seed spitting distances for females versus males.

```
Two-sample T for females vs males

          N    Mean    StDev    SE Mean
females   10   47.80    9.02       2.9
males     10   56.50    8.45       2.7


Difference = mu (females) - mu (males)
Estimate for difference:   -8.70000
95% CI for difference:   (-16.90914, -0.49086)
T-Test of difference = 0 (vs not =): T-Value = -2.23 P-Value = 0.039 DF = 18
```

# Evaluating More Means with ANOVA

Now that you can compare two independent populations inside and out, at some point two populations will not be enough. Suppose you want to compare more than two populations regarding some response variable *(y)*. This idea kicks the *t*-test up a notch into the territory of ANOVA. The ANOVA procedure is built around a hypothesis test called the *F-test,* which compares how much the groups differ from each other, compared to how much variability is in each group. In this section, I set up an example of when to use ANOVA and show you the steps involved in the ANOVA process. You can then apply the ANOVA steps to the following example throughout the rest of the chapter.

## Spitting seeds: A situation just waiting for ANOVA

Before you can jump into using ANOVA, you must figure out what question you want answered and collect the necessary data.

Suppose you want to compare the watermelon seed spitting distances for four different age groups: 6–8, 9–11, 12–14, and 15–17. The hypotheses for this example are Ho: $\mu_1 = \mu_2 = \mu_3 = \mu_4$ versus Ha: At least two of these means are different, where the population means μ represent those from the age groups, respectively. Over the years of this contest, you have collected data on 200 children from each age group, so you have some prior ideas about what the distances typically look like. This year, you have 20 entrants, 5 in each age group. You can see the data from this year, in inches, in Table 9-1.

**Table 9-1 Watermelon Seed Spitting Distances for Four Child Age Groups (Measured in Inches)**

| *6–8 Years* | *9–11 Years* | *12–14 Years* | *15–17 Years* |
|---|---|---|---|
| 38 | 38 | 44 | 44 |
| 39 | 39 | 43 | 47 |
| 42 | 40 | 40 | 45 |
| 40 | 44 | 44 | 45 |
| 41 | 43 | 45 | 46 |

Do you think you see a difference in distances for these age groups based on this data? If you just combined all the data, you would see quite a bit of difference (the range of the combined data goes from 38 inches to 47 inches). Perhaps accounting for which age groups each contestant is in does explain at least some of what's going on. But don't stop there. In the next section, you see the official steps you need to do to answer your question.

## Walking through the steps of ANOVA

You have decided on the quantitative response variable *(y)* you want to compare for your *k* various population (or treatment) means, and you collected a random sample of data from each population. Now you're ready to conduct ANOVA on your data to see whether the population means are different for your response variable, *y*.

REMEMBER

The characteristic that defines these populations is called the *treatment variable, x*. Statisticians use the word *treatment* in this context because one of the biggest uses of ANOVA is for designed experiments where subjects are randomly assigned to treatments, and the responses are compared for the various treatment groups. So statisticians oftentimes use the word *treatment* even when the study isn't an experiment, and they're comparing regular populations. Hey, don't blame me! I'm just following the proper statistical terminology.

Just to get a feeling for what an ANOVA procedure involves and to give you a quick reference for a later time, here are the general steps in a one-way ANOVA:

1. **Check the ANOVA conditions, using the data collected from each of the *k* populations.**

   See the next section, "Checking the conditions," for the specifics on these conditions.

2. **Set up the hypotheses Ho: $\mu_1 = \mu_2 = \ldots = \mu_k$ versus Ha: At least two of the population means are different.**

   Another way to state your alternative hypothesis is by saying Ha: At least two of $\mu_1, \mu_2, \ldots \mu_k$ are different.

3. **Collect data from *k* random samples, one from each population.**

4. **Conduct an *F*-test on the data from step three, using the hypotheses from step two, and find the *p*-value.**

   See the section "Doing the *F*-test" later in this chapter for these instructions.

5. **Make your conclusions: If you reject Ho (when your *p*-value is less than 0.05 or your prespecified $\alpha$ level), you conclude that at least two of the population means are different; otherwise, you conclude that you didn't have enough evidence to reject Ho (you can't say the means are different).**

If these steps look like a foreign language to you, don't fear — I describe each of these steps in detail in the sections to follow.

# Checking the Conditions

Step one of ANOVA is checking to be sure all necessary conditions are met before diving into the data analysis. The conditions for using ANOVA are just an extension of the conditions for a *t*-test (see the section "Comparing Two Means with a *t*-Test"). The following conditions all need to hold in order for ANOVA to be conducted:

- The *k* populations are independent (in other words, their outcomes don't affect each other).
- The *k* populations each have a normal distribution.
- The variances of the *k* normal distributions are equal.

I go into more detail about these conditions in the following sections.

## Checking off independence

To check the first condition, examine how the data was collected from each of the separate populations. In order to maintain independence, the outcomes from one population can't affect the outcomes of the other populations. If the data has been collected by using a separate random sample from each population (*random* here meaning that each individual in the population had an equal chance of being selected), this factor ensures independence at the strongest level.

In the watermelon seed spitting data (see Table 9-1), the data aren't randomly sampled from each age group because the data represents everyone who participated in the contest. But, you can argue that the seed spitting distances from one age group don't affect the seed spitting distances from the other age groups, so the independence assumption is okay here also.

## Looking for what's normal

The second ANOVA condition is that each of the $k$ populations has a normal distribution. To check this condition, make a separate histogram of the data from each group and see whether it resembles a normal distribution. Data from a normal distribution should look symmetric (in other words, if you split the histogram down the middle, it looks the same on each side) and have a bell-shape. Don't expect the data in each histogram to follow a normal distribution exactly (remember it's only a sample), but it shouldn't be extremely different from a normal, bell-shaped distribution.

Since the data contains only five children per age group, checking conditions can be iffy. But in this case, you have past data for 200 children in each age group, so you can use that to check the conditions. The histograms and descriptive statistics of the seed spitting data for the four age groups are shown in Figure 9-2, all in one panel, so you can easily compare them to each other on the same scale. Looking at the four histograms in Figure 9-2, you can see that each graph resembles a bell shape; the normality condition isn't being violated here. (Red flags should come up if you see two peaks in the data, or a skewed shape where the peak is off to one side, or if the histogram is flat, for example.)

You can use Minitab to make histograms for each of your samples and have all of them appear on one large panel, all using the same scale. To do this, go to Graph>Histogram and click OK. Choose the variables that represent data from each sample by highlighting them in the left-hand box and clicking Select. Then click on Multiple Graphs, and a new window opens. Under the Show Graph Variables option, check the following box: In separate panels of the same graph. On the Same Scales for Graphs option, check the box for $x$ and the box for $y$. This option gives you the same scale on both the $x$ and $y$ axes for all the histograms. Then click OK.

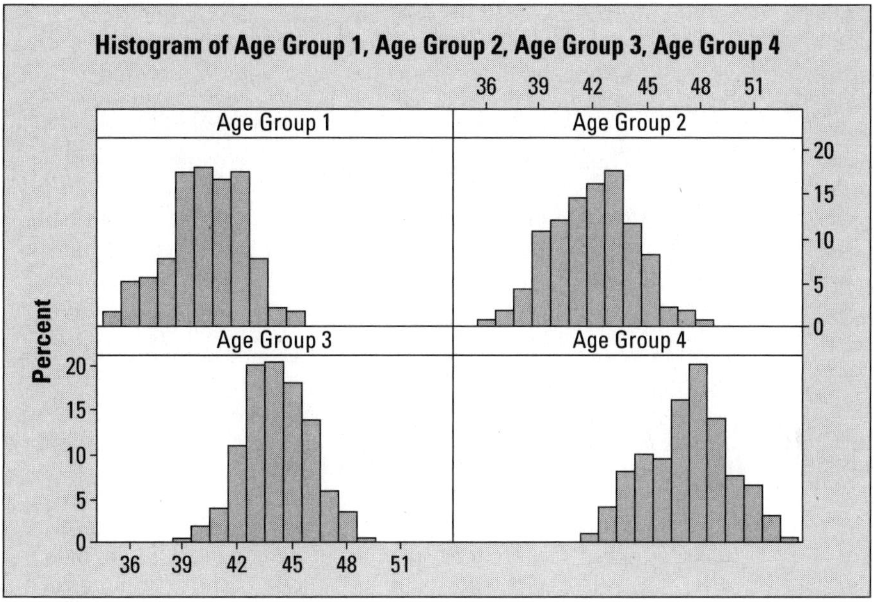


**Figure 9-2:** Checking ANOVA conditions by using histograms and descriptive statistics.

**Descriptive Statistics: Age Group 1, Age Group 2, Age Group 3, Age Group 4**

| Variable | Total Count | Mean | Variance |
|---|---|---|---|
| Age Group 1 | 200 | 40.116 | 4.256 |
| Age Group 2 | 200 | 41.880 | 4.994 |
| Age Group 3 | 200 | 44.165 | 3.249 |
| Age Group 4 | 200 | 47.405 | 5.154 |

## Taking note of spread

The third condition for ANOVA is that the variance in each of the *k* populations is the same. To check this out on your data, use Minitab to find the variance in each sample and compare them. The variances for each sample should be close. What does *close* mean? A hypothesis test can handle this question; however, it falls outside the scope of most intermediate statistics courses. So you are left with a judgment call. Compare all the variances as a group and look for any glaring differences. If a difference is large enough for you to write home about (say 10 percent or more), this variance indicates a problem. (Not only do you have a problem with the ANOVA conditions, but if you're writing your mom about your stats problems you might need to get a bit of a life.) If no big differences exist in the variances, you can say that the equal variance condition is met. The variances for the seed spitting data are shown in Figure 9-2 for each age group. They are quite close, so this condition is met.

To find descriptive statistics for each sample, go to Stat>Basic Statistics> Display Descriptive Statistics. Click on each variable in the left-hand box for which you want the descriptive statistics and then click Select. Click on the Statistics option, and a new window appears with tons of different types of statistics. Click on the ones you want and click off the ones you don't want. Click OK. Then click OK again. Your descriptive statistics are calculated.

Note that you don't need the sample sizes in each group to be equal to carry out ANOVA; however, in intermediate stats, you'll typically see what statisticians call a *balanced design,* where each sample from each population has the same sample size. (For more precision in your data, the larger the sample sizes, the better; see Chapter 3.)

# Setting Up the Hypotheses

Step two of ANOVA is setting up the hypotheses to be tested. You're testing to see whether or not all the population means can be deemed equal to each other. The null hypothesis for ANOVA is that all the population means are equal. That is, Ho: $\mu_1 = \mu_2 = \ldots = \mu_k$, where $\mu_1$ is the mean of the first population, $\mu_2$ is the mean of the second population, and so on until you reach $\mu_k$ (the mean of the $k^{th}$ population).

Now what appears in the alternative hypothesis (Ha) must be the opposite of what is in the null hypothesis (Ho). What's the opposite of having all *k* of the population's means equal to each other? You may think the opposite is that they're all different. But that's not the case. In order to blow Ho wide open, all you need is for at least two of those means to not be equal. The alternative hypothesis, Ha, is that at least two of the population means are different from each other. That is, Ha: At least two of $\mu_1, \mu_2, \ldots \mu_k$ are different.

Note that Ho and Ha for ANOVA are an extension of the hypotheses for a two sample *t*-test (which only compares two independent populations). And while the alternative hypothesis in a *t*-test may be that one mean is greater than, less than, or not equal to the other, you don't consider any alternative other than $\neq$ in ANOVA. You only want to know whether or not the means are equal — at this stage of the game anyway. After you reach the conclusion that Ho is rejected in ANOVA, you can proceed to figure out how the means are different, which ones are bigger than others, and so on, using multiple comparisons. Those details appear in Chapter 10.

# Doing the F-Test

Step three, collecting the data, includes taking *k* random samples, one from each population. Step four of ANOVA is doing the *F*-test on this data, which is the heart of the ANOVA procedure. This test is the actual hypothesis test of Ho: $\mu_1 = \mu_2 = \ldots = \mu_k$ versus Ha: At least two of $\mu_1, \mu_2, \ldots \mu_k$ are different.

You have to carry out three major steps in order to complete the *F*-test (don't get these steps confused with the main ANOVA steps; consider the *F*-test a few steps within a step):

1. **Break down the variance of *y* into sums of squares.**
2. **Find the mean sums of squares.**
3. **Put the mean sums of squares together to form the *F*-statistic.**

I describe each step of the *F*-test in detail and apply it to the example of comparing watermelon seed spitting distances (see Table 9-1) in the following sections.

Because data analysts rely heavily on computer software to conduct each step of the *F*-test, you can do the same. All computer software packages organize and summarize the important information from the *F*-test into a table format for you. This table of results for ANOVA is called (what else?) the *ANOVA table*. Because the ANOVA table is a critical part of the entire ANOVA process, I start the following sections out by describing how to run ANOVA in Minitab to get the ANOVA table, and I continue to reference this section as I describe each step of the ANOVA process.

## Running ANOVA in Minitab

Using Minitab to run ANOVA, you first have to enter the data from the *k* samples. You can enter the data one in of two ways:

- **Stacked data** means that you enter all the data into two columns. Column one includes the number indicating what sample the data value is from (1 to *k*), and the responses *(y)* are in column two. To analyze this data, go to Stat>ANOVA>One-Way Stacked. Highlight the response *(y)* variable and click Select. Highlight the factor (population) variable and click Select. Click OK.
- **Unstacked** is the other method of entering data: a separate column for the data in each sample. To analyze the data entered this way, go to Stat>ANOVA>One-Way Unstacked. Highlight the names of the columns where your data are located. Click OK.

I typically use the unstacked version just because I think it helps visualize the data. However, the choice is up to you, and the results come out the same no matter which one you choose.

## Breaking down the variance into sums of squares

The first step of the *F*-test is splitting up the variability in the *y* variable into portions that define where the variability is coming from. The term *analysis of variance* is a great description for exactly how you conduct a test of *k* population means. With the overall goal of testing whether *k* population (or treatment) means are equal, you take a random sample from each of the *k* populations. You first put all the data together into one big group and measure how much total variability there is; this variability is called the *sums of squares total,* or SSTO. If the data are really diverse, SSTO is large. If the data are very similar, SSTO is small.

Now the total variability in the combined data set (SSTO) can be split into two parts:

- **SST:** The variability between the groups, known as the sums of squares for treatment
- **SSE:** The variability within the groups, known as the sum of squares for error

This splitting up of the variability in your data results in one of the most important equalities in ANOVA. That equality is SSTO = SST + SSE.

The formula for SSTO is the numerator of the formula for $s^2$, the variance of a single data set, so $SSTO = \Sigma\Sigma\left(x_{ij} - \overline{x}\right)^2$, where *i* and *j* represent the $j^{th}$ value in the sample from the $i^{th}$ population. SSTO represents the total squared distance between the data values and their overall mean. The formula for SST is $SST = n_i \Sigma\left(\overline{x}_i - \overline{x}\right)^2$, where $n_i$ is the size of the sample coming from the $i^{th}$ population. SST represents the total squared distance between the means from each sample and the overall mean. The formula for SSE is $SSE = \Sigma\Sigma\left(x_{ij} - \overline{x}_i\right)^2$, where $x_{ij}$ is the $j^{th}$ value in the sample from the $i^{th}$ population and $\overline{x}_i$ is the mean of the sample coming from the $i^{th}$ population. This formula represents the total squared distance between the values in each sample and their corresponding sample means. Using algebra, you can show (with some serious elbow grease) that SSTO = SST + SSE.

The Minitab output for the watermelon seed spitting contest for the four age groups is shown in Figure 9-3. Under the Source column of the ANOVA table, you see *Factor* listed in row one. The factor variable (as described by Minitab) represents the treatment or population variable. In column three of the Factor row, you see the SST, which is equal to 89.75. In the Error row (row two), you locate the SSE in column three, which equals 56.80. In row three (Total), column three, you see the SSTO, which is 146.55. Using the values of SST, SSE, and SSTO from the Minitab output, you can verify that SST + SSE = SSTO.

**Figure 9-3:** ANOVA Minitab output for the watermelon seed spitting example.

**One-Way ANOVA: Age Group 1, Age Group 2, Age Group 3, Age Group 4**

| Source | DF | SS | MS | F | P |
|---|---|---|---|---|---|
| Factor | 3 | 89.75 | 29.92 | 8.43 | 0.001 |
| Error | 16 | 56.80 | 3.55 | | |
| Total | 19 | 146.55 | | | |

S = 1.884   R-Sq = 61.24%   R-Sq(adj) = 53.97%

Now you're ready to use these sums of squares to complete the next step of the *F*-test (keep reading).

## Locating those mean sums of squares

After you have the sums of squares for treatment, SST, and the sums of squares for error, SSE (see preceding section for more on these), you want to compare them to see whether the variability in the *y*-values that is due to the model (SST) is large compared to the amount of error left over in the data after the groups have been accounted for (SSE). So you ultimately want a ratio comparing SST to SSE somehow. To make this ratio form a statistic that statisticians know how to work with (in this case, an *F*-statistic), they decided to find the mean of each of SST and SSE and work with that. Finding the mean sums of squares is the second step of the *F*-test.

*MST* is the mean sums of squares for treatments, which measures the mean variability that occurs between the different treatments (the different samples in the data). What you're looking for is the amount of variability in the data as you move from one sample to another. A great deal of variability between samples (treatments) may indicate that the populations are different as well. You can find MST by taking SST and dividing by $k - 1$ (where $k$ is the number of treatments).

*MSE* is the mean sums of squares for error, which measures the mean within-treatment variability. The *within-treatment variability* is the amount of variability that you see within each sample itself, due to chance and/or other factors not included in the model. You can find MSE by taking SSE divided by $n - k$ (where $n$ is the total sample size and $k$ is the number of treatments). The values of $k - 1$ and $n - k$, respectively, are called the *degrees of freedom* for SST and SSE. Minitab calculates and posts the degrees of freedom for SST and SSE, as well as the values of MST and MSE, in the ANOVA table in columns two and four, respectively.

From the ANOVA table for the seed spitting data in Figure 9-3, you can see that column two has the heading *DF,* which stands for degrees of freedom. You can find the degrees of freedom for SST in the Factor row (row two); this value is equal to $k - 1 = 4 - 1 = 3$. The degrees of freedom for SSE is found to be $n - k = 20 - 4 = 16$. (Remember you have four age groups and five children in each group for a total of $n = 20$ data values.) The degrees of freedom for SSTO is $n - 1 = 20 - 1 = 19$ (found in the Total row under DF.) You can verify that the degrees of freedom for SSTO = degrees of freedom for SST + degrees of freedom for SSE.

The values of MST and MSE are shown in column four of Figure 9-3, with the heading *MS.* You can see the MST in the Factor row, which is 29.92. This value was calculated by taking SST = 89.75, and dividing it by degrees of freedom, 3. You can see MSE in the Error row, equal to 3.55. MSE is found by taking SSE = 56.80 and dividing that value by its degrees of freedom, 16.

By finding the mean sums of squares, you've completed step two of the *F*-test, but don't stop here! You need to continue to the next section if you want to complete the process.

# Figuring the F-statistic

The test statistic for the test of the equality of the *k* population means is $F = \frac{MST}{MSE}$. The result of this formula is called the *F-statistic.* The *F*-statistic has an *F*-distribution, which is equivalent to the square of a *t*-test (when the numerator degrees of freedom is 1). All *F*-distributions start at zero and are skewed to the right. The degree of curvature and the height of the curvature of each *F*-distribution is reflected in two *degrees of freedom,* represented by $k - 1$ and $n - k$. (These come from the denominators of MST and MSE, respectively, where *n* is the total sample size and *k* is the total number of treatments or populations.) A shorthand way of denoting the *F*-distribution for this test is $F_{(k-1,\, n-k)}$.

In the watermelon seed spitting example, you're comparing four means and have a sample of size five from each population. Figure 9-4 shows the corresponding *F*-distribution, which has degrees of freedom $4 - 1 = 3$ and $20 - 4 = 16$; in other words $F_{(3,\, 16)}$.

You can see the *F*-statistic on the Minitab ANOVA output (see Figure 9-3) in the Factor row, under the column indicated by *F.* For the seed spitting example, the value of the *F*-statistic is 8.43. This number was found by taking MST = 29.92 divided by MSE = 3.55. You can then locate 8.43 on the *F*-distribution in Figure 9-4 to see where it stands. (More on that in the next section.)

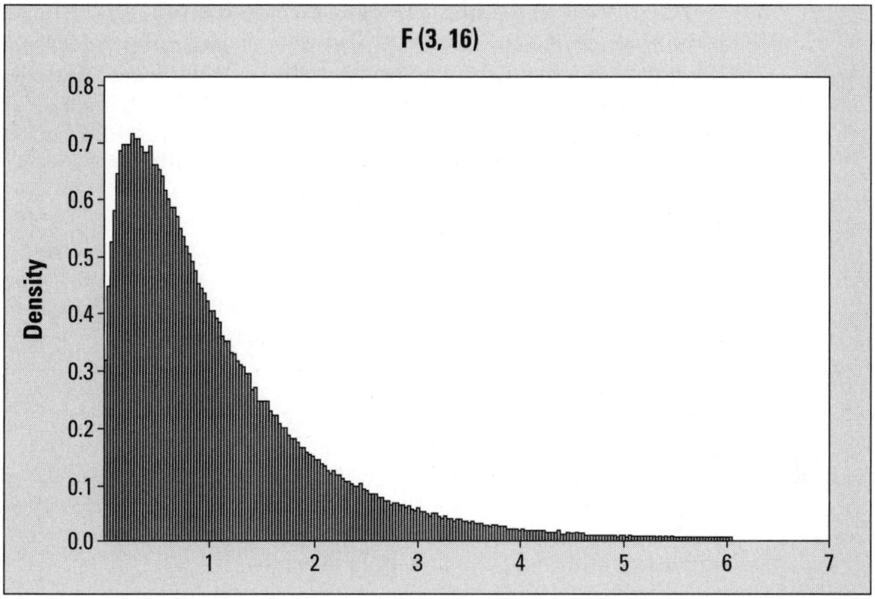


**Figure 9-4:** *F*-distribution with (3, 16) degrees of freedom.

Be sure to not to exchange the order of the degrees of freedom for the *F*-distribution. The difference between $F_{(3, 16)}$ and $F_{(16, 3)}$ is big.

## Making conclusions from ANOVA

If you've completed the *F*-test and found your *F*-statistic (step four in the ANOVA process), you're ready for step five of ANOVA: making conclusions for your hypothesis test of the *k* population means. If you haven't already, you can compare the *F*-statistic to the corresponding *F*-distribution with $k - 1$, $n - k$ degrees of freedom, to see where it stands and make a conclusion. You can make the conclusion in one of two ways: the *p*-value approach or the critical-value approach. (The approach you use depends primarily on whether you have access to a computer, especially during exams.) I describe these two approaches in the following sections.

### Using the p-value approach

On Minitab ANOVA output (see Figure 9-3), the value of the *F*-statistic is located in the Factor row, under the column noted by *F*. The associated *p*-value for the *F*-test is located in the Factor row under the column headed by *P*. The *p*-value tells you whether or not you can reject Ho. If the *p*-value is less than your prespecified $\alpha$ (typically 0.05), reject Ho. Conclude that the *k* population means aren't all equal and that at least two of them are different. If the *p*-value is greater than $\alpha$, then you can't reject Ho. You don't have enough evidence in your data to say the *k* population means have any differences.

The *F*-statistic for comparing the mean watermelon seed spitting distances for the four age groups is 8.43. The *p*-value as indicated in Figure 9-3 is 0.001. That means the results are highly statistically significant. You reject Ho and conclude that at least one pair of age groups differ in its mean watermelon seed spitting distances. (You would hope that a 17-year-old could do a lot better than a 6-year-old, but maybe those 6-year-olds have a lot more spitting going on in their lives than 17-year-olds do.)

Using Figure 9-4, you see how the *F*-statistic of 8.43 stands on the *F*-distribution with (4 – 1, 20 – 4) = (3, 16) degrees of freedom. You can see it's way off to the right, out of sight. It makes sense that the *p*-value, which measures the probability of being beyond that *F*-statistic, is 0.001.

### If you've gotta use critical values . . .

If you're in a situation where you don't have access to a computer (as is still the case in many statistics courses today when it comes to taking exams), finding the exact *p*-value for the *F*-statistic isn't possible. However, statistical software packages automatically calculate all *p*-values exactly (so on any computer output you can see them as such).

To approximate the *p*-value from your *F*-statistic (in the event you don't have a computer or computer output available), you find a cutoff value on the *F*-distribution with ($k - 1$, $n - k$) degrees of freedom that draws a line in the sand between rejecting Ho and not rejecting Ho. This cutoff (also known as the *critical value*) is determined by your prespecified $\alpha$ (typically 0.05). You choose the critical value so that the area to its right on the *F*-distribution is equal to $\alpha$.

Table A-5 in the Appendix shows the critical values of the *F*-distribution with various degrees of freedom, all using $\alpha = 0.05$. Other *F*-distribution tables are available in various statistics textbooks and Internet links for other values of $\alpha$; however, $\alpha = 0.05$ is by far the most common $\alpha$ level used for the *F*-distribution and is sufficient for your purposes.

This table of values for the *F*-distribution is called the *F-table* (students are typically given these with their exams). For the seed spitting example, the *F*-statistic has an *F*-distribution with degrees of freedom (3, 16), which I calculate in a previous section. To find the critical value, go to Table A-5 in the Appendix. Because the degrees of freedom are (3, 16), go to column 3 and row 16 on the *F*-table. The critical value is 3.2389 (or 3.24). Your *F*-statistic for the seed spitting example is 8.43, which is well beyond this critical value (you can see how 8.43 compares to 3.24 by looking at Figure 9-4). Your conclusion is to reject Ho at level $\alpha$. At least two of the age groups differ on mean seed spitting distances.

With the critical value approach, any *F*-statistic that lies beyond the critical value results in rejecting Ho, no matter how far or close to the line it is. If your *F*-statistic is beyond the value found in Table A-5, then you reject Ho and say at least two of the treatments (or populations) have different means.

### What's next?

After you've rejected Ho in the *F*-test and concluded that not all the populations means are the same, your next question may be: Which ones are different? You can answer that question by using a statistical technique called *multiple comparisons.* Statisticians use many different multiple comparison procedures to further explore the means themselves after the *F*-test has been rejected. I discuss and apply some of the more common multiple comparison techniques in Chapter 10.

## Checking the Fit of the ANOVA Model

As with any other model, you must determine how well the ANOVA model fits before you can use its results with confidence. In the case of ANOVA, the model basically boils down to a treatment variable (also known as the population you're in) plus an error term. To assess how well that model fits the data, see the values of $R^2$ and $R^2$ adjusted on the last line of the ANOVA output below the ANOVA table. For the seed spitting data, you see those values at the bottom of Figure 9-3.

The value of $R^2$ measures the percentage of the variability in the response variable *(y)* explained by the explanatory variable *(x).* In the case of ANOVA, the *x* variable is the factor due to treatment (where the treatment can represent a population being compared). A high value of $R^2$ (say above 80 percent) means this model fits well. The value of $R^2$ adjusted, the preferred measure, takes $R^2$ and adjusts it for the number of variables in the model. In the case of one-way ANOVA, you have only one variable, the factor due to treatment so $R^2$ and $R^2$ adjusted won't be very far apart. For more on $R^2$ and $R^2$ adjusted, see Chapter 5.

For the watermelon seed spitting data, the value of $R^2$ adjusted (as found in the last row of Figure 9-3) is only 53.97 percent. That means age group (while shown to be statistically significant by the *F*-test; see the section "Making conclusions from ANOVA") explains just over half of the variability in the watermelon seed spitting distances. Because age group alone explains only a little over half of what's going on in the seed spitting distances, you may find other variables you can examine in addition to age group, making an even better model.

The results of the *t*-test done to compare the spitting distances of males and females in the section "Comparing Two Means with a *t*-Test" (see Figure 9-1) showed that males and females were significantly different on mean seed spitting distances. So I would venture a guess that if you include gender as well as age group thereby creating what statisticians call a *two-factor ANOVA* (or *two-way ANOVA*), the resulting model would fit the data even better, resulting in higher values of $R^2$ and $R^2$ adjusted. (See Chapter 11 for two-way ANOVA.)

## Up-front rejection the best policy for most refusal letters

Many medical and psychological studies use designed experiments to compare the responses of several different treatments, looking for differences. A *designed experiment* is a study in which subjects are randomly assigned to treatments (experimental conditions) and their responses are recorded. The results are used to compare treatments to see which one(s) work best, which ones work equally well, and so on.

One example of one such experiment that employs ANOVA is from The Ohio State University research press release Web site. The experiment tested three traditional principles of writing refusal letters:

- Using a buffer — a neutral or positive sentence that delays the negative information
- Placing the reason before the refusal
- Ending the letter on a positive note as a way of reselling the business

Subjects were randomly assigned to treatments, and their responses to the rejection letters were compared (likely on some sort of scale such as 1 = very negative to 7 = very positive with 4 being a neutral response).

This scenario can be analyzed by using ANOVA. It compares three treatments (forms of the rejection letters) on some quantitative variable (response to the letter). You can argue that this isn't a continuous variable, because it has enough possible values that ANOVA isn't unreasonable. The data were also shown to have a bell shape.

The null hypothesis would be Ho: Mean responses to the three types of rejection letters are equal, versus Ha: At least two forms of the rejection letter resulted in different mean responses.

In the end, the researcher did find some significant results. In other words, the different ways the rejection letter was written affected the participants in different ways. Using multiple comparison procedures (see Chapter 10), you would be able to go in and determine which forms of the rejection letters gave different responses and how the responses differed.

So in case you have to write a rejection letter at some point, the researcher recommends the following guidelines for writing it:

- Don't use buffers to begin negative messages.
- Give a reason for the refusal when it makes the sender's boss look good.
- Present the negative positively but clearly; offer an alternative or compromise if possible.
- A positive ending isn't necessary.

# Chapter 10

# Pairing Things Down with Multiple Comparisons

### In This Chapter

- When and how to follow up ANOVA with multiple comparisons
- Comparing two well-known multiple comparison procedures

You're comparing the means of not two, but *k* independent populations, and you find out (using ANOVA — see Chapter 9) that you reject Ho: All the population means are equal, and you conclude Ha: At least two of the population means are different. Now you gotta know — which of those populations are different? Answering this question requires a follow-up procedure to ANOVA called *multiple comparisons*, which makes sense because you want to compare the multiple means you have and see which ones are different.

In this chapter, you figure out when you need to use a multiple comparison procedure. You see two of the most well-known multiple comparison procedures: Fisher's LSD (least significant difference) and Tukey's test. They can help you answer that burning question: So some of the means are different, but which ones are different?

## Following Up after ANOVA

This section runs through the ANOVA procedure in the case where Ho is rejected and leads you to the next step: multiple comparisons.

Suppose you want to compare the average number of cell-phone minutes used per month for children and young adults, where the age groups are the following:

- Group 1: 19 years old and under
- Group 2: 20-39 years old

- Group 3: Adult males 40-59 years old
- Group 4: Adult females 60 years old and over

You collect data on a random sample of 10 people from each group (where no one knows anyone else to keep independence), and you record the number of minutes each person used their cell phone in one month. The first ten lines of a hypothetical data set are shown in Table 10-1.

**Table 10-1 One Month's Cell Phone Minutes for Four Age Groups**

| *19 and Under (Group 1)* | *20–39 (Group 2)* | *40–59 (Group 3)* | *60 and Over (Group 4)* |
|---|---|---|---|
| 800 | 250 | 700 | 200 |
| 850 | 350 | 700 | 120 |
| 800 | 375 | 750 | 150 |
| 650 | 320 | 650 | 90 |
| 750 | 430 | 550 | 20 |
| 680 | 380 | 580 | 150 |
| 800 | 325 | 700 | 200 |
| 750 | 410 | 700 | 130 |
| 690 | 450 | 590 | 160 |
| 710 | 390 | 650 | 30 |

The means and standard deviations of the sample data are shown in Figure 10-1, as well as confidence intervals for each of the population means separately (see Chapter 3 for info on confidence intervals). Looking at Figure 10-1, it appears that all four means are different, with 19 and under heading the pack, with 40- to 59-year-olds not far behind, and with 20- to 39-year-olds and those over 60 bringing up the rear (in that order).

Knowing that man can't live by sample results alone, you decide that ANOVA is needed to see whether any differences that appear in the samples can be extended to the population (see Chapter 9). By using the ANOVA procedure, you test whether the average cell minutes used is the same across all groups. The results of the ANOVA, using the data from Table 10-1, are shown in Figure 10-2.

**Figure 10-1:** Basic statistics and confidence intervals for the cell-phone data.

```
                          Individual 95% CIs For Mean Based on
                                       Pooled StDev
Level     N     Mean      StDev    ------+---------+---------+---------+---
Group 1   10   748.00     64.60                                    (-*-)
Group 2   10   368.00     59.08                     (-*-)
Group 3   10   657.00     64.99                                (-*-)
Group 4   10   125.00     62.41    (-*-)
                                   ------+---------+---------+---------+---
                                        200       400       600       800
```

Looking at Figure 10-2, the *F*-test for equality of all four population means has a *p*-value of 0.000, meaning it is less then 0.001. That says at least two of these groups have a significant difference in their cell-phone use (see Chapter 9 for info on the *F*-test and its results).

**Figure 10-2:** ANOVA results for comparing cell-phone use for four age groups.

**One-way ANOVA: Group 1, Group 2, Group 3, Group 4**

```
Source    DF        SS      MS       F      P
Factor     3   2416010  805337  204.13  0.000
Error     36    142030    3945
Total     39   2558040

S = 62.81   R-Sq = 94.5%   R-Sq(adj) = 93.99%
```

Okay, so what's your next question? You just found out that the average number of cell-phone minutes per month isn't the same across these four groups. Remember, this doesn't mean all four groups are different (see Chapter 9). However, it does mean that at least two groups are significantly different in their cell-phone use. So your questions are: Which groups are different, and how are they different?

Determining which populations have differing means after ANOVA has been rejected involves a new data-analysis technique called *multiple comparisons.* While many different multiple comparison procedures are out there, statisticians have their favorites, which I present in the next section.

Don't attempt to explore the data with a multiple comparison procedure if the test for equality of the populations isn't rejected. In this case, you must conclude that you don't have enough evidence to say the population means aren't equal, so you must stop there. Always look at the *p*-value of the *F*-test on the ANOVA output before moving on to conduct any multiple comparisons.

# Pinpointing Differing Means with Fisher and Tukey

You've conducted ANOVA to see whether a group of *k* populations have the same mean, and you rejected Ho. You conclude that at least two of those populations have different means. But you don't have to stop there; you can go on to find out how many and which means are different by conducting multiple comparison tests.

In this section, you see two of the most well-known multiple comparison procedures: *Fisher's paired differences* (also known as *Fisher's test* or *Fisher's LSD*) and *Tukey's simultaneous confidence intervals* (also known as *Tukey's test*).

Although I only discuss two procedures in this section, tons of other multiple comparison procedures are out there. Although the other procedures' methods differ a great deal, their overall goal is the same: to figure out which population means differ by comparing their sample means.

## Fishing for differences with Fisher's LSD

In this section, I outline Fisher's LSD and apply it to the cell-phone example.

### Examining Fisher's LSD procedure

Suppose you're comparing *k* population means. Fisher's LSD (short for *least significant difference*) conducts a *t*-test on each of the $\frac{k(k-1)}{2}$ pairs of populations in the study, each one at level $\alpha = 0.05$. For example, if you have four populations labeled A, B, C, D, you would have $\frac{4(4-1)}{2} = 6$ *t*-tests to perform: A versus B; A versus C; A versus D; B versus C; B versus D; and C versus D.

The number of tests is calculated by knowing that you have *k* possible means for the first one in the pair, then $k - 1$ left for the second one in the pair. Because the order of the means doesn't matter, you can divide by 2 to avoid overcounting.

Fisher's LSD is very straightforward, easy to conduct, and easy to understand. However, Fisher's LSD has some issues. Because each *t*-test is conducted at $\alpha$ level 0.05, each test done has a 5 percent chance of making a Type I error (rejecting Ho when you shouldn't have — see Chapter 3). Although a 5-percent error rate for each test doesn't seem too bad, the errors have a multiplicative effect as the number of tests increases. For example, the chance of making at least one Type I error with six *t*-tests, each at level $\alpha = 0.05$, is 26.50 percent, which would be your *overall error rate* for the procedure.

You could help lower the error rate for Fisher's test if you lower the value of $\alpha$ for each test from 0.05 to, say, 0.01. However, doing so makes it harder to reject Ho for each pair of means. A lower value of $\alpha$ also doesn't solve the error-rate problem; it just slows it down for a bit, until the number of tests gets larger, and the error rate goes back up again.

If you want or need to know how I arrived at the number 26.50 percent as the overall error rate in that last example, here it goes: The probability of making a Type I error for each test is 0.05. The chance of making at least one error in six tests equals one minus the probability of making no errors in six tests. The chance of not making an error in one test is $1 - \alpha = 0.95$. The chance of no error in six tests is this quantity times itself six times, or $(0.95)^6$, which equals 0.735. Now take one minus this quantity to get $1 - 0.735 = 0.2650$ or 26.50 percent.

To conduct Fisher's LSD, go to Stat>ANOVA>One-way or One-way unstacked. (If your data appear in two columns with Column 1 representing the population number and Column 2 representing the response, just click One-way because your data is stacked. If your data is shown in *k* columns, one for each of the *k* populations, click One-way unstacked.) In either case, the next step is to highlight the data for the groups you're comparing and click Select. Then click on Comparisons. Click on Fisher's. The individual error rate is listed at 5 (percent), which is typical. If you want to change it, type in the desired error rate (between 0.5 and 0.001) and click OK. You may type in your error rate as a decimal, 0.05, or as a number greater than one, such as 5. Numbers greater than one are interpreted as a percentage.

### Applying Fisher's LSD to cell phones

An ANOVA procedure was done on the cell-phone data presented in Table 10-1 to compare the mean number of minutes used for four age groups. Looking at Figure 10-2, you see Ho (all the populations means are equal) was rejected. The next step is to conduct multiple comparisons by using Fisher's LSD to see which population means differ. Figure 10-3 shows the Minitab output.

The first block of results shows "Group 1 subtracted from" where Group 1 = age 19 and under. Each line after that represents the other age groups (Group 2 = 20- to 39-year-olds, Group 3 = 40- to 59-year-olds, and Group 4 = 60 and over). Each line shows the results of comparing the mean for the other group minus the mean for Group 1. For example, the first line shows Group 2 being compared with Group 1.

Moving to the right in that same row, you see the confidence interval for the difference in these two means, which turns out to be –436.97 to –323.03. Because 0 isn't contained in this interval, you conclude that these two means are different in the populations also. You can also say, because this difference $\mu_2 - \mu_1$ is negative, that $\mu_2$ is less than $\mu_1$. Or, a better way to think of it may be that $\mu_1$ is greater than $\mu_2$. That is, Group 1's mean is greater than Group 2's mean.

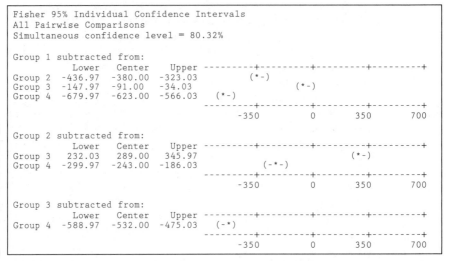

```
Fisher 95% Individual Confidence Intervals
All Pairwise Comparisons
Simultaneous confidence level = 80.32%

Group 1 subtracted from:
           Lower   Center    Upper  --------+---------+---------+---------+
Group 2  -436.97  -380.00  -323.03           (*-)
Group 3  -147.97  -91.00   -34.03                    (*-)
Group 4  -679.97  -623.00  -566.03   (*-)
                                    --------+---------+---------+---------+
                                         -350         0       350       700

Group 2 subtracted from:
           Lower   Center    Upper  --------+---------+---------+---------+
Group 3   232.03   289.00   345.97                              (*-)
Group 4  -299.97  -243.00  -186.03              (-*-)
                                    --------+---------+---------+---------+
                                         -350         0       350       700

Group 3 subtracted from:
           Lower   Center    Upper  --------+---------+---------+---------+
Group 4  -588.97  -532.00  -475.03   (-*)
                                    --------+---------+---------+---------+
                                         -350         0       350       700
```

**Figure 10-3:** Output showing Fisher's LSD applied to the cell-phone data.

Each subsequent row in the "Group 1 subtracted from" section of Figure 10-3 shows similar results. None of the confidence intervals contain 0, so you conclude that the mean cell-phone use for Group 1 isn't equal to the mean cell-phone use for any other group. Moreover, because all confidence intervals are in negative territory, you can conclude that the mean cell-phone use time for those 19 and under is greater than all the others. This process continues as you move down through the output until all six pairs of means are compared. Then you put them all together into one conclusion.

For example, in the second portion of the output, Group 2 is subtracted from Groups 3 and 4. You see the confidence interval for the "Group 3" line is 232.03, 345.97; this gives possible values for Group 3's mean minus Group 2's mean. The interval is entirely positive, so conclude that Group 3's mean is greater than Group 2's mean (according to this data). On the next line, the interval for Group 4 minus Group 2 is –299.97 to –186.03. All these numbers are negative, so conclude Group 4's mean is less than Group 2's. Combine conclusions to say that Group 3's mean is greater than Group 2's, which is greater than Group 4's.

In the cell-phone example, none of the means are equal to each other, and based on the signs of confidence intervals and the results of all the individual pairwise comparisons, the following order of cell-phone mean usage prevails: $\mu_1 > \mu_3 > \mu_2 > \mu_4$. (Hypothetical data aside, it might be the case that 40- to 59-year-olds use a lot of cell phone time because of their jobs.)

Notice near the top of Figure 10-3 that you see "simultaneous confidence level = 80.32 percent." That means the overall error rate for this procedure is 1 – 0.8032 = 0.1968, which is close to 20 percent.

## Separating the turkeys with Tukey's test

This section dives into Tukey's test and applies it to the cell-phone example.

### Setting up Tukey's test

The basic idea behind Tukey's test is to provide a series of simultaneous confidence intervals for the differences in the means. It still examines all possible pairs of means and keeps the overall error rate (also known as the *familywise error rate*) at $\alpha$ (like Fishers LSD), but it also keeps the individual Type I error rate for each pair of means at $\alpha$ as well. This difference takes care of a lot of issues raised with Fisher's LSD procedure (refer to the preceding section).

Although the details of the formulas used for Tukey's test are beyond the scope of this book, they're not based on the *t*-test, but rather something called a *studentized range statistic,* which is based on the highest and lowest means in the group, and their difference. The individual error rates are held at 0.05 because Tukey developed a cutoff value for his test statistic, which is based on all pairwise comparisons (no matter how many means are in each group).

If you calculate the results by hand, you can look at tables to make your conclusions. However, all applications I have ever seen both in the classroom and outside of it use a computer for these calculations. (For sanity's sake, I suggest you do the same.)

To conduct Tukey's test, go to Stat>ANOVA>One-way or One-way unstacked. (If your data appears in two columns with Column 1 representing the population number and Column 2 representing the response, just click One-way because your data is stacked. If your data is shown in *k* columns, one for each of the *k* populations, click One-way unstacked.) The next step is to highlight the data for the groups you're comparing and click Select. Then click on Comparisons. Click on Tukey's. The familywise (overall) error rate is listed at 5 (percent), which is typical. If you want to change it, type in the desired error rate (between 0.5 and 0.001) and click OK. You may type in your error rate as a decimal, such as 0.05, or as a number greater than one, such as 5. Numbers greater than one are interpreted as a percentage.

### Doing Tukey's test on the cell phone data

The Minitab output for comparing the groups regarding cell-phone use by using Tukey's test appears in Figure 10-4. Looking at Figure 10-4, you see that its results can be interpreted in the same was as for Figure 10-3. Some of the numbers in the confidence intervals are different, but in this case, the main conclusions are the same: Those 19 and under use their cell phones most, followed by 40- to 59-year-olds, then 20- to 39-year-olds, and finally those 60 and over.

The results of Fisher and Tukey don't always agree, usually because the overall error rate of Fisher's procedure is larger than Tukey's (except when only two means are involved). Most statisticians I know prefer Tukey's procedure over Fisher's. That doesn't mean they don't have other procedures they like even better than Tukey's, but Tukey's is the most common procedure, and many people like to use it.

```
Tukey 95% Simultaneous Confidence Intervals
All Pairwise Comparisons

Individual confidence level = 98.93%

Group 1 subtracted from:

           Lower    Center     Upper  +---------+---------+---------+---------
Group 2  -455.68   -380.00   -304.32           (-*-)
Group 3  -166.68   -91.00    -15.32                   (-*-)
Group 4  -698.68   -623.00   -547.32  (-*-)
                                      +---------+---------+---------+---------
                                   -700      -350         0       350


Group 2 subtracted from:

           Lower    Center     Upper  +---------+---------+---------+---------
Group 3   213.32    289.00    364.68                                (-*-)
Group 4  -318.68   -243.00   -167.32              (-*-)
                                      +---------+---------+---------+---------
                                   -700      -350         0       350


Group 3 subtracted from:

           Lower    Center     Upper  +---------+---------+---------+---------
Group 4  -607.68   -532.00   -456.32     (-*-)
                                      +---------+---------+---------+---------
                                   -700      -350         0       350
```

**Figure 10-4:** Output for Tukey's test used to compare cell-phone usage.

Another multiple comparison procedure is listed on Minitab's repertoire after you ask it to do multiple comparisons. This procedure is called *Dunnett's test.* Dunnett's test is a special multiple comparison procedure used in a designed experiment that contains a control group. The test compares each treatment group to the control group and determines which treatments do better than others that way. Dunnett's test is better able to find real differences in this situation than other multiple comparison procedures, because it focuses only on the differences between each treatment and the control — not the differences between every single pair of treatments in the entire study.

# Chapter 11

# Getting a Little Interaction with Two-Way ANOVA

### In This Chapter

- Building and carrying out ANOVA with two factors
- Getting familiar with (and looking for) interaction effects and main effects
- Putting the terms to the test
- Demystifying the two-way ANOVA table

*Analysis of variance* (ANOVA) is often used in experiments to see whether different levels of an explanatory variable $(x)$ get different results on some quantitative variable $y$. (See Chapter 9.) The $x$ variable in this case is called a *factor,* and it has certain levels to it, depending on how the experiment is set up. For example, say you want to compare the average reduction in blood pressure on certain dosages of a drug. The factor is drug dosage. Suppose it has three levels: 10mg per day, 20mg per day, or 30mg per day. Suppose someone else studies the response to that same drug and examines whether the times taken per day (one time or two times) has any effect on blood pressure. In this case, the factor is number of times per day, and it has two levels: once and twice.

Suppose you want to study the effects of dosage *and* number of times taken together, because you believe both may have an affect on the response. So what you have is called a *two-way ANOVA,* using two factors together to compare the average response. So it's an extension of one-way ANOVA (refer to Chapter 9) with a twist, because the two factors you use may operate on the response differently together than they would separately.

In this chapter, you examine two-way ANOVA — setting up the model, making your way through the ANOVA table, taking the $F$-tests, and drawing the appropriate conclusions.

# Setting Up the Two-Way ANOVA Model

The two-way ANOVA model extends the ideas of the one-way ANOVA model and adds an interaction term to examine how various combinations of the two factors affect the response. In this section, you see the building blocks of a two-way ANOVA: the treatments, main effects, the interaction term, and the sums of squares equation that puts everything together.

## Determining the treatments

The two-way ANOVA model contains two factors, $A$ and $B$, and each factor has a certain number of levels (say $i$ levels of factor A and $j$ levels of factor B). In the drug study example from the chapter intro, you have A = drug dosage with $i = 1, 2,$ or 3 and B = number of times taken per day with $j = 1$ or 2. Each person involved in the study is subject to one of the three different drug dosages and will take the drug in one of the two methods given. That means you have $3 * 2 = 6$ different combinations of factors A and B that you can apply to the subjects, and you can study it in the two-way ANOVA model.

Each different combination of levels of factors A and B is called a *treatment* in the model. Table 11-1 shows the six treatments in the drug study. For example, Treatment 4 is the combination of 20mg of the drug taken in two doses of 10mg each per day.

**Table 11-1 The Six Treatment Combinations for the Drug Study**

| *Amount* | *One Time/Day* | *Two Times/Day* |
|---|---|---|
| 10mg | Treatment 1 | Treatment 2 |
| 20mg | Treatment 3 | Treatment 4 |
| 30mg | Treatment 5 | Treatment 6 |

If factor A has $i$ levels and factor B has $j$ levels, you have $i * j$ different combinations of treatments in your two-way ANOVA model.

## Stepping through the sums of squares

The two-way ANOVA model contains three terms:

- **The main effect A:** A term for the effect of factor A on the response
- **The main effect B:** A term for the effect of factor B on the response
- **The interaction of A and B:** The effect of the combination of factors A and B (denoted AB)

The sums of squares equation for the one-way ANOVA (see Chapter 9) is SSTO = SST + SSE, where SSTO is the total variability in the response variable, *y*; SST is the variability explained by the treatment variable (call it factor A); and SSE is the variability left over as error. The purpose of this model is to test to see whether the different levels of factor A produce different responses in the *y* variable. The way you do it is by using Ho: $\mu_1 = \mu_2 = \ldots = \mu_i$, where *i* is the number of levels of factor A (the treatment variable). If you reject Ho, then factor A (which separates the data into the groups being compared) is significant. If you can't reject Ho, you can't conclude that factor A is significant.

In the two-way ANOVA, you add another factor to the mix (B) plus an interaction term (AB). The sums of squares equation for the two-way ANOVA model is SSTO = SSA + SSB + SSAB + SSE. Here SSTO is the total variability in the *y*-values; SSA is the sums of squares due to factor A (representing the variability in the *y*-values explained by factor A.); and similarly for SSB and factor B. SSAB is the sums of squares due to the interaction of factors A and B, and SSE is the amount of variability left unexplained, and deemed error. (While the mathematical details of all the formulas for these terms are unwieldy and beyond the focus of this book, they just extend the formulas for one-way ANOVA found in Chapter 9. ANOVA handles the calculations for you, so you don't have to worry about that part.)

To carry out a two-way ANOVA in Minitab, enter your data in three columns. Column 1 contains the responses (the actual data). Column 2 represents the level of factor A (Minitab calls it the row factor). Column 3 represents the level of factor B (Minitab calls it the *column factor*). Go to Stat>Anova>Two-way. Click on Column 1 in the left-hand box and it appears in the Response box on the right-hand side. Click on Column 2 and it appears in the row factor box; click on Column 3 and it appears in the column factor box. Click OK.

For example, suppose you have six data values in Column 1: 11, 21, 38, 14, 15, 62. Suppose Column 2 contains 1, 1, 1, 2, 2, 2, and Column 3 contains 1, 2, 3, 1, 2, 3. This means that factor A has two levels (1, 2), and factor B has three levels (1, 2, 3). The number 11 was the response when Level 1 of factor A and Level 1 of factor B were applied. The second data value, 21, came from Level 1 of A and Level 2 of B. The third value, 38, came from Level 1 of A and Level 3 of B. The fourth number, 14, came from Level 2 of A and Level 1 of B. The number 15 is the response from Level 2 of A and Level 2 of B, and finally, the number 62 corresponds to the result of Level 2 of A and Level 3 of B.

Suppose factor A has $i$ levels and factor B has $j$ levels, with a sample of size $m$ collected on each combination of A and B. The degrees of freedom for factor A, factor B, and the interaction term AB are $(i - 1)$, $(j - 1)$, and $(i - 1) * (j - 1)$ respectively. This formula is just an extension of the degrees of freedom for the one-way model for factors A and B. The degrees of freedom for SSTO is $i * j * m - 1$, and the degrees of freedom for SSE is $i * j * (m - 1)$.

# Understanding Interaction Effects

The interaction effect is the heart of the two-way ANOVA model. Knowing that the two factors may act together in a different way than they would separately is important and must be taken into account. In this section, you see the many ways in which the interaction term AB and the main effects of factors A and B affect the response variable in a two-way ANOVA model.

## What is interaction anyway?

Interaction is when two factors meet, or interact with each other, on the response in a way that's different from how each factor affects the response separately. For example, before you can test to see whether dosage of medicine (factor A) or number of times taken (factor B) are important in explaining changes in blood pressure, you have to look at how they operate together to affect blood pressure. That is, you have to examine the interaction term.

Suppose you're taking one type of medicine for cholesterol and one medicine for a heart problem. Suppose researchers only looked at the effects of each drug alone, saying each one was good for managing the problem for which it was designed, with little to no side effects. Now you come along and mix the two drugs in your system. As far as the individual study results are concerned, all bets are off. With only those separate studies to go on, they will have no idea how the drugs will interact with each other, and you can be in a great deal of trouble very quickly. Fortunately, drug companies and medical researchers do a great deal of work studying drug interactions, and your pharmacist knows which drugs interact as well. You can bet a statistician was involved in this work from day one!

Baking is another good example of how interaction works. Slurp down one raw egg, drink a cup of milk, and eat a cup of sugar, a cup of flour, and a stick of margarine. Then eat a cup of chocolate chips. Each one of these items has a certain taste, texture, and affect on your taste buds that, in most cases, won't be all that great. But mix them all together in a bowl and voilà! You have a batch of chocolate chip cookie dough, thanks to the magic effects of interaction.

## Interacting with interaction plots

In the two-way ANOVA model, you have two factors and their interaction. A number of results could come out of this model in terms of significance of the individual terms, as you can see in the following:

- Factors A and B are both significant.
- Factor A is significant but not factor B.
- Factor B is significant but not factor A.
- Neither factors A nor B are significant.
- The interaction term AB is significant.

Figure 11-1 depicts each of these five situations, respectively, in terms of a diagram, using the drug-study example. Plots that show how factors A and B react separately and together on the response variable *y* are called *interaction plots.* In the following sections, I describe each of these five situations in detail in terms of what the plots are telling you and what the results mean in the context of an example.

### Factors A and B are significant

Figure 11-1a shows the situation when both A and B are significant in the model (no interaction present). The lines represent the levels of the times-per-day factor (B); the *x*-axis represents the levels of the dosage factor (A); and the *y*-axis represents the average value of the response variable *y*, change in blood pressure, at each combination of treatments.

The top line moving across Figure 11-1a shows that when the drug is taken two times per day, the change in blood pressure increases with dosage level. The bottom line shows the same thing happens when the drug is taken once per day, except that the effects on blood pressure are lower overall than the effects of taking the drug twice a day. That means factor A is significant because blood pressure changes across dosage levels, and factor B is significant because blood pressure is different from one line to another. (Assume the difference is large enough to be significant.) Here the different combinations of factors A and B don't affect the overall trends, so there's no interaction effect.

REMEMBER

Two parallel lines in an interaction plot means a lack of an interaction effect. In the drug-study example, the levels of A don't change blood pressure differently for different levels of B.

### Factor A is significant but not factor B

Figure 11-1b shows that blood pressure changes across dosage levels for taking the drug once or twice a day. However, the two lines are so close together that whether you take the drug once or twice a day has no effect. So factor A (dosage) is significant, and factor B (times per day) isn't. Parallel lines again means no interaction effect.

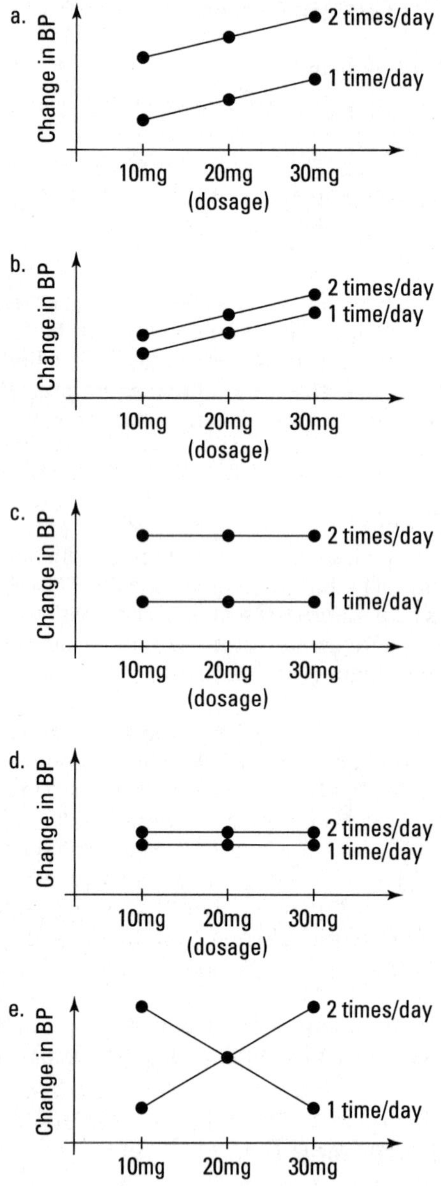


**Figure 11-1:** Five examples of the results from a two-way ANOVA with interaction.

### Factor B is significant but not factor A

Figure 11-1c shows where factor B is significant but A isn't. The lines are flat across dosage levels indicating that dosage has no effect on blood pressure. However, the two lines for times per day are spread apart, so their effect on blood pressure is significant. Parallel lines mean no interaction effect.

### Neither factor is significant

Figure 11-1d shows two flat lines that are very close to each other. By the previous discussions about Figures 11-1b and 11-1c, you can guess that this figure represents the case where neither factor A nor factor B are significant, and you don't have an interaction effect because the lines are parallel.

### Interaction term is significant

Finally you get to Figure 11-1e, the most interesting interaction plot of all. The big picture is that because the two lines cross, then factors A and B interact with each other in the way that they operate on the response. If they didn't interact, then the lines would be parallel.

Start with the top line of Figure 11-1e. When you take the drug two times per day at the low dose, you get a low change in blood pressure; as you increase dosage, blood pressure increases also. But when you take the drug once per day, the opposite result happens.

If you didn't look for a possible interaction effect before you examined the main effects, you may have thought no matter how many times you take this drug per day, the effects will be the same. Not so! Always check out the interaction term first in any two-way ANOVA. If the interaction term is significant, you have no way to pull out the effects due to just factor A or just factor B; they're moot. Checking the main effects of factor A or B without checking out the interaction AB term is considered a no-no in the two-way ANOVA world. Another taboo is examining the factors individually (also known as the main effect) if the interaction term is significant.

## Testing the Terms in Two-Way ANOVA

In a one-way ANOVA, you have only one hypothesis test. You use an *F*-test to determine whether the means of the *y* values are the same or different as you go across the levels of the one factor. In two-way ANOVA you have more items to test besides the overall model. You have the interaction term AB and possibly the main effects of A and B. Each test in a two-way ANOVA is an *F*-test based on the ideas of one-way ANOVA (see Chapter 9 for more on this).

First, you test whether the interaction term AB is significant. To do this, you use the test statistic $F = \frac{MS_{AB}}{MSE}$, which has an *F*-distribution with $(i-1) * (j-1)$ degrees of freedom from $MS_{AB}$ (mean sum of squares for the interaction term of A and B) and $i * j * (m - 1)$ degrees of freedom from MSE (mean sum of squares for error), respectively. (Recall that $i$ and $j$ are the number of levels of A and B, and $m$ is the sample size at each combination of A and B.) You basically want to see whether more of the total variability in the $y$'s can be explained by the AB term compared to what is left in the error term. A large value of $F$ means that the AB term is significant, and you leave it in the model.

If the interaction term isn't significant, you take the AB term out of the model, and you can explore the effects of factors A and B separately regarding the response variable $y$. The test for Factor A uses the test statistic $F = \frac{MS_A}{MSE}$, which has an $F$-distribution with $i - 1$ degrees of freedom from $MS_A$ (mean sum of squares for factor A) and $i * j * (m - 1)$ degrees of freedom from MSE (mean sum of squares for error), respectively. Testing for factor B uses $F = \frac{MS_B}{MSE}$, which has an $F$-distribution with $j - 1$ and $i * j * (m - 1)$ degrees of freedom.

The results you can get from testing the terms of the ANOVA model are the same as those represented in Figure 11-1. They're all provided in Minitab output outlined in the next section, including their sum of squares, degrees of freedom, mean sum of squares, and $p$-values for their appropriate $F$-tests.

# Running the Two-Way ANOVA Table

The ANOVA table for two-way ANOVA includes the same elements as the ANOVA table for one-way ANOVA (see Chapter 9). But where in the one-way ANOVA you had one line for Factor A's contributions, now you add lines for the effects of Factor B and the interaction term AB. Minitab calculates the ANOVA table for you as part of the output from running a two-way ANOVA.

In this section, you can figure out how to interpret the results of a two-way ANOVA, assess the model's fit, and use a multiple comparisons procedure, using the drug-data study.

## Interpreting the results: Numbers and graphs

The drug-study example has, say, four people in each treatment combination of three possible dosage levels (10, 20, 30mg per day) and two possible times for taking the drug (one time per day and two times per day). The total sample size is $4 * 3 * 2 = 24$. I made up five different data sets in which the analyses represent each of the five scenarios shown in Figure 11-1. Their ANOVA tables, as created by Minitab, are shown in Figure 11-2.

The order of the graphs in Figure 11-1 and the ANOVA tables in Figure 11-2 isn't the same. Can you match them up? (I promise to give you the answers, so keep reading.)

a

Two-way ANOVA: BP versus Dosage, Times

| Scarce | DF | SS | MS | F | P |
|---|---|---|---|---|---|
| Dosage | 2 | 56.3333 | 28.1667 | 112.67 | 0.000 |
| Times | 1 | 4.1667 | 4.1667 | 16.67 | 0.001 |
| Interaction | 2 | 0.3333 | 0.1667 | 0.67 | 0.526 |
| Error | 18 | 4.5000 | 0.2500 | | |
| Total | 23 | 65.3333 | | | |

S = 0.5  R-Sq = 93.11%  R-Sq(adj) = 91.20%

b

Two-way ANOVA: BP versus Dosage, Times

| Source | DF | SS | MS | F | P |
|---|---|---|---|---|---|
| Dosage | 2 | 0.0833 | 0.04167 | 0.16 | 0.855 |
| Times | 1 | 0.3750 | 0.37500 | 1.42 | 0.249 |
| Interaction | 2 | 16.7500 | 8.37500 | 31.74 | 0.000 |
| Error | 18 | 4.7500 | 0.26389 | | |
| Total | 23 | 21.9583 | | | |

S = 0.5137  R-Sq = 78.37%  R-Sq(adj) = 72.36%

c

Two-way ANOVA: BP versus Dosage, Times

| Source | DF | SS | MS | F | P |
|---|---|---|---|---|---|
| Dosage | 2 | 0.0833 | 0.041667 | 0.08 | 0.926 |
| Times | 1 | 0.3750 | 0.375000 | 0.69 | 0.416 |
| Interaction | 2 | 0.7500 | 0.375000 | 0.69 | 0.513 |
| Error | 18 | 9.7500 | 0.541667 | | |
| Total | 23 | 10.9583 | | | |

S = 0.7360  R-Sq = 11.03%  R-Sq(adj) = 0.00%

d

Two-way ANOVA: BP versus Dosage, Times

| Source | DF | SS | MS | F | P |
|---|---|---|---|---|---|
| Dosage | 2 | 36.7500 | 18.3750 | 47.25 | 0.000 |
| Times | 1 | 0.6667 | 0.6667 | 1.71 | 0.207 |
| Interaction | 2 | 0.0833 | 0.0417 | 0.11 | 0.899 |
| Error | 18 | 7.0000 | 0.3889 | | |
| Total | 23 | 44.5000 | | | |

S = 7.6236  R-Sq = 84.27%  R-Sq(adj) = 79.90%

e

Two-way ANOVA: BP versus Dosage, Times

| Source | DF | SS | MS | F | P |
|---|---|---|---|---|---|
| Dosage | 2 | 0.0833 | 0.0417 | 0.16 | 0.855 |
| Times | 1 | 12.0417 | 12.0417 | 45.63 | 0.000 |
| Interaction | 2 | 0.0833 | 0.0417 | 0.16 | 0.855 |
| Error | 18 | 4.7500 | 0.2639 | | |
| Total | 23 | 16.9583 | | | |

S = 0.5137  R-Sq = 71.99%  R-Sq(adj) = 64.21%

**Figure 11-2:** ANOVA tables for the interaction plots from Figure 11-1.

Notice that each ANOVA table in Figure 11-2 shows the degrees of freedom for dosage is 3 – 1 = 2; the degrees of freedom for times per day is 2 – 1 = 1; the degrees of freedom for the interaction term is (3 – 1)(2 – 1) = 2; the degrees of freedom for total is 3 * 2 * 4 – 1 = 23; and degrees of freedom for error is 3 * 2 * (4 – 1) = 18.

Here are the answers to match the graphs from Figure 11-1 with the output from Figure 11-2:

- In the ANOVA table for Figure 11-2a, you see that the interaction term isn't significant (*p*-value = 0.526), so the main effects can be studied. The *p*-values for dosage and times taken are 0.000 and 0.001, indicating both factors A and B respectively are significant; this matches the plot in Figure 11-1a.
- In Figure 11-2b, you see that the *p*-value for interaction is significant (*p*-value = 0.000) so you can't examine the main effects of factors A and B (in other words, don't look at their *p*-values). This represents the situation in Figure 11-1e.
- Figure 11-2c shows nothing is significant (*p*-value for interaction term is 0.513; *p*-values for main effects of A (dosage) and B (times taken) are 0.926 and 0.416, respectively). These results coincide with Figure 11-1d.
- Figure 11-2d matches Figure 11-1b, with no interaction effect (*p*-value = 0.899), dosage (factor A) is significant (*p*-value = 0.000), and times per day (factor B) isn't (*p*-value = 0.207).
- Figure 11-2e matches Figure 11-c. Dosage * times per day is not significant (*p*-value = 0.855); times per day is significant with *p*-value 0.000 but not dosage level (*p*-value = 0.855).

### Assessing the fit

To assess the fit of the two-way ANOVA models, you can use the $R^2$ adjusted (see Chapter 5). The higher this number is, the better (the maximum is 100 percent or 1.00). Notice that all the ANOVA tables in Figure 11-2 show a fairly high $R^2$ adjusted except for Figure 11-2c. In this table, none of the terms was significant.

### Multiple comparisons

In the case where you find that an interaction effect is statistically significant, you can conduct multiple comparisons to see which combinations of factors A and B create different results in the response. The same ideas hold here as those for Chapter 10 on multiple comparisons, except the tests are performed on all $i * j$ interactions.

To perform multiple comparisons for a two-way ANOVA by using Minitab, enter your responses (data) in Column 1, your levels of Factor A in Column 2, and your levels of Factor B in Column 3. Choose Stat>ANOVA>General Linear Model. In the Responses box, enter your Column 1 variable. In Model, enter 1 <space> 2 <space> 1*2 (for the main effects and the interaction effect, respectively; here <space> means leave a space where indicated). Click on Comparisons. In Terms, enter columns 2 and 3. Check the Method you want to use for your multiple comparisons (see Chapter 10). Click OK.

# Chapter 12

# Rock My World: Relating Regression to ANOVA

### In This Chapter

- Relating the formulas and procedures for one-way ANOVA and regression
- Making the connection between these two seemingly unrelated procedures

So you're motoring on in your intermediate stat course, working your way through regression (where you estimate *y*, using one or more *x* variables — see Chapter 4). Then you hit a new topic, ANOVA, which stands for *analysis of variance* — comparing the means of several populations (see Chapter 9). That seems to be no problem. But wait a minute; now your professor starts talking about how ANOVA is related to regression — suddenly everything starts to spin out of control. How do you reconcile two techniques that appear to be as different as apples and oranges? That's what this chapter is all about.

Think of this chapter as your bridge across the gap that lies between regression and ANOVA, allowing you to walk smoothly across, answering any questions that a professor may throw into your path. You don't apply these two techniques in this chapter (you can find that information in Chapters 4 and 9). The goal of this chapter is to determine and describe the relationship between regression and ANOVA so they don't look quite so much like an apple and an orange.

## Seeing Regression through the Eyes of Variation

Every statistical model tries to explain why the different outcomes *(y)* are what they are. It tries to figure out what factors or explanatory variables *(x)* can help explain that variability in those *y*'s. In this section, you start with the *y*-values by themselves and see how their variability plays a central role in the regression model. This is the first step toward applying ANOVA (the analysis of *variance*) to the regression model.

## Verifying variability in the y's and looking at x to explain it

No matter what *y* variable you're interested in predicting, you will always have variability in those *y*-values. If you want to predict the length of a fish, you may notice that fish have many different lengths (indicating a great deal of variability). Even if you put all the fish of the same age and species together, you still have some variability in their lengths (it will be less than before, but still there nonetheless). The first step to understanding the basic ideas of regression and ANOVA is to understand that variability in the *y*'s is to be expected, and your job is to try to figure out what can explain most of it. This section deals with seeing and explaining variability in the *y*-values.

## Seeing the variability in Internet use

Both regression and ANOVA work to get a handle on explaining the variability in the *y* variable using an *x* variable. After you collect your data, you can find the standard deviation in the *y* variable to get a sense of how much the data varies within the sample. From there, you collect data on an *x* variable and see how much it contributes to explaining that variability.

Suppose you notice that people spend different amounts of time on the Internet, and you want to explore why that may be. You start by taking a small sample of 20 people and record how many hours per month they spend on the Internet. The results (in hours) are 20, 20, 22, 39, 40, 19, 20, 32, 33, 29, 24, 26, 30, 46, 37, 26, 45, 15, 24, and 31. The first thing you notice about this data is the large amount of variability in it. The *standard deviation* (average distance from the data values to their mean) of this data set is 8.93, which is quite large given the size of the numbers in the data set.

## Finding an "x-planation" for Internet use

So you figure out that the *y*-values (such as amount of time someone uses the Internet from the preceding section) have a great deal of variability in them. What can help explain this? Part of the variability is due to chance. But you suspect some variable is out there (call it $x$) that has some connection to the $y$ variable, and that variable can help you make more sense out of this seemingly wide range of $y$-values.

For example, if you record the calories for five types of candy bars as 100, 200, 300, 400, and 500, you would say "Wow, that's a lot of variation in calories; I wonder why that is?" Then you notice that the weights of the candy bars are 1, 2, 3, 4, and 5 ounces, respectively. This relationship can be expressed as $y = 100x$, where $y$ equals calories and $x$ equals weight.

Now you can look at what before was a bunch of variability in the $y$-values and say, "Hey, that's not just random variability; the differing $y$-values can be explained by the weight of candy bar ($x$)." You can now use $x$ in a nice regression model to estimate $y$. Notice that you're talking about splitting the total variability in the $y$'s into the part due to $x$ and the part due to chance (error). That's ANOVA language! Hey, perhaps regression and ANOVA are related after all . . .

To continue with the Internet use example, suppose you have a brainstorm that number of years of education could possibly be related to Internet use. In this case, the explanatory variable (input variable, $x$) is years of education, and you want to use it to try to estimate $y$, the number of hours on the Internet in a month. You take a larger random sample of 250 Internet users and ask them how many years of education they had (so $n = 250$). You can check out the first ten observations from your data set containing the $(x, y)$ pairs in Table 12-1. If a significant connection of some sort exists between the $x$-values and the $y$-values, then you can say that $x$ is helping to explain some of the variability in the $y$'s. If it explains enough variability, you can place $x$ into a simple regression model and use it to estimate $y$.

**Table 12-1 First Ten Observations from the Education and Internet Use Example**

| *Years of Education* | *Hours on Internet (For One Month)* |
|---|---|
| 15 | 41 |
| 15 | 32 |
| 11 | 33 |
| 10 | 42 |
| 10 | 28 |
| 10 | 21 |

*(continued)*

**Table 12-1 *(continued)***

| *Years of Education* | *Hours on Internet (For One Month)* |
|---|---|
| 10 | 17 |
| 10 | 14 |
| 9 | 18 |
| 9 | 14 |

## Getting results with regression

After you have a possible *x* variable picked, you collect pairs of data *(x, y)* on a random sample of individuals from the population, and you look for a possible linear relationship between them. To do this, use Minitab to make a scatterplot of the data and calculate the correlation *(r)*. If the data appear to follow a straight line (as shown on the scatterplot), you go ahead and perform a simple linear regression of the response variable *y* based on the *x* variable. The *p*-value of the *x* variable in the simple linear regression analysis tells you whether or not the *x* variable does a significant job in predicting *y*. Some of the details of getting the regression results are described below (for full information, see Chapter 4).

Looking at the small snippet of 10 out of the 250 person data set in Table 12-1, you can begin to see that you may have a pattern between education and Internet use. It looks like as education increases so does Internet use.

To do a simple linear regression using Minitab, enter your data in two columns: the first column for your *x* variable and the second column for your *y* variable (as in Table 12-1). Go to Stat>Regression>Regression. Click on your *y* variable in the left-hand box; the *y* variable then appears in the Response box on the right-hand side. Click on your *x* variable in the left-hand box; the *x* variable then appears in the Predictor box in the right-hand side. Click OK, and your regression analysis is done. As part of every regression analysis, Minitab also provides you with the corresponding ANOVA results, found at the bottom of the output.

The simple linear regression output that Minitab gives you for the education and Internet example is in Figure 12-1. (Notice the ANOVA output at the bottom; you can see the connection in the upcoming section "Regression and ANOVA: A Meeting of the Models.")

**Regression Analysis: Internet versus Education**

```
The regression equation is
Internet = -8.29 + 3.15 Education

Predictor      Coef  SE Coef       T      P
Constant     -8.290    2.665   -3.11  0.002
Education    3.1460   0.2387   13.18  0.000

S = 7.23134      R-Sq = 41.2%      R-Sq(adj) = 41.0%

Analysis of Variance

Source           DF       SS      MS       F      P
Regression        1   9085.6  9085.6  173.75  0.000
Residual Error  248  12968.5    52.3
Total           249  22054.0
```

**Figure 12-1:** Output for simple linear regression applied to education and Internet use data.

Looking at Figure 12-1, you see that the *p*-value on the row marked *Education* is 0.000, which means the *p*-value's less than 0.001. Therefore the relationship between years of education and Internet use is statistically significant. A scatterplot of the data (not shown here) also indicates that the data appear to have a positive linear relationship. That means as you increase number of years of education, Internet use also tends to increase (on average).

## Assessing the fit of the regression model

Before you go ahead and use a regression model to make predictions for $y$ based on an $x$ variable, you must first assess the fit of your model. One way to get a rough idea of how well your regression model fits is by using a *scatterplot* (a graph showing all the pairs of data plotted in the $x$-$y$ plane). Use the scatterplot to see whether the data appears to fall in the pattern of a line. If the data appears to follow a straight-line pattern (or even something close to that — anything but a curve or a scattering of points that has no pattern at all), you calculate the correlation, $r$, to see how strong the linear relationship between $x$ and $y$ is (the closer $r$ is to +1 or –1, the stronger the relationship; the closer $r$ is to zero, the weaker the relationship). Minitab can do scatterplots and correlations for you; see Chapter 4 for more on simple linear regression, including making a scatterplot and finding the value of $r$.

If the data doesn't have a significant correlation, stop the analysis; you can't go further to find a line that fits a relationship that doesn't exist.

Next you come to the more general way of assessing not only the fit of a simple linear regression model, but many other models too (for example: multiple, nonlinear, and logistic regression models in Chapters 5, 7, and 8, to name a few). In simple linear regression, the value of $R^2$, as indicated by Minitab and statisticians as a capital *R* (squared), is equal to the square of the Pearson correlation coefficient, *r* (indicated by Minitab and statisticians by a small *r*). In all other situations, $R^2$ provides a more general measure of model fit. (Note that *r* only measures the fit of a straight-line relationship between one *x* variable and one *y* variable; see Chapter 4.) Finally, $R^2$ adjusted modifies $R^2$ to account for the number of variables in the model. $R^2$ is what statisticians use to assess model fit (see Chapter 5 for more).

The value of $R^2$ adjusted for the model of using education to estimate Internet use (Figure 12-1) is equal to 41 percent. This value reflects the percentage of variability in Internet use that can be explained by a person's years of education. This number isn't great, but it's not terrible either. Note the square root of 41 percent is 0.64 for *r* itself, which in the case of linear regression indicates a moderate relationship.

This evidence gives you the green light to use the results of the regression analysis to estimate number of hours of Internet use in a month by using years of education. The regression equation as it appears in the top part of the Figure 12-1 output is Internet = –8.29 + 3.15 * 16 = 42.11. So if you have 16 years of education, for example, your estimated Internet use is 42.11, or about 42 hours per month (about 10.5 hours per week).

But wait! Look again at Figure 12-1 and zoom in on the bottom part. I didn't ask for anything special to get this info on the Minitab output, but you can see an ANOVA table there. That seems like a fish out of water doesn't it? But in the next section you see how an ANOVA table can describe regression results (albeit it in a different way).

# Regression and ANOVA: A Meeting of the Models

Okay, here it comes. You've already broken down the regression output into all its pieces and parts. The next step toward understanding the connection between regression and ANOVA is to apply the sums of squares from ANOVA to regression (something that is typically not done in a regression analysis). Before you start, think of this process as going to a 3-D movie, where you have to wear special glasses in order to see all the special effects!

In this section, you see the sums of squares in ANOVA applied to regression and how the degrees of freedom work out. You build an ANOVA table for regression and discover how the *t*-test for a regression coefficient is related to the *F*-test in ANOVA. I know you can hardly wait, so I won't keep you in suspense any longer.

## Comparing sums of squares

*Sums of squares* is a term you may remember from ANOVA (see Chapter 9), but it certainly isn't a term you normally use when talking about regression (as in Chapter 4). Yet, both types of models can be broken down into sums of squares, and that similarity gets at the true connection between ANOVA and regression. In step-by-step terms, you first partition out the variability in the *y* variable by using formulas for sums of squares from ANOVA (sums of squares for total, treatment, and error). Then you find those same sums of squares for regression — this is the twist on the process because you typically don't find sums of squares for regression. You compare the two procedures through their sums of squares. This section shows you the details of how this comparison is done.

### Partitioning variability by using SSTO, SSE, and SST for ANOVA

ANOVA is all about partitioning the total variability in the *y*-values into sums of squares (see all the info you ever need on one-way ANOVA in Chapter 9). The key idea is that SSTO = SST + SSE, where SSTO is the total variability in the *y*-values; SST measures the variability explained by the model (also known as the treatment, or *x* variable in this case); and SSE measures the variability due to error (what's left over after the model is fit).

The corresponding formulas for SSTO, SSE, and SST are $\Sigma\left(y_i-\bar{y}\right)^2$, $\Sigma\left(y_i-\hat{y}_i\right)^2$, and $\Sigma\left(\hat{y}_i-\bar{y}\right)^2$ respectively, where $\bar{y}$ is the mean of the *y*'s, $y_i$ is each observed value of *y*, and $\hat{y}_i$ is each predicted value of *y* from the ANOVA model. Use these formulas to calculate the sums of squares for ANOVA (Minitab does this for you when it performs ANOVA). Keep these values of SSTO, SST, and SSE. You will use them to compare to the results from regression.

### Finding sums of squares for regression

In regression, you measure the deviations in the *y*-values by taking each $y_i$ minus its mean, $\bar{y}$. Square each result and add them all up, and you have SSTO. Next, take the residuals, which represent the difference between each $y_i$ and it's estimated value from the model, $\hat{y}_i$. Square the residuals and add them up, and you get the formula for SSE.

Now that you have calculated SSTO and SSE, you need the bridge between them. That is, you need a formula that connects the variability in the $y_i$'s (SSTO) and the variability in the residuals after fitting the regression line (SSE). That bridge is SSR (equivalent to SST in ANOVA). In regression, $\hat{y}_i$ represents the predicted value of $y_i$ based on the regression model. These are the values on the regression line. To assess how much this regression line helps to predict the *y*-values, you compare it to the model you would get without any *x* variable in it.

Without any other information, the only thing you can do to predict *y* is look at the average, $\bar{y}$. So, SST compares the predicted value from the regression line to the predicted value from the flat line (the mean of the *y*'s) by subtracting them. The result is $\left(\hat{y}_i - \bar{y}\right)$. Square each result and sum them all up, and you get the formula for SST.

Now for one last hoop to jump through (as if you haven't had enough already). Instead of calling the sum of squares for the regression model SST as is done in ANOVA, statisticians call it *SSR* for *sum of squares regression.* Consider SSR from regression to be equivalent to the SST from ANOVA. The reason this is important is because computer output lists the sums of squares for the regression model as SSR not SST.

To summarize the sums of squares as they apply to regression, you have SSTO = SSR + SSE where

- SSTO measures the variability in the observed *y*-values around their mean. This value represents the variance of the *y*-values.
- SSE represents the variability between the predicted values for *y* (the values on the line) and the observed *y*-values. SSE represents the variability left over after the line has been fit to the data.
- SSR measures the variability in the predicted values for *y* (the values on the line) from the mean of *y*. SSR is the sum of squares due to the regression model (the line) itself.

Minitab calculates all the sums of squares for you as part of the regression analysis. You can see this calculation in the section "Bringing regression to the ANOVA table."

## Dividing up the degrees of freedom

In ANOVA, you test a model for the treatment (population) means by using an *F*-test, which is $F = \frac{MST}{MSE}$. To get MST (the mean sum of squares for treatment), you take SST (the sum of squares for treatment) and divide by its degrees of freedom. You do the same with MSE (that is, take SSE, the sum of squares for error, and divide by its degrees of freedom). The question now is, what do those degrees of freedom represent and how do they relate to regression? This section addresses that issue.

### Degrees of freedom in ANOVA

In ANOVA, the degrees of freedom for SSTO is $n - 1$, which represents the sample size minus one. In the formula for SSTO, $\Sigma\left(y_i - \bar{y}\right)^2$, you see there are $n$ observed $y$-values minus one mean. That in a very general way is where the $n - 1$ comes from.

Note that if you divide SSTO by $n - 1$, you get $\frac{\Sigma\left(y_i - \bar{y}\right)^2}{n-1}$, the variance in the $y$-values. This calculation makes good sense because the variance also measures the total variability in the $y$-values.

The degrees of freedom for SSE is $n - k$. In the formula for SSE, $\Sigma\left(\hat{y}_i - \bar{y}\right)^2$, you see there are $n$ observed $y$-values, and $k$ is the number of treatments in the model. In regression, the number of coefficients in the model is $k = 2$ (the slope and the $y$-intercept). So you have degrees of freedom $n - 2$ associated with SSE when you're doing regression.

### Degrees of freedom in regression

The degrees of freedom for SST in ANOVA equals the number of treatments minus one. How does the degrees of freedom idea relate to regression? The number of treatments in regression is equivalent to the number of parameters in a model (a parameter being an unknown constant in the model that you're trying to estimate).

When you test a model you're always comparing it to a different (simpler) model to see whether it fits the data better. In linear regression you compare your regression line $y = b_0 + b_1x$, to the horizontal line $y = \bar{y}$. This second, simpler model just uses the mean of $y$ to predict $y$ all the time, no matter what $x$ is. In the regression line, you have two coefficients: one to estimate the parameter for the $y$-intercept ($b_0$) and one to estimate the parameter for slope ($b_1$) in the model. In the second, simpler model, you have only one parameter: the value of the mean. The degrees of freedom for SSR in simple linear regression is the difference in the parameters of the two models: $2 - 1 = 1$.

Putting all this together, the degrees of freedom for regression must add up for the equation SSTO = SSR + SSE. The degrees of freedom corresponding to this equation are $(n - 1) = (2 - 1) + (n - 2)$, which is true if you do the math. So the degrees of freedom for regression, using the ANOVA approach, all check out. Whew!

In Figure 12-1, you can see the degrees of freedom for each sums of squares listed under the *DF* column of the ANOVA part of the output. You see SSR has $2 - 1 = 1$ degree of freedom, SSE has $250 - 2 = 248$ degrees of freedom (because $n = 250$ observations were in the data set and $k = 2$ and you find $n - k$ to get degrees of freedom for SSE). The degrees of freedom for SSTO is $250 - 1 = 249$.

## Bringing regression to the ANOVA table

In ANOVA, you test your model Ho: All $k$ population means are equal versus Ha: At least two population means are different by using a *F*-test. You build your *F*-test statistic by relating the sums of squares for treatment to the sum of squares for error. To do this, you divide SSE and SST by their degrees of freedom ($n - k$ and $k - 1$, respectively, where $n$ is the sample size and $k$ is the number of treatments) to get the mean sums of squares for error (MSE) and mean sums of squares for treatment (MST). In general, you want MST to be large compared to MSE, which would indicate that the model fits well. The results of all these statistical gymnastics are summarized by Minitab in a table called (cleverly) the ANOVA table.

The ANOVA table shown in the bottom part of Figure 12-1 for the Internet-use data represents the ANOVA table you get from using the regression line as your model. Under the Source column, you may be used to seeing treatment, error, and total. For regression, the treatment is the regression line, so you see *regression* instead of treatment. The error term in ANOVA is labeled *residual error,* because in regression, you measure error in terms of residuals. Finally you see *total,* which is the same the world around.

The SS column represents the sums of squares for the regression model. The three sums of squares listed in the SS column are SSR (for regression), SSE (for residuals), and SST (total). These sums of squares are calculated using the formulas from the previous section; the degrees of freedom, *DF* in the table, are found by using the formulas from the previous section also.

The MS column takes the value of SS "whatever"(you fill in the blank) and divides it by the respective degrees of freedom, just like ANOVA. For example in Figure 12-1, SSE is 12,968.5, and the degrees of freedom is 248. Take the first value divided by the second one to get 52.29 or 52.3, which is listed in the ANOVA table for MSE.

The value of the *F*-statistic, using the ANOVA method, is $F = \frac{MST}{MSE} = \frac{9{,}085.6}{52.3} =$ 173.7 in the Internet example, which you can see in column five of the ANOVA part of Figure 12-1 (subject to rounding). The *F*-statistics's *p*-value is calculated based on an *F*-distribution with $2 - 1 = 1$ and $250 - 2 = 248$ degrees of freedom, respectively. (In the Internet example, the *p*-value listed in the last column of the ANOVA table is 0.000, meaning the regression model fits.) But remember, in regression you don't use an *F*-statistic and an *F*-test. You use a *t*-statistic and a *t*-test. What gives? The next section explains.

## Relating the F- and t-statistics: The final frontier

In regression, one way of testing whether the best-fitting line is statistically significant is to test Ho: slope = 0 versus Ha: slope ≠ 0. To do this, you use a *t*-test (see Chapter 3). The slope is the heart and soul of the regression line, because it describes the main part of the relationship between $x$ and $y$. If the slope of the line equals zero (you can't reject Ho), you're just left with $y = b_1$, a horizontal line, and your model $y = b_0 + b_1x$ isn't doing anything for you.

In ANOVA, you test to see whether the model fits by testing Ho: The means of the populations are all equal, versus Ha: At least two of the population means aren't equal. To do this you use an *F*-test (taking MST and dividing it by MSE; see Chapter 10).

The sets of hypotheses in regression and ANOVA seem totally different, but in essence, they're both doing the same general thing: testing whether a certain model fits. In the regression case, the model you want to see fit is the straight line, and in the ANOVA case, the model of interest is a set of (normally distributed) populations with at least two different means (and the same variance). Here each population is labeled as a treatment by ANOVA.

But more than that, you can think of it this way: Suppose you took all the populations from the ANOVA and lined them up side by side on an *x-y* plane (see Figure 12-2). If the means of those distributions are all connected by a flat line (representing the mean of the *y*'s), then you would have no evidence against Ho in the *F*-test, so you can't reject it — your model isn't doing anything for you (it doesn't fit). This idea is similar to the idea of fitting a flat horizontal line through the *y*-values in regression; a straight-line model with a nonzero slope doesn't work in that case.

The big thing is that statisticians can prove (so you don't have to) that an *F*-statistic is equivalent to the square of a *t*-statistic, and the *F*-distribution is equivalent to the square of a *t*-distribution when the SSR has df = 2 – 1 = 1. And when you have a simple linear regression model, the degrees of freedom is exactly one! (Note that *F* is always greater than or equal to zero, which is needed if you're making it the square of something.) So there you have it! The *t*-statistic for testing the regression model is equivalent to an *F*-statistic for ANOVA when the ANOVA table is formed for the simple regression model.

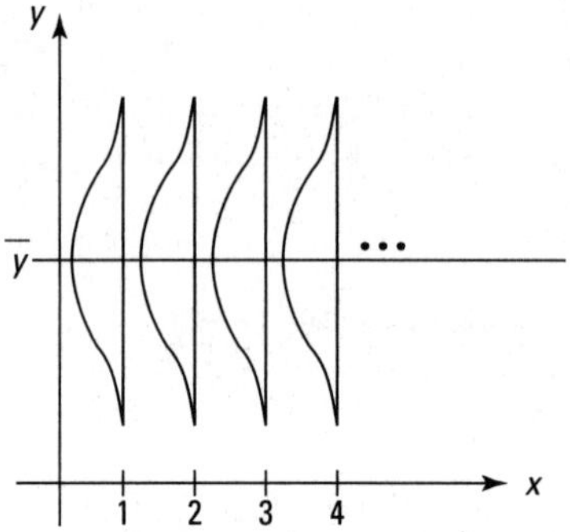


**Figure 12-2:** Connecting means of populations to the slope of a line.

Indeed (the stat professor's way of saying "and this is the *really* cool part. . ."), if you look at the value of the *t*-statistic for testing the slope of the education variable in Figure 12-1, you see that it's 13.18 (look at the row marked *Education* and the column marked *T*). Square that value, and you get 173.71. The *F*-statistic in the ANOVA table of Figure 12-1 is equal to 173.75. The *F*-statistic from ANOVA and the *t*-statistic from regression are equal to each other in Figure 12-2, subject to a little round-off error done by Minitab on the output. (Just like magic! I still get chills just thinking about it.)

# Part IV

# Building Strong Connections with Chi-Square Tests

The 5th Wave By Rich Tennant

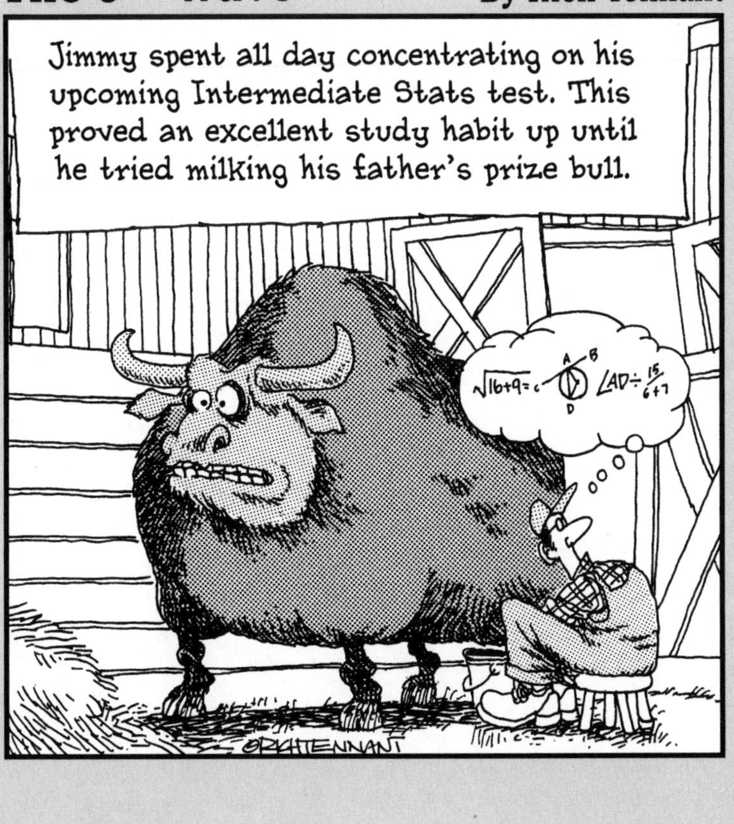
Jimmy spent all day concentrating on his upcoming Intermediate Stats test. This proved an excellent study habit up until he tried milking his father's prize bull.

## In this part . . .

Have you ever wondered if the percentage of M&Ms of each color is the same in every bag? Or whether someone's vote in an election is related to gender? Have you ever wondered if banks really have a case for denying loans based on a low credit score? This part answers all of those questions and more, using the Chi-square distribution.

# Chapter 13

# Forming Associations with Two-Way Tables

### In This Chapter

- Reading and interpreting two-way tables
- Figuring probabilities and checking for independence
- Watching out for Simpson's Paradox

Looking for relationships between two categorical (qualitative) variables is a very common goal for researchers. For example, many medical studies center on how some characteristic about a person either raises or lowers his chance of getting some disease. Marketers ask questions like, "Who is more likely to buy our product: males or females?" Sports stat freaks wonder about things like "Does winning the coin toss at the beginning of a football game increase your team's chance of winning the game?"

To answer each of the above questions, you must first collect data (from a random sample) on the two categorical variables being compared — call them $x$ and $y$. Then you organize that data into a table that contains columns and rows, showing how many individuals from the sample appear in each combination of $x$ and $y$. Finally, you use the information in the table to conduct a hypothesis test (called the Chi-square test). Using the Chi-square test, you can determine whether you can see a relationship between $x$ and $y$ in the population from which the data was drawn. This last step needs the machinery from Chapter 14 to accomplish it. The goals of this chapter are to understand what it means for two qualitative variables ($x$ and $y$) to be associated and to discover how to use percentages to determine whether a sample data set appears to show a relationship between $x$ and $y$.

Suppose you're collecting data on cell-phone users, and you want to find out whether more females use cell phones than males. A study of 508 randomly selected male cell-phone users and 508 randomly selected female cell-phone users conducted by a wireless company found that women tend to use their phones for personal calls more than men (big shocker). The survey showed that 427 of the women said they used their wireless phones primarily to talk with friends and family, while only 325 of the men admitted to doing so.

But you can't stop there. You need to break down this information, calculate some percentages, and compare them to see how close they really are. Sample results vary from sample to sample, and differences can appear by chance.

In this chapter, you find out how to organize data from qualitative variables (data based on categories rather than measurements) into a table format. This skill is especially useful when you're trying to look for relationships between two qualitative variables, such as using a cell phone for personal calls (a yes or no category) and gender (male or female). You also summarize the data to answer your questions. And, finally, you get to figure out, once and for all, what's going on with that Simpson's Paradox thing.

# Breaking Down a Two-Way Table

A *two-way table* is a table that contains rows and columns, which help you organize data from categorical (qualitative) variables in the following ways:

- The rows represent the possible categories for one categorical variable, such as males and females.
- The columns represent the possible categories for a second categorical variable, such as using your cell phone for personal calls, or not.

Here I review the basic ideas of organizing and filling in a two-way table.

## Organizing data into a two-way table

To organize your data into a two-way table, first set up the rows and columns. Table 13-1 shows the setup for the cell-phone data (refer to the example I give at the beginning of the chapter).

**Table 13-1 Two-Way Table Set Up for the Cell-Phone Data**

| | *Personal Calls: Yes* | *Personal Calls: No* |
|---|---|---|
| Males | | |
| Females | | |

Notice that Table 13-1 has four empty cells inside of it (not counting the empty space in the upper-left corner). Because gender has two choices (male or female), and personal cell-phone use has two choices (yes or no), the resulting two-way table has $2 * 2 = 4$ cells.

TIP

To figure out the number of cells in any two-way table, multiply the number of possible categories for the row variables times the number of possible categories for the column variable.

## Filling in the cell counts

After you set up the table with the appropriate number of rows and columns, you need to fill in the appropriate numbers in each of the cells of the two-way table. The number in each cell of a two-way table is called the *cell count* for that cell. The upper-left cell in the two-way table shown in Table 13-1 represents the number of males who use their cell phones for personal calls. With the information you have in the cell-phone problem, the cell count for this cell is 325. Because you know that 427 females use their cell phones for personal calls, this number goes into the lower-left cell.

Now, to figure out the numbers in the remaining two cells, you do a bit of subtraction. You know from the information given that the total number of male cell-phone users in the survey is 508. Each male either uses his cell phone for personal calls (falling into the *yes* group), or he doesn't (falling into the *no* group). Because 325 males fall into the *yes* group, and you have 508 males total, 183 males ($508 - 325 = 183$) don't use their cell phones for personal calls. This number is the cell count for the upper-right cell of the two-way table. Finally, because 508 females took the survey, and 427 of them use their cell phones for personal calls, you know that the rest of them ($508 - 427 = 81$) don't. Therefore, 81 is the cell count for the lower-right cell of the table. Table 13-2 shows the completed table for the cell-phone user problem, with the four cell counts filled in.

**Table 13-2 Completed Two-Way Table for the Cell-Phone Data**

| | *Personal Calls: Yes* | *Personal Calls: No* |
|---|---|---|
| Males | 325 | 183 (508 – 325) |
| Females | 427 | 81 (508 – 427) |

Just to save you a little time, if you have the total number in a group and how many of those individuals fall into one of the categories of the two-way table, you can determine the number falling into the remaining category by subtracting the total number in the group minus the number in the given category. You can complete this process for each remaining group in the table.

## Making marginal totals

One of the most important aspects of a two-way table is to have easy access to all the pertinent totals. Because every two-way table is made up of rows and columns, you can imagine that the totals for each row and the totals for each column are important. Also, the grand total is important to know.

If you take a single row and add up all the cell counts in the cells of that row, you get what is called a *marginal row total* for that row. Where does this marginal row total go on the table? You guessed it — out in the margin at the end of that row. You can find the marginal row totals for every row in the table and put them into the margins at the end of the rows. This group of marginal row totals for each row represents what statisticians call the *marginal distribution* for the row variable. The marginal row totals should add up to the *grand total,* which is the total number of individuals in the study. (The individuals may be people, cities, dogs, companies, and so on, depending on the scenario of the problem at hand.)

Similarly, if you take a single column and add up all the cell counts in the cells of that column, you get the *marginal column total* for that column. This number goes in the margin at the bottom of the column. Follow this pattern for each column in the table, and you have the marginal distribution for the column variable. Again, the sum of all the marginal column totals equals the grand total. The grand total is always located in the lower-right corner of the two-way table.

The marginal row total, marginal column totals, and the grand total for the cell-phone example are shown in Table 13-3.

**Table 13-3 Marginal and Grand Totals for the Cell Phone Data**

| | *Personal Calls: Yes* | *Personal Calls: No* | *Marginal Row Totals* |
|---|---|---|---|
| Males | 325 | 183 (508 – 325) | 508 |
| Females | 427 | 81 (508 – 427) | 508 |
| **Marginal Column Totals** | 752 | 264 | 1,016 **(Grand Total)** |

The marginal row totals add the cell counts in each row; yet the marginal row totals show up as a column in the two-way table. This phenomenon occurs because when summing the cell counts in a row, you put the result in the margin at the end of the row, and when you do this for each row, you're stacking the row totals into a column. Similarly, the marginal column totals add the cell counts in each column; yet they show up as a row in the two-way table. Don't let this be a source of confusion when you're trying to navigate or set up a two-way table. It's always a good idea to label your totals as marginal row, marginal column, or grand total to help keep it clear.

# Breaking Down the Probabilities

A percentage, when applied to a two-way table, represents the portion of the individuals in the sample falling into a certain group. This idea can be expanded to a probability, which gives the chance that an individual person selected from this group falls into a certain category.

A two-way table gives you the opportunity to find many different kinds of probabilities to help you find the answers to different questions about your data or to look at the data another way. In this section, I cover the three most important types of probabilities found in a two-way table: marginal probabilities, joint probabilities, and conditional probabilities. (If you need more info on these terms, check out *Probability For Dummies* [Wiley].)

When you find probabilities based on a sample, as you do in this chapter, you have to realize that those probabilities pertain to that sample only. They do not transfer automatically to the population being studied. For example, if you take a random sample of 1,000 adults and find that 55 percent of them watch reality TV, this study doesn't mean that 55 percent of all adults in the entire population watch reality TV. (The media makes this mistake every day.) You need to take into account the fact that sample results vary. In Chapters 14 and 15, you do just that. But this chapter zeros in on summarizing the information in your sample, which is the first step toward that end (but not the last step in terms of making conclusions about your corresponding population).

## Marginal probabilities

A *marginal probability* makes a probability out of the marginal total, for either the rows or the columns. A marginal probability represents the proportion of the entire group that belongs in that single row or column category. Each marginal probability represents only one category for only one variable — it doesn't consider the other variable at all. In the cell-phone example, you have four possible marginal probabilities (refer to Table 13-3):

- Marginal probability of female ($508/1,016 = 0.50$). That means, 50 percent of all the cell-phone users in this sample were females.
- Marginal probability of male ($508/1,016 = 0.50$). That means, 50 percent of all the cell-phone users in this sample were males.
- Marginal probability of using a cell phone for personal calls ($752/1,016 = 0.74$). Therefore, 74 percent of all cell-phone users in this sample make personal calls with their cell phones.
- Marginal probability of not using a cell phone for personal calls ($264/1,016 = 0.26$). In other words, 26 percent of all the cell-phone users in this sample don't make personal calls with their cell phones.

Statisticians use shorthand notation for all probabilities. If you let M = male, F = female, Yes = personal cell-phone use, and No = no personal cell-phone use, then each of the preceding marginal probabilities is written this way:

- P(F) = 0.50
- P(M) = 0.50
- P(Yes) = 0.74
- P(No) = 0.26

TIP

Notice that P(F) and P(M) add up to 1.00. This result is no coincidence, because these two categories make up the entire gender variable. Similarly, P(Yes) and P(No) sum up to 1.00 because those choices are the only two for the personal cell-phone use variable. Everyone has to be classified somewhere.

WARNING!

Be advised that some probabilities aren't useful in terms of discovering information about the population in general. For example, P(F) = 0.50 in the previous example because the researchers determined ahead of time that they wanted exactly 508 females and exactly 508 males. The fact that 50 percent of the sample is female and 50 percent of the sample is male doesn't mean that in the entire population of cell-phone users 50 percent are males and 50 percent are females. The sample was just set up that way. If you want to study what proportion of cell-phone users are females and males, you need to take a combined sample instead of two separate ones, and see how many males and females appear in the combined sample.

## Joint probabilities

A *joint probability* gives the probability of the intersection of two categories, one from the row variable and one from the column variable. It's the probability that someone selected from the whole group has two particular characteristics at the same time. A joint probability is found by taking the cell count for those having both characteristics and dividing by the grand total. In other words, both characteristics happen jointly, or together.

The cell-phone example has four joint probabilities:

- The probability that someone from the entire group is male and uses his cell phone for personal calls. This probability is $325/1{,}016 = 0.32$, meaning that 32 percent of all the cell-phone users in this sample are males using their cell phones for personal calls.
- The probability that someone from the entire group is male and doesn't use his cell phone for personal calls is $183/1{,}016 = 0.18$.
- The probability that someone from the entire group is female and makes personal calls with her cell phone is $427/1{,}016 = 0.42$.
- The probability that someone from the entire group is female and doesn't make personal calls with her cell phone is $81/1{,}016 = 0.08$.

The notation for the joint probabilities previously listed is as follows, where $\cap$ represents the intersection of the two categories listed:

- $P(M \cap Yes) = 0.32$
- $P(M \cap No) = 0.18$
- $P(F \cap Yes) = 0.42$
- $P(F \cap No) = 0.08$

TIP

The sum of all the joint probabilities for any two-way table should be 1.00, unless you have a little round-off error, which makes it very close to, but not exactly, 1.00. The sum is 1.00, because everyone in the group is classified somewhere with respect to both variables. It's like dividing the entire group into four parts and showing which proportion falls into each part.

## Conditional probabilities

A *conditional probability* is what you use if you want to compare subgroups in the sample. In other words, if you want to break down the table further, a conditional probability is what you use. Each row has a conditional probability for each cell within the row, and each column has a conditional probability for each cell within that column.

***Note:*** Because conditional probability is one of the sticking points for a lot of students, I want to spend extra time on it. My goal in this section is for you to have a good understanding of what a conditional probability really means and how you can use it in the real world (something many statistics textbooks neglect to mention, I have to say).

### *Figuring conditional probabilities*

Consider the cell-phone example in Table 13-3. Suppose you want to look at just the males who took the survey. The total number of males is 508. You can break this group down into two subgroups by using conditional probability. You can find the probability of using cell phones for personal calls (males only), and you can find the probability of not using cell phones for personal calls (males only). Similarly, you can break down the females by those females who use cell phones for personal calls and those females who don't.

In each case, to find a conditional probability, you first look at a single row or column of the table that represents the known characteristic about the individuals. The marginal total for that row or column now represents your new grand total, because this group becomes your entire universe when you examine it. Then take the cell counts from that row or column and divide the sum by that row or column's marginal total.

In the cell-phone example, you have the following conditional probabilities when you break the table down by gender:

- The conditional probability that a male uses a cell phone for personal calls is $^{325}/_{508} = 0.64$.
- The conditional probability that a male doesn't use a cell phone for personal calls is $^{183}/_{508} = 0.36$.
- The conditional probability that a female uses a cell phone for personal calls is $^{427}/_{508} = 0.84$.
- The conditional probability that a female doesn't use a cell phone for personal calls is $^{81}/_{508} = 0.16$.

To interpret these results, you say that within this sample if you're male, you're more likely than not to use your cell phone for personal calls (64 percent compared to 36 percent). However, the percentage of personal-call makers is higher for females (84 percent versus 16 percent).

WARNING! The conclusions you can make from two-way tables in this chapter must refer only to the sample, not the population it came from. Before going on to make general statements about the conditional probability within a population, you need to conduct a confidence interval for a population proportion (which is equivalent to a probability). See Chapter 3 or your intro stats book for information on a hypothesis test for a population proportion.

Notice that for the males in the previous example, the two probabilities (0.64 and 0.36) add up to 1.00. This is no coincidence. The males have been broken down by cell-phone use for personal calls, and because everyone in the study is a cell-phone user, each male has to be classified in one group or the other. Similarly, the two probabilities for the females sum to 1.00.

### *Notation for conditional probabilities*

Conditional probabilities are denoted by a straight up-and-down line that lists and separates the event that is known to have happened (what's given) and the event for which you want to find the probability. You can write the notation like this: P(XX | XX). You place the given event to the right of the line and the event for which you want to find the probability to the left of the line. For example, suppose you know someone is female (F) and you want to find out the chance she is a Democrat (D). In this case, you're looking for P(D | F). On the other hand, say you know a person is a Democrat and you want the probability that person is female — you're looking for P(F | D).

The straight up-and-down line in the conditional probability notation isn't a division sign; the line is just a line separating events A and B. Also, be careful of the order in which you place A and B into the conditional probability notation. In general, $P(A \mid B) \neq P(B \mid A)$.

Following is the notation used for the conditional probabilities in the cell-phone example:

- **P(Yes | M) = 0.64.** You can say it this way: "The probability of Yes given Male is 0.64."
- **P(No | M) = 0.36.** In human terms, say "The probability of No given Male is 0.36."
- **P(Yes | F) = 0.84.** Say this one with gusto: "The probability of Yes given Female is 0.84."
- **P(No | F) = 0.16.** You translate this notation by saying "The probability of No given Female is 0.16."

You can see that P(Yes | M) + P(No | M) = 1.00 because you're breaking all males into two groups: those using cell phones for personal calls (Y) and those not (N). Notice however, that P(Yes | M) + P(Yes | F) doesn't sum to 1.00. In the first case, you're looking only at the males, and in the second case, only at the females.

### *Comparing two groups with conditional probabilities*

One of the most common questions regarding two categorical (qualitative) variables is this: Are they related? To answer this question, you use conditional probabilities. You set up and find the conditional probabilities you need to see whether two variables are related.

To compare the conditional probabilities, take one variable and find the conditional probabilities based on the other variable. Do this for each category of the first variable. Compare those conditional probabilities (you can even graph them for the two groups) and see whether they're different or the same. (If the conditional probabilities are the same for each group, the variables aren't related in the sample. If they're different, the variables are related in the sample.) To be able to generalize the results, you need to use the sample results to draw a conclusion from the overall population involved by doing a Chi-square test (see Chapter 14).

Revisiting the cell-phone example from the previous section, you can ask specifically: Is personal use related to gender? You know that you want to compare cell-phone use for males and females to find out whether use is related to gender. However, it's very difficult to compare cell counts — for example, 325 males use their phones for personal calls, compared to 427 females. In fact, it's impossible to compare these numbers without using some total for perspective. Three hundred twenty-five out of what?

You have no way of comparing the cell counts in two groups without creating percentages (dividing each cell count by the appropriate total). Percentages give you a means of comparing two numbers on equal terms. For example, suppose you give a one-question opinion survey (yes, no, no opinion) to a random sample of 1,099 people; 465 respondents said yes, 357 said no, and 277 had no opinion. To truly interpret this information, you're probably in your head trying to compare these numbers to each other. That's what percentages do for you. Showing the percentage in each group in a side-by-side fashion gives you a relative comparison of the groups with each other.

But first, you need to bring conditional probabilities into the mix. In the cell-phone example, if you want the percentage of females who use their cell phones for personal calls, you take 427 divided by the total number of females (508) to get 84 percent. Similarly, to get the percentage of males who use their cell phones for personal calls, take the cell count (325) and divide it by that row total for males (508), which gives you 64 percent. This percentage is the conditional probability of using a cell phone for personal calls, given the person is male.

Now you're ready to compare the males and females by using conditional probabilities. Take the percentage of females who use their cell phones for personal calls and compare it to the percentage of males who use their cell phones for personal calls. By finding these conditional probabilities, you can easily compare the two groups and say that in this sample at least, more females use their cell phones (84 percent) for personal calls than men (64 percent).

### Using graphs to display conditional probabilities

One way to highlight conditional probabilities as a tool for comparing two groups is to use graphs such as a pie chart comparing the results of the other variable for each group or a bar chart comparing the results of the other variable for each group.

Figures 13-1a and 13-1b use two pie charts to compare males and females on cell-phone use. Figure 13-1a shows cell-phone use for only the males; this pie chart shows the conditional distribution of use for (given) males. Figure 13-1b shows the conditional distribution of cell phone use for (given) females. A comparison of Figures 13-1a and 13-1b shows the slices for cell-phone use aren't equal (or even close) for males compared to females. That result means that gender and cell-phone use for personal calls are dependent in this sample.

You may be wondering how close the two pie charts need to look (in terms of how close the slice amounts are for one pie compared to the other) in order to say the variables are independent. This question isn't one you can answer completely until you conduct a hypothesis test for the proportions themselves (see the Chi-square test in Chapter 14). For now, with respect to your sample data, if the difference in the appearance of the slices for the two graphs is enough that you would write a newspaper article about it, then I'd go for dependence. Otherwise, conclude independence.

You can also make a bar chart to show the same idea. (For more info on pie charts and bar charts, see *Statistics For Dummies* [written by me and published by Wiley] or your intro stats textbook.)

Another way you can make comparisons is to break down the two-way table by the column variable. (You don't always have to use the row variable for comparisons.) In the cell-phone example (Table 13-3), you can compare the group of personal-call makers to the group of no-personal-call makers and see what percentage in each group is male and female. This type of comparison puts a different spin on the information, because you're comparing the behaviors to each other, in terms of gender.

With this new breakdown of the two-way table, you get the following:

- The conditional probability of being male, given you use your cell phone for personal calls, is $P(M \mid Yes) = {}^{325}\!/_{752} = 0.43$. ***Note:*** The denominator is 752, the total number of people who make personal calls with their cell phones.
- The conditional probability of being female, given you use your cell phone for personal calls, is $P(F \mid Yes) = {}^{427}\!/_{752} = 0.57$.

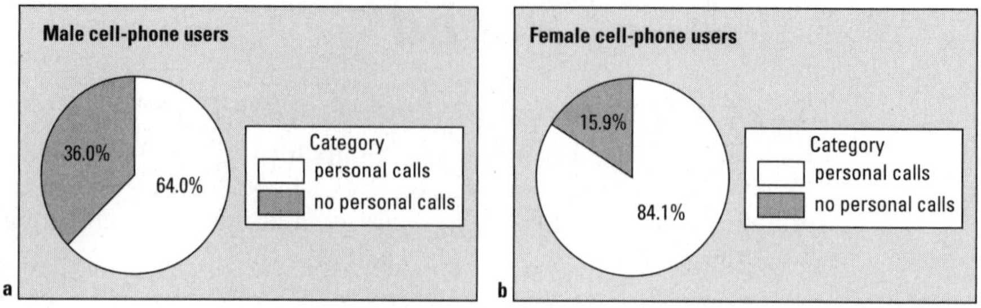


**Figure 13-1:** Pie charts comparing male versus female personal cell-phone use.

Again, these two probabilities add up to 1.00, because you're breaking down the personal-call makers according to gender (male or female), and the last two probabilities sum to 1.00, because you're breaking down the non-personal-call makers by gender (male and female).

The overall conclusions are similar to those found in the previous section, but the specific percentages and the interpretation are different. Interpreting the data this way, if you use your cell phone for personal calls, you're more likely to be female than male (57 percent compared to 43 percent). And if you don't use your cell phone to make personal calls, you're more likely to be male (69 percent versus 31 percent).

## What should you divide by? That is the question!

To get the correct answer for any probability in a two-way table, here's the trick: Always be sure to identify the group that is being examined. What is the probability "out of"? In the cell phone example (refer to Table 13-3):

- If you want the percentage *of all users* who are males using their phones for personal calls, then you take the cell count 325, and divide by 1,016, the grand total.
- If you want the percentage *of males* who are using their cell phones for personal calls, you take 325 divided by 508, the total number of males.
- If you want the percentage *of personal-call makers* who are male, you take 325 divided by 752 (the total number of people who make personal calls with their cell phones).

In each of these three cases, the numerator is the same, but the denominators are different, leading you to very different answers. Deciding which number to divide by is a very common source of confusion for people, and this trick can really help give you an edge on keeping it straight.

# Trying to be Independent

Independence is a big deal in statistics. The term generally means that two items have outcomes whose probabilities don't affect each other. The items could be events A and B, variables $x$ and $y$, or survey results from two people selected at random from a population, and so on. If the outcomes of the two items do affect each other, statisticians call those two items *dependent* (or not independent). In this section, you check for and interpret independence of two categories of qualitative variables in a sample, and you check for and interpret independence of two qualitative variables in a sample.

## Checking for independence between two categories

Statistics instructors often have students check to see whether two categories (one from a qualitative variable $x$ and the other from a qualitative variable $y$) are independent. I prefer to just compare the two groups and talk about how similar or different the percentages are, broken down by another variable. However, to cover all the bases and make sure you can answer this very popular question, here's the official definition of independence, straight from the statistician's mouth: Two categories are *independent* if their joint probability equals the product of their marginal probabilities. The only caveat here is that neither of the categories can be completely empty.

For example, if being female is independent of being a Democrat, then $P(F \cap D) = P(F) * P(D)$, where D = Democrat and F = Female. So, to show that two categories are independent, find the joint probability and compare it to the product of the two marginal probabilities. If you get the same answer both times, the categories are independent. If not, then the categories are not independent, but rather, they are dependent.

You may be wondering: Don't all probabilities work this way, where the joint probability equals the product of the marginals? No, they don't. For example, if you draw a card from a standard 52-card deck, you get a red card with probability ½. You draw a black card with probability ½. The chance, though, of drawing both a black and red card with one draw is 0, while the product of the probabilities for black times red comes out to $\frac{1}{2} * \frac{1}{2} = \frac{1}{4}$.

Now, if you look at a red card that is a two, the joint probability of a red two, which is $\frac{2}{52} = \frac{1}{26}$, equals the probability of a red card ($\frac{1}{2}$) times the probability of a two, which is $\frac{4}{52}$ (because $\frac{1}{2} * \frac{4}{52} = \frac{1}{26}$).

Another way to check for independence is to compare the conditional probability to the marginal probability. Specifically, if you want to check whether being female is independent of being Democrat, check either of the following two situations (they'll both work if the variables are independent):

- **Is P(F | D) = P(F)?** That is, if you know someone is a Democrat, does that affect the chance that they will also be female? If yes, F and D are independent. If not, F and D are dependent.
- **Is P(D | F) = P(D)?** This question is asking whether being female changes your chances of being a Democrat. If yes, D and F are independent. If not, D and F are dependent.

Is knowing that you're in one category going to change the probability of being in another category? If so, the two categories aren't independent. If it doesn't affect the probability, then the two categories are independent.

## *Checking for independence between two variables*

The discussion in the previous section focuses on checking if two specific categories are independent in a sample. If you want to extend this idea to showing that two entire categorical variables are independent, you must check the independence conditions for every combination of categories in those variables. All of them must work, or independence is lost. The first case where dependence is found between two categories means that the two variables are dependent. If you find that the first case shows independence, you must continue checking all the combinations before declaring independence.

Suppose a doctor's office wants to know whether calling patients to confirm their appointments is related to whether they actually show up. The variables are $x$ = called the patient (called or didn't call) and $y$ = patient showed up for their appointment (showed or didn't show). Here are the four conditions that need to hold before you declare independence:

1. P(showed) = P(showed | called)
2. P(showed) = P(showed | didn't call)
3. P(didn't show) = P(didn't show | called)
4. P(didn't show) = P(didn't show | didn't call)

If any one of these conditions isn't met, you stop there and declare the two variables to be dependent in the sample. If (and only if) all the conditions are met, you declare the two variables independent in the sample.

You can see the results of a sample of 100 randomly selected patients in Table 13-4.

**Table 13-4 Confirmation Calls Related to Showing Up for the Appointment**

| | *Called* | *Didn't Call* | *Row totals* |
|---|---|---|---|
| Showed | 57 | 33 | 90 |
| Didn't Show | 3 | 7 | 10 |
| Column Totals | 60 | 40 | 100 |

Checking the conditions for independence, you can start at the first condition and check to see whether P(showed) = P(showed | called). From the last column of Table 13-4, you can see that P(showed) is equal to $^{90}/_{100}$ = 0.90, or 90 percent. Next, you can find P(showed | called) by looking at the first column of Table 13-4. This probability is $^{57}/_{60}$ = 95 percent. Because these two probabilities aren't equal (although they're close), then you say that showing up and calling first are dependent. In other words, people come a little more often when you call them first. (To determine whether these sample results carry through to the population, which also takes care of the question of how close the probabilities need to be in order to conclude independence, see Chapter 14.)

# Demystifying Simpson's Paradox

*Simpson's Paradox* is a phenomenon where results appear to be in direct contradiction to one another, which can make even the best student's heart race. This situation can go unnoticed unless three variables (or more) are examined, in which case you organize the results into a *three-way table,* with columns within columns or rows within rows.

Simpson's Paradox is a favorite among statistics instructors (because it's so mystical and magical — and the numbers get so gooey and complex) but Simpson's Paradox is a nonfavorite among many students, mainly because of the following two reasons (in my opinion):

- Due to the way Simpson's Paradox is presented in most statistics courses, you can easily get buried in the details and have no hope of seeing the big picture: Simpson's Paradox presents a big problem in terms of interpreting data, and you need to understand it fully in order to avoid it.
- Most textbooks do a good job of showing you examples of Simpson's Paradox, but they do a not-so-good job of explaining why it occurs (some even neglect to explain the why part at all).

My goals in this section are for you to know what Simpson's Paradox is, to be able to understand and explain why and how it happens, and to know how to be watchful for it. This is a tall order, I know, but stick with me.

# Experiencing Simpson's Paradox

Simpson's Paradox was discovered in 1951 by an American Statistician named E. H. Simpson. He realized that if you analyze some data sets one way, by breaking them down by two variables only, you can get one result, but when you break the data down further by a third variable, the results switch direction. That's why his result is called *Simpson's Paradox* — a paradox being an apparent contradiction in results.

In the following sections, you can see Simpson's Paradox play out in an example and all the details in between.

## Simpson's Paradox in action: Video games and the gender gap

Suppose I am interested in finding out who is better at playing video games, men or women. I watch males and females choose and play a variety of video games, and each time someone plays a video game, I record whether he or she wins or loses. Suppose I record the results of 200 video games, as seen in Table 13-5. (Note that the females played 120 games, and the males played 80 games.)

**Table 13-5 Video Games Won and Lost for Males versus Females**

| *All Games* | *Won* | *Lost* | *Marginal Row Totals* |
|---|---|---|---|
| Males | 44 | 36 | 80 |
| Females | 84 | 36 | 120 |
| **Marginal Column Totals** | 128 | 72 | 200 **(Grand Total)** |

Looking at Table 13-5, you see the proportion of males who won their video games, P(Won | Male), is $44/80 = 0.55$. The proportion of females who won their video games, P(Won | Female), is $84/120 = 0.70$. So overall, the females won more of their video games than the males did. Does this finding mean that women are better than men at video games in general in the sample?

Not so fast, my friend. Notice that the people in the study were allowed to choose the video games they played. This factor blows the study wide open. Suppose females and males choose different types of video games: Can this affect the results? The answer may be *yes.* Considering other variables that could be related to the results but weren't included in the original study (or at least not in the original data analysis) is important. These additional variables that cloud the results are called *confounding variables.*

## Factoring in difficulty level

Many people may expect the video game results from the previous section to be turned around, that men are better at playing video games than women. According to the research, men spend more time playing video games, on average, and are by far the primary purchaser of video games, compared to women. So what explains the eyebrow-raising results in this study? Is there another possible explanation? Is important information missing that is relevant to this case?

One of the variables that wasn't considered when I made Table 13-5 was the difficulty level of the video game being played. Suppose I go back and include the difficulty level of the chosen game each time, along with each result (won or lost). Level one indicates easy video games, comparable to the level of Ms. Pac Man (games that are my speed), and level two means more challenging video games (like war games or sophisticated strategy games).

Table 13-6 represents the results with this new information added on difficulty level of games played. You have three variables now: level of difficulty (one or two); gender (male or female); and outcome (won or lost). Statisticians therefore call Table 13-6 a three-way table.

**Table 13-6 A Three-Way Table for Gender, Game Level, and Game Outcome**

| | ***Level-One Games*** | | ***Level-Two Games*** | |
|---|---|---|---|---|
| | ***Won*** | ***Lost*** | ***Won*** | ***Lost*** |
| Males | 9 | 1 | 35 | 35 |
| Females | 72 | 18 | 12 | 18 |

Note in Table 13-6 that the number of level-one video games chosen was 9 + 1 + 72 + 18 = 100, and the number of level-two video games chosen was 35 + 35 + 12 + 18 = 100. But now you need to look at who chose which level of game. The next section probes this very issue.

### Comparing success rates with conditional probabilities

To compare the success rates for males versus females using Table 13-6, you can figure out the appropriate conditional probabilities, first for level-one games and then for level-two games.

For level-one games (only), the conditional probability of winning given male is $P(\text{Won} \mid \text{Male}) = 9/10 = 0.90$. So for the level-one games, males won 90 percent of the games they played. For level-one games, the percentage of games won by the females is $P(\text{Won} \mid \text{Female}) = 72/90 = 0.80$, or 80 percent. These results mean that at level one, the males did 10 percent better than the females at winning their games. But this percentage appears to contradict the results found in Table 13-5. (Just wait — the contradictions don't end here!)

Now figure the conditional probabilities for the level-two video games won. For the men, the percentage of males winning level-two games was $35/70 = 0.50$, or 50 percent. For the ladies, the percentage of women winning level-two games was $12/30 = 0.40$, or 40 percent. Once again, the males outdid the females!

Step back and think about this scenario for a minute. Table 13-5 shows that females won a higher percentage of the video games they played overall. But Table 13-6 shows that males won more of the level-one games and that males won more of the level-two games. What's going on? No need to check your math. No mistakes were made — no tricks were pulled. This inconsistency in results happens in real life from time to time in situations where an important third variable is left out of a study, a situation aptly named *Simpson's Paradox.* (See why it's called a paradox?)

## Asking why: Simpson's Paradox

REMEMBER

Confounding variables are the underlying cause of Simpson's Paradox. (A *confounding variable* is a third variable that's related to each of the other two variables and can affect the results if not accounted for.)

In the video game example, when you look at the video game outcomes (won or lost) broken down by gender only (Table 13-5), females won a higher percentage of their overall games than males (70 percent overall winning percentage for females compared to 55 overall winning percentage for males). Yet, when you split up the results by the level of the video game (level one or level two; see Table 13-6), the results reverse themselves, and you see that males did better than females on the level-one games (90 percent to 80 percent), and males also did better on the level-two games (50 percent versus 40 percent).

To see why this seemingly impossible result happens, take a look at the marginal row *probabilities* versus the marginal row *totals* in Table 13-6 (for the level-one games). The percentage of times a male won when he played an easy video game was 90 percent. However, males chose level-one video games only 10 times (out of 80 total level-one games played by men. That's only 12.5 percent).

To break this idea down further, the males' non-stellar performance on the challenging video games (50 percent — but still better than the females) coupled with the fact that the males chose challenging video games 70 out of 80 = 87.5 percent of the time really brought down that overall winning percentage (55 percent). And even though the men did really well on the level-one video games, they didn't play many of them (compared to the females), so their high winning percentage on level-one video games (90 percent) didn't count much toward their overall winning percentage.

Meanwhile, in Table 13-6, you see that females chose level-one video games 90 times (out of 120). Even though the females only won 72 out of the 90 games (80 percent, a lower percentage than the males), they chose to play many more of the level-one games, boosting their overall winning percentage.

Now the opposite situation happens when you look at the level-two video games in Table 13-6. The males chose the harder video games 70 times (out of 80), while the females only chose the harder ones 30 times out of 120. The males did better than the females on level-two video games (winning 50 percent of them versus 40 percent for the females). However, level-two video games are harder to win than level-one video games. This factor means that the males' winning percentage on level-two video games, being only 50 percent, doesn't contribute much to their overall winning percentage. However, the low winning percentage for females on level-two video games doesn't hurt them much, because they didn't play many level-two video games.

The bottom line is that the occurrence or non-occurrence of Simpson's Paradox is a matter of weights. In the overall totals from Table 13-5, the males don't look as good as the females. But when you add in the difficulty of the games (shown in Table 13-6), you see that most of the males' wins came from harder games (which have a lower winning percentage). The females played many more of the easier games on average, and easy games have a higher chance of winning no matter who plays them. So it all boils down to this: Which games did the males choose to play, and which games did the females choose to play? The males chose harder games, which contributed in a negative way to their overall winning percentage and made the females look better than they actually were.

Level of game wasn't included in the original summary, Table 13-5, but it should have been included because it's a variable that affected the results. Level of game, in this case, was the confounding variable.

## Keeping one eye open for Simpson's Paradox

Simpson's Paradox shows you the importance of including data about possible confounding variables when attempting to look at relationships between qualitative variables.

In the video game example I use in previous sections, level of difficulty of the game was a confounding variable; more men chose to play the more difficult games, which are harder to win, thereby lowering their overall success rate.

You can avoid Simpson's Paradox by making sure that obvious confounding variables are included in a study; that way, when you look at the data you get the relationships right the first time, and no room exists for misconstruing the results. And as with all other statistical results, if it looks too good to be true, or too simple to be correct, it probably is! Beware of someone that tried to oversimplify any result. While three-way tables are more difficult to examine, they are often worth using.

# Chapter 14

# Being Independent Enough for the Chi-Square Test

## *In This Chapter*

- Testing for independence in the population (not just the sample)
- Using the Chi-square distribution
- Discovering the connection between the *Z*-test and the Chi-square test

You've seen these hasty judgments before — people who collect one sample of data and try to use it to make conclusions about the whole population. When it comes to two qualitative variables (where data falls into categories and don't represent measurements), the problem seems to be even more widespread.

For example, a TV news show finds that out of 1,000 presidential voters, 200 females are voting Republican, 300 females are voting Democrat, 300 males are voting Republican, and 200 males are voting Democrat. The news anchor shows the data and then states that 30 percent ($^{300}/_{1,000}$) of all voters are females voting Democrat (and so on for the other counts). This conclusion is misleading. It is true that in this sample of 1,000 voters, 30 percent of them are females voting Democrat. However, this result doesn't automatically mean that 30 percent of the entire population of voters are females voting Democrat. Results change from sample to sample.

People often understand that they can expect sample results to change, yet they don't seem to realize that some conclusions come out differently due to even small changes in the sample results. For example, if you ask ten people about their views on an issue, you may get six people in favor (the majority) and four against. But the next time you take a sample of ten people, the results may reverse, and you'll have four people in favor and six people against (the majority). This inconsistency is especially prone to happening if the sample size is small.

In this chapter, you see how to move beyond just summarizing the sample results from a two-way table (discussed in Chapter 13) to using those results in a hypothesis test to make conclusions about an entire population. This process requires a new probability distribution called the *Chi-square distribution,* which you get very familiar with in this chapter. You also find out how to answer a very popular question among researchers: Are these two categorical (qualitative) variables independent (not related to each other) in the entire population?

# A Hypothesis Test for Independence

A recent survey conducted by American Demographics asked men and women about the color of their next house. The results showed that 36 percent of the men wanted to paint their houses white, and 25 percent of the women wanted to paint their houses white. Table 14-1 illustrates the results from a sample of 1,000 people (500 men and 500 women).

**Table 14-1 Gender and House-Paint Preference: Observed Cell Counts**

| | *White Paint* | *Nonwhite Paint* | *Marginal Row Totals* |
|---|---|---|---|
| Men | 180 | 320 | 500 |
| Women | 125 | 375 | 500 |
| **Marginal Column Totals** | 305 | 695 | 1,000 **(Grand Total)** |

The *marginal row totals* represent the total number in each row; the *marginal column totals* represent the total number in each column (see Chapter 13 for more information on row and column marginal totals). Notice that of the males, the percentage who want to paint their houses white is $180/500 = 0.36$, or 36 percent, as stated previously. And the percentage of females who want to paint their houses white is $125/500 = 0.25$, or 25 percent. (Both of these percentages represent conditional probabilities as explained in Chapter 13.)

The American Demographics report concluded from this data that ". . . men and women agree on exterior house paint colors; the main exception being the top male choice, white (36 percent would paint their next house white versus 25 percent of women)." This type of conclusion is commonly formed, but it's an overgeneralization of the results at this point. You know that in this sample, more men wanted to paint their houses white than women, but is 180 really that different from 125, with a sample size of 1,000 people whose results will vary the next time you do the survey? How do you know these results carry over to the population of all men and women? That question can't be answered without a formal statistical procedure called a *hypothesis test* (see Chapter 3 for the basics on hypothesis tests).

To show that men and women in the population differ according to favorite house color, first note that you have two qualitative variables — gender (male or female) and paint color (white or nonwhite). What you really want to know is whether these two variables are related to each other or not. If they are related, then favorite paint color depends on gender, which means these two variables are dependent. If they aren't related, then favorite paint color doesn't depend on gender, and the two variables are independent.

To test whether two qualitative variables are independent, you need a Chi-square test. The steps for the Chi-square test are the following, with full details supplied in the next sections (note that Minitab can conduct this test for you also, from step three on down):

1. **Collect your data and summarize it in a two-way table.**

   These numbers represent the observed cell counts. (For more on two-way tables, see Chapter 13.)

2. **Set up your null hypothesis, Ho: Variables are independent; and the alternative hypothesis, Ha: Variables are dependent.**

3. **Calculate the expected cell counts under the assumption of independence.**

   The expected cell count for a cell is the row total times the column total divided by the grand total.

4. **Check the conditions of the Chi-square test before proceeding; each expected cell count must be greater than or equal to five.**

5. **Figure the Chi-square test statistic.**

   This statistic finds the observed cell count minus the expected cell count, squares the difference, and divides it by the expected cell count. Do these steps for each cell and then add them all up.

6. **Look up your test statistic on the Chi-square table (Table A-3 in the Appendix) and find the *p*-value (or one that's close).**

7. **If your result is less than your prespecified cutoff ( the $\alpha$ level), usually 0.05, reject Ho and conclude dependence of the two variables.**

   If your result is greater than the $\alpha$ level, fail to reject Ho; the variables can't be deemed dependent.

To conduct a Chi-square test in Minitab, enter your data in the spreadsheet exactly as it appears in your two-way table (see Chapter 13 for setting up a two-way table for qualitative data). Go to Stat>Tables>Chi-Square Test. Click on the two variable names in the left-hand box corresponding to your column variables in the spreadsheet. They appear in the Columns Contained in the Table box. Then click on OK.

## Collecting and organizing the data

The first step toward any data analysis is collecting your data. In the case of two categorical (qualitative) variables, you collect data on the two variables at the same time for each person. In the house-color example from the previous section, you note each person's gender, and then ask each person his or her preference for exterior house color. Keeping the data together in pairs (for example: male, white paint; female, nonwhite paint), you then organize it into a two-way table where the rows represent the categories of one qualitative variable (for example, males and females for gender), and the columns represent the categories of the other qualitative variable (for example, white paint and nonwhite paint).

The data for the house-paint example is organized in Table 14-1. You can see by looking at the grand total in the lower-right-hand corner of the table that 1,000 people participated in the survey; you see by the row totals that the 1,000 people were comprised of 500 men and 500 women. The connection between the two pieces of information collected is kept by organizing the data into one two-way table versus two individual tables, one for gender and one for house-paint preference. That way, you can look at the relationship between the two variables. (For the full details on organizing and interpreting the results from a two-way table, see Chapter 13.)

## Determining the hypotheses

Every hypothesis test (whether it be a Chi-square test or some other test) has two hypotheses:

- A *null hypothesis,* which you have to believe unless someone showed you otherwise. The notation for this hypothesis is Ho.
- An *alternative hypothesis,* which you want to conclude in the event that you can't support the null hypothesis anymore. The notation for this hypothesis is Ha.

For a full discussion of hypothesis testing, see my other book *Statistics For Dummies* (Wiley) or your intro stats textbook. For a quick review, see Chapter 3 of this book.

In the case where you're testing for the independence of two qualitative variables, the null hypothesis is when no relationship exists between them. In other words, they're independent. The alternative hypothesis is when the two variables are related, or dependent.

For the paint color example from the previous section, you write Ho: gender and paint color are independent versus Ha: gender and paint color are dependent. You have now completed step two of the Chi-square test.

## Figuring expected cell counts

When you've collected your data and set up your two-way table (for example, see Table 14-1), you already know what the observed values are for each cell in the table. Now you need something to compare them to. You're now ready for step three of the Chi-square test — finding expected cell counts. The null hypothesis says that the two variables $x$ and $y$ are independent. That's the same as saying $x$ and $y$ have no relationship. Assuming independence, you can determine which numbers should be in each cell of the table by using a formula for what is called the expected cell counts. (Each individual square in a two-way table is called a *cell,* and the number that falls into each cell is called the *cell count;* see Chapter 13 for more information.)

### Standing alone: Independent data

In general, *independence* means that you can find no major difference in the way the rows look, as you move down a column. That is, the proportion of the data falling into each column across the row is about the same for each row. So to find the expected cell counts for any two-way table, take the row total times the column total divided by the grand total, and do this process for each cell in the table.

Table 14-2 shows an example of independent data from a two-way table. Suppose that in this case the table represents data collected from men and women regarding whether they agree with a certain policy (yes or no). The proportion of all men who said yes is $\frac{10}{60} = 0.17$, or 17 percent. When you look at the same percentage for the women, you get the same number, 0.17. For both males and females, you get $\frac{50}{60} = 0.83$, or 83 percent, for the No group. Because males and females voted exactly the same way, these variables are likely going to be independent in the population as well as the sample.

**Table 14-2 Gender and Opinion: Observed Cell Counts = Expected Cell Counts (Independent)**

| | *Yes* | *No* | *Marginal Row Totals* |
|---|---|---|---|
| Men | 10 | 50 | 60 |
| Women | 10 | 50 | 60 |
| **Marginal Column Totals** | 20 | 100 | 120 **(Grand Total)** |

To get the expected cell counts for the upper-left cell in Table 14-2, take 60 (row one total) times 20 (column one total) divided by 120 (grand total) = 10. For the next cell in the first row, you multiply 60 by $\frac{100}{120} = 50$. The same results occur in row two, because the numbers are all the same as in row one. Because Table 14-2 represents two independent variables, you get the same expected cell counts for each row.

Under independence, you can find no difference between what you observed and what you expected.

The expected cell-count formula can actually make sense if you look at it the right way. That is, if the two variables are independent, the proportion of the data falling into each column across the row is about the same for each row. So to find the expected cell count for any cell, you take the row total for the row that cell is in, and you multiply that total by the proportion of the table that falls into the column that cell is in (that is, the column total divided by the grand total).

### Tying the knot: Dependent data

If two variables are dependent, then the value of one variable affects the value of the other variable. For example, suppose you believe women chew gum more than men. Then gender and gum chewing would be dependent, because if you knew someone's gender, that would change the probability of them being a gum chewer. Dependent variables affect each other's probabilities. In the end, the cell counts you actually observe from variables that are dependent won't match what you expected the cell counts to look like under Ho: The variables are independent. Big differences between observed and expected cell counts means that the variables are dependent.

Table 14-3 shows some data that is dependent because the relationship isn't the same for each row. More men in the sample said no to gum chewing ($^{35}/_{60}$ = 58 percent) than women in this sample ($^{25}/_{60}$ = 42 percent). However, this may not hold for all men and women in the population.

**Table 14-3 Gum Chewing: Observed Cell Counts**

| | *Yes* | *No* | *Marginal Row Totals* |
|---|---|---|---|
| Men | 25 | 35 | 60 |
| Women | 35 | 25 | 60 |
| **Marginal Column Totals** | 60 | 60 | 120 **(Grand Total)** |

Making conclusions about the population based on the sample (observed) data in a two-way table is taking too big of a leap. You need to conduct a Chi-square test in order to broaden your conclusions to the entire population. Ignoring the fact that sample results vary is where the media, and even some researchers, can get into trouble. Stopping with the sample results only and going merrily on your way can lead to conclusions that others can't confirm when they take new samples.

To check whether a two-way table is dependent, you first find the expected cell counts by taking the row total times the column total divided by the grand total and do this for each cell in the table. For Table 14-3, the expected cell count for the males who chew gum is $60 * {}^{60}/_{120} = 30$. The expected cell count for the males who don't chew gum is $60 * {}^{60}/_{120} = 30$. For the females who chew gum, you take $60 * {}^{60}/_{120} = 30$, and the same for females who don't chew gum. If gender and gum chewing are independent, you should expect to observe 30 in each cell (on average).

Next you compare the expected cell counts to the actual observed cell counts by looking at their differences (see Table 14-3 for the observed cell counts and Table 14-4 for the expected cell counts for the gum chewing example). You can see by Table 14-3 that the observed cell counts are 25, 35, 35, and 25. The expected cell count is 30 for each cell, as you can see in Table 14-4. The differences between the observed and expected cell counts are $25 - 30 = -5$; $35 - 30 = 5$; $35 - 30 = 5$; and $25 - 30 = -5$. These differences appear to be small with the naked eye, which may indicate gum chewing preference knows no gender. However, until you do a Chi-square test for independence (Chapter 15), you can never really know for sure.

**Table 14-4 Gum Chewing: Expected Cell Counts**

| | *Yes* | *No* | *Marginal Row Totals* |
|---|---|---|---|
| Men | $60 * ({}^{60}/_{120}) = 30$ | $60 * ({}^{60}/_{120}) = 30$ | 60 |
| Women | $60 * ({}^{60}/_{120}) = 30$ | $60 * ({}^{60}/_{120}) = 30$ | 60 |
| **Marginal Column Totals** | 60 | 60 | 120 **(Grand Total)** |

## *Checking the conditions for the test*

The time has come for step four of the Chi-square test: checking conditions. The Chi-square test has one main condition that must be met in order to test for independence on a two-way table: The expected count for each cell must be at least five, that is, greater than or equal to five. Expected cell counts that fall below five aren't reliable in terms of the variability that can take place. This problem is similar to trying to predict the outcome of only five flips of a coin — almost anything can happen. But if you flip the coin more times, you have a better idea of what you can expect to flip.

If you're analyzing data and you find that your data set doesn't meet the expected cell count of at least five for one or more cells, you can combine some of your rows and/or columns. This combination makes your table smaller, but it increases the cell counts for the cells that you do have, and that helps.

# Calculating the Chi-square test statistic

Every hypothesis test uses data to make the decision about whether or not to reject Ho in favor of Ha. In every hypothesis test, you take information from the data and put it together into a test statistic. The *test statistic,* in general, finds the distance between your observed results (your data) and the results you expect if Ho were true. If that difference is large, then you reject Ho in favor of Ha. If that difference is small, you fail to reject Ho. (For more information on test statistics, see another book I wrote, *Statistics For Dummies* [Wiley], or your intro stats book.)

In the case of testing for independence in a two-way table, you use a hypothesis test based on the Chi-square test statistic. In the following sections, you can see the steps for calculating and interpreting the Chi-square test statistic, which is step five of the Chi-square test.

## Working out the formula

A major component of the Chi-square test statistic is the expected cell count for each cell in the table. The formula for finding the expected cell count, $e_{ij}$, for the cell in row $i$, column $j$ is $e_{ij} = \frac{\text{row } i \text{ total} * \text{column } j \text{ total}}{\text{grand total}}$. Note that the values of $i$ and $j$ vary for each cell in the table. In a two-way table, the upper-left cell of the table is in row one, column one. The cell in the upper-right corner is in row one, column two. The cell in the lower-left corner is in row two, column one, and the lower-right-hand cell is in row two, column two.

The formula for the Chi-square test statistic is $\chi^2 = \sum_i \sum_j \frac{\left(o_{ij} - e_{ij}\right)^2}{e_{ij}}$, where $o_{ij}$ is the observed cell count for the cell in row $i$, column $j$, and $e_{ij}$ is the expected cell count for the cell in row $i$, column $j$.

When you calculate the expected cell count for some cells, you typically get a number that has some digits after the decimal point (in other words, the number isn't a whole number). Don't round this number off, despite the temptation to do so. This expected cell count is actually an overall-average expected value, and you can keep the count as it is, with decimal included.

Here are the major steps in how the Chi-square test statistic is calculated (Minitab does these steps for you as well):

1. **Subtract the observed cell count from the expected cell count for the upper-left-hand cell in the table.**
2. **Square the result from step one to make the number positive.**

3. **Divide the result from step two by the expected cell count.**
4. **Repeat this process for all the cells in the table and add up all the results.**

   The final sum that you get is your Chi-square test statistic.

The reason you divide by the expected cell count in the Chi-square test statistic is to account for cell-count sizes. If you expect a big cell count, say 100, and are off by only 5 for the observed count of that cell, that difference shouldn't count as much as if you expected a small cell count (like 10) and the observed cell count was off by 5. Dividing by the expected cell count puts a more fair weight on the differences that go into the Chi-square test statistic.

To perform a Chi-square test in Minitab, enter the raw data (the data on each person) in two columns. The first column is the values of your first variable in your data set. (For example, if your first variable is gender, go down the column entering the gender of each person.) Then enter your second variable in the second column, using the same row to represent each person in the data set. (If your second variable is paint preference, for example, enter each person's house-paint preference in column two, keeping the data from each person together in each row.) Go to Stat>Tables>Cross-tabulation and $\chi$-square. (But don't stop here: Keep reading.)

On the left-hand side, click on the variable that you wish to be in the rows of your two-way table (you may click on the first variable if you wish). Click Select, and the variable name appears in the row variable portion of the table on the right. Then go to the column variable blank on the right-hand side and click on it. You will be asked to choose your column variable. Go to the left-hand side and click on the name of your second variable. Click Select. Then click on the Chi-square button and choose Chi-square analysis by checking the box. If you want the expected cell counts included, check that box also. Then click OK, and OK.

The Chi-square test statistic can never be negative, because it's built on sums of squares of differences in the numerator and expected cell counts in the denominator (which are always positive).

The Minitab output for the Chi-square analysis for the house-paint example (from Table 14-1) is shown in Figure 14-1. You can pick out quite a few numbers from the output in Figure 14-1 that are especially important. First, you see three numbers listed in each cell. The first (top) number is the observed cell count for that cell; this matches the observed cell count for each cell shown in Table 14-1. (Notice the marginal row and column totals of Figure 14-1 also match those from Table 14-1.)

The second number in each cell of Figure 14-1 is the expected cell count for that cell; you find it by taking the row total times the column total divided by the grand total (see the section "Figuring the expected cell counts"). For example, the expected cell count for the upper-left cell (males who prefer white house paint) is $500 * {}^{305}\!/\!_{1,000} = 152.50$.

The third number in each cell of Figure 14-1 is that part of the Chi-square test statistic that comes from that cell. (See steps one through three of the previous section, "Working out the formula.") The sum of the third numbers in each cell equals the value of the Chi-square statistic listed in the last line of the output. (For the house-paint example, the Chi-square test statistic is 14.27.)

Interpreting the Chi-square test statistic is step six of the Chi-square test; you work through that process in the next section.

**Chi-Square Test: Gender, House-Paint Preference**

```
Expected counts are printed below observed counts
Chi-Square contributions are printed below expected counts

        White Paint  Nonwhite Paint    Total
    M           180             320      500
             152.50          347.50
              4.959           2.176

    F           125             375      500
             152.50          347.50
              4.959           2.176

Total           305             695     1000

Chi-Sq = 14.271, DF = 1, P-Value = 0.000
```

**Figure 14-1:** Minitab output for the house-paint data.

## Finding your results on the Chi-square table

The only way to be able to make an assessment about your Chi-square test statistic is to compare it to all the possible Chi-square test statistics you would get if you had a two-way table with the same row and column totals, yet you distributed the numbers in the cells in every way possible. (You can do that in your sleep, right?) Some resulting tables give large Chi-square test statistics, and some give small Chi-square test statistics.

Putting all these Chi-square test statistics together gives you what's called a *Chi-square distribution.* You find your particular test statistic on that distribution (step six of the Chi-square test), and see where it stands compared to the rest. If your test statistic is large enough that it appears way out on the right tail of the Chi-square distribution (boldly going where no test statistic has gone before), you reject Ho. If the test statistic isn't that far out, then you can't reject Ho.

In the next sections, you find out more about the Chi-square distribution and how it behaves, so you can make a decision about the independence of your two variables based on your Chi-square statistic.

### Determining degrees of freedom

Each type of two-way table has its own Chi-square distribution, depending on the number of rows and columns it has, and each Chi-square distribution is identified by its *degrees of freedom*. In general, a two-way table with $r$ rows and $c$ columns uses a Chi-square distribution with $(r - 1) * (c - 1)$ degree of freedom. A two-way table with two rows and two columns uses a Chi-square distribution with one degree of freedom. Notice that $1 = (2 - 1) * (2 - 1)$. A two-way table with three rows and two columns uses a Chi-square distribution with $(3 - 1) * (2 - 1) = 2$ degrees of freedom.

TECHNICAL STUFF

Understanding *why* degrees of freedom are calculated this way is likely to be beyond the scope of your statistics class. But if you really want to know, the degrees of freedom represents the number of cells in the table that are flexible, or "free," given all the marginal row and column totals. For example, suppose that a two-way table has all row and column totals equal to 100 and the upper-left cell is 70. Then the upper-right cell must be 100 (row total) – 30 = 70. Because the column one total is 100, and the upper-left cell count is 70, the lower-left cell count must be 100 – 70 = 30. Similarly, the lower-right cell count must be 70.

So you have only one free cell in a two-way table after you have the marginal totals set up. That's why the degree of freedom for a two-way table is 1. In general, you always lose one row and one column because of knowing the marginal totals, because these last row and column values can be calculated through subtraction. That's where the formula $(r - 1) * (c - 1)$ comes from. (That's more than you wanted to know, isn't it?)

### Discovering how Chi-square distributions behave

Figure 14-2 shows pictures of Chi-square distributions with one, two, four, six, eight, and ten degrees of freedom, respectively. Here are some important points about Chi-square distributions:

- For one degree of freedom, the distribution looks like a hyperbola (see Figure 14-2, top left); for more than one degree of freedom, it looks like a mound that has a long right tail (see Figure 14-2, lower right).
- All the values are greater than or equal to zero.

- The shape is always skewed to the right (tail going off to the right).
- As the number of degrees of freedom increases, the mean (the overall average) increases (moves to the right) and the variances increase (resulting in more spread).
- No matter what the degree of freedom is, the values on the Chi-square distribution (known as the *density*) approaches zero for increasingly larger Chi-square values. That means that larger and larger Chi-square values are less and less likely to happen.

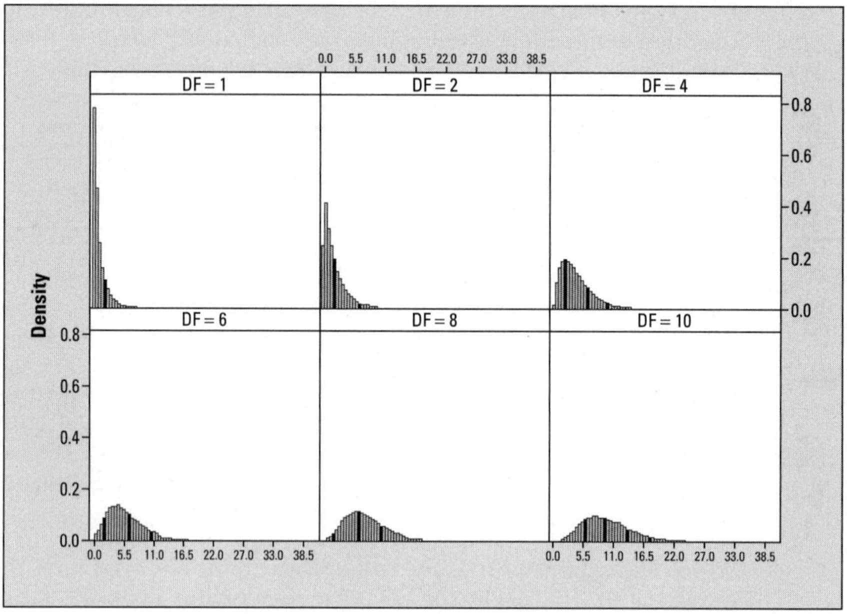


**Figure 14-2:** Chi-square distributions with 1, 2, 4, 6, 8, and 10 degrees of freedom (moving from upper left to lower right).

### *Using the Chi-square table*

After you find your Chi-square test statistic and its degrees of freedom, you want to determine how large your statistic is, relative to its corresponding distribution. (You're now venturing into step seven of the Chi-square test.) If you think about it graphically, you want to find the probability of being beyond (getting a larger number than) your test statistic. If that probability is small, your Chi-square test statistic is something unusual — it's out there — and you can reject Ho. You then conclude that your two variables are not independent (they are related somehow).

REMEMBER

In case you're following along at home, the Chi-square test statistic for the independent data from Table 14-2 is zero, because the observed cell counts are equal to the expected cell counts for each cell, and their differences are always equal to zero. (This result never happens in real life!) This scenario represents a *perfectly independent* situation and results in the smallest possible value of a Chi-square test statistic.

If the probability of being to the right of your Chi-square test statistic (on a graph) isn't small enough, you don't have enough evidence to reject Ho. You then stick with Ho; you can't reject it. You conclude that your two variables are independent (unrelated).

How small of a probability do you need to reject Ho? For most hypothesis tests, statisticians generally use 0.05 as the cutoff. (For more information on cutoff values, also known as $\alpha$ levels, flip to Chapter 3, or check out my other book *Statistics For Dummies* [Wiley].)

Your job now is to find the probability of being beyond your Chi-square test statistic on the corresponding Chi-square distribution with $(r - 1) * (c - 1)$ degrees of freedom. Each Chi-square distribution is different, and because the number of possible degrees of freedom is infinite, showing every single value of every Chi-square distribution isn't possible. In Table A-3 (in the Appendix in the back of this book), you see some of the most important values on each Chi-square distribution with degrees of freedom from 1 to 50.

To use the Chi-square table (Table A-3 in the Appendix), you find the row that represents your degrees of freedom (abbreviated *df*). Move across that row until you reach the value that is closest to your Chi-square test statistic, without going over. (It's like a game show, when you're trying to win the showcase by guessing the price.) Then go to the top of the column you're in. That number represents the area to the right (above) of the Chi-square test statistic you saw in the table. The area above your particular Chi-square test statistic is less than or equal to this number. This result is the approximate *p*-value of your Chi-square test.

Using the house-paint example (see Figure 14-1), the Chi-square test statistic was 14.27. You have $(2 - 1) * (2 - 1) = 1$ degree of freedom. On Table A-3 (in the Appendix), you go to the row for $df = 1$, and go across to the number closest to 14.27 (without going over). That number is 7.88, in the last column. (This number is much less than 14.27, but it's the biggest number on the table for that row.) The number at the top of that column is 0.005.

## Drawing your conclusions

You have two alternative ways to draw conclusions from the Chi-square test statistic. You can look up your test statistic on the Chi-square table (located in Table A-3 in the Appendix) and see the probability of being greater than that. This method is known as *approximating the p-value.*(The *p-value* of a test statistic is the probability of being at or beyond your test statistic on the distribution to which the test statistic is being compared — in this case, the Chi-square distribution.) Or you can have the computer calculate the exact *p*-value for your test. (For more on *p*-values and $\alpha$ levels, see my other book *Statistics For Dummies.* For a quick review on these topics, see Chapter 3 of this book.)

Before you do anything though, set your $\alpha$, the cutoff probability for your *p*-value, in advance. If your *p*-value is less than your $\alpha$ level, reject Ho. If it is more, you can't reject Ho.

### Approximating p-value from the table

For the house-paint example (see Figure 14-1), the Chi-square test statistic was 14.27 with 1 df (degree of freedom). The closest number in row one of Table A-3 (in the Appendix), without going over, is 7.88 (in the last column). The number at the top of that column is 0.005. This number is less than your typical $\alpha$ level of 0.05, so you reject Ho. You know that your *p*-value is less than 0.005 because your test statistic was more than 7.88. In other words, if 7.88 is the minimum evidence you need to reject Ho, you have more evidence than that with a value of 14.28. More evidence against Ho means a smaller *p*-value. However, because Table A-3 only gives a few values for each Chi-square distribution, the best you can say using this table is that your *p*-value for this test is less than 0.005.

Here's the big news: Because your *p*-value is less than 0.05, you can conclude based on this data that gender and house-paint color are likely to be related in the population (dependent), like the Demographics Survey said (located at the beginning of this chapter). Only now, you have a formal statistical analysis that says this result found in the sample is also likely to occur in the entire population. This statement is much stronger!

REMEMBER

If your data shows you can reject Ho, you only know at that point that the two variables have some relationship. The Chi-square test statistic doesn't tell you what that relationship is. In order to explore the relationship between the two variables, you find the conditional probabilities in your two-way table (see Chapter 13). You can use those results to give you some ideas as to what may be happening in the population. For example, in the house-paint data (because paint preference is related to gender), you can examine the relationship further by first finding the percentage of men that prefer white houses, which comes out to $^{180}/_{500} = 0.36$, or 36 percent, calculated from Table 14-1. Now compare this result to the percentage of women who prefer white houses: $^{125}/_{500} = 0.25$, or 25 percent. You can now conclude that in this population (not just the sample), men prefer white houses more than women do.

### Extracting the p-value from computer output

After Minitab calculates the test statistic for you, it reports the exact *p*-value for your hypothesis test. The *p*-value measures the likelihood that your results were found just by chance while Ho is still true. It tells you how much strength you have against Ho. If the *p*-value is 0.001, for example, you have much more strength against Ho than if the *p*-value, say, is 0.10.

Looking at the Minitab output for the house-paint data in Figure 14-1, the *p*-value is reported to be 0.000. This means that the *p*-value is smaller than 0.001; for example, it may be 0.0009. That's a very small *p*-value! (Minitab only reports results to three decimal points, which is typical of many statistical software packages.)

The Chi-square test for the gum-chewing data from Table 14-3 results in a *p*-value of 0.068. This calculation is what statisticians call a *marginal result,* because it's just on the other side of 0.05. (The test statistic turned out to be only 3.33, and that didn't seem to be very large.) This *p*-value is larger than the typical $\alpha$ of 0.05, but not a lot larger. Technically speaking, you can't reject Ho at level $\alpha = 0.05$. In practical terms, even though gum chewing and gender seem to be dependent in the sample, you can't say that you can expect to find this relationship in the population.

I've seen situations where people who get a result that isn't quite what they want (like a *p*-value of 0.068) do some tweaking to get what they want. What they do is change their $\alpha$ level from 0.05 to 0.10 after the fact. This change makes the *p*-value less than the $\alpha$ level, and they feel they can reject Ho and say that a relationship exists. But what's wrong with this? They changed the $\alpha$ after they looked at the data, which isn't allowed. That's like changing your bet in blackjack after you find out what the dealer's cards look like. (Tempting, but a serious no-no.) Always be wary of large $\alpha$ levels, and make sure that you always choose your $\alpha$ before collecting any data — and stick to it. The good news is that when *p*-values are reported, anyone reading them can make his own conclusion; no cut-and-dry rejection and acceptance region is set in stone. But setting an $\alpha$ level once, then changing it after the fact to get a better conclusion is never good!

## Comparing Two Tests for Comparing Two Proportions

You can use the Chi-square test to check whether two population proportions are equal (for example, is the proportion of female cell-phone users the same as the proportion of male cell-phone users?). Now you may be thinking, "But wait a minute, don't statisticians already have a test for two proportions? I seem to remember it from my intro stats course . . . I'm thinking . . . yeah, it's the *Z*-test for two proportions. What's that test got to do with a Chi-square test?" In this section, you answer that question, and use both methods to investigate a possible gender gap in cell-phone use.

## Getting reacquainted with the Z-test for two population proportions

The way that most people figure out how to test the equality of two population proportions is to use a *Z-test for two population proportions* (where you collect a random sample from each of the two populations, find and subtract their two sample proportions, and divide by their pooled standard error; see your intro stats book for details on this particular test). This test is possible to do as long as the sample sizes from the two populations are large — at least five successes and five failures in each sample.

The null hypothesis for the *Z*-test for two population proportions is Ho: $p_1 = p_2$, where $p_1$ is the proportion of the first population that falls into the category of interest and $p_2$ is the proportion of the second population that falls into the category of interest. And as always, the alternative hypothesis is one of the following choices Ha: not equal to, greater than, or less than.

Suppose you want to compare the proportion of cell-phone users for men versus women. You make $p_1$ be the proportion of males who own a cell phone, and $p_2$ is the proportion of all females who own a cell phone. You collect data, find the sample proportions from each group, $\hat{p}_1$ *and* $\hat{p}_2$, take their difference and make a *Z*-statistic out of it using the formula $Z = \dfrac{\hat{p}_1 - \hat{p}_2}{\sqrt{\hat{p}\left(1-\hat{p}\right)}\sqrt{\dfrac{1}{n_1}+\dfrac{1}{n_2}}}$, where $\hat{p} = \dfrac{x_1+x_2}{n_1+n_2}$. Here, $x_1$ and $x_2$ are the number of individuals from samples one and two, respectively, with the desired characteristic; $n_1$ and $n_2$ are the two sample sizes.

Suppose that you collect data on 100 men and 100 women and find 45 male cell-phone owners and 55 female cell-phone owners,. This means that $\hat{p}_1$ equals $^{45}/_{100} = 0.45$, and $\hat{p}_2$ equals $^{55}/_{100} = 0.55$. Your samples have at least five *successes* (having the desired characteristic; in this case, cell-phone ownership) and five *failures* (not having the desired characteristic, which is cell-phone ownership.) So you go ahead and compute the *Z*-statistic for comparing the two population proportions (males versus females) based on this data is –1.41, as shown on the last line of the Minitab output in Figure 14-3.

**Figure 14-3:** Minitab output comparing proportion of male and female cell-phone owners.

**Test Cell Phone for Two Proportions**

```
Sample    X    N   Sample p
M        45  100   0.450000
F        55  100   0.550000


Difference = p (1) - p (2)
Estimate for difference: -0.1
95% CI for difference:(-0.237896, 0.0378957)
Test for difference = 0 (vs not = 0): Z = -1.41 P-Value = 0.157
```

The *p*-value for the test statistic of $Z = -1.41$ is 0.157 (calculated by Minitab, or by looking at the area below the *Z*-value of –1.41 on a *Z*-table; see your intro stats text for one of those). This *p*-value (0.157) is greater than the typical $\alpha$ level (prespecified cutoff) of 0.05, so you can't reject Ho. You can't say that the two population proportions aren't equal. That is, you must conclude that the proportion of cell-phone owners for males is no different than for females. Even though the sample seemed to have evidence for a difference (after all, 45 percent isn't equal to 55 percent), you don't have enough evidence in the data to say that this same difference carries over to the population. So you can't lay claim to a gender gap in cell-phone use, at least with this sample.

## Equating Chi-square tests and Z-tests for a two-by-two table

Here's the key to relating the *Z*-test to a Chi-square test for independence. If you use the *Z*-test to see whether the proportion of male cell-phone owners is equal to the proportion of female cell-phone owners, you're really looking at whether you can expect the same proportion of cell-phone owners despite gender (after you take the sample sizes into account). And that means you are testing whether gender (male or female) is independent of cell-phone ownership (yes or no).

If the proportion of female cell-phone owners equals the proportion of male cell-phone owners, then the proportion of cell-phone owners is the same regardless of gender, so gender and cell-phone ownership are independent. On the other hand, if you find the proportion of male cell-phone owners to be unequal to the proportion of female cell phone owners, then you can say that cell-phone use differs by gender — so gender and cell-phone ownership are dependent.

Therefore, the *Z*-test for two proportions and the Chi-square test for independence in a two-by-two table (one with two rows and two columns) are equivalent if the sample sizes from the two populations are large enough; that is, when the number of successes and the number of failures in each cell of the two samples is at least five.

With the cell-phone data from the previous section, you have 45 males using cell phones (out of 100 males) and 55 females using cell phones (out of 100 females). The Minitab output for the Chi-square test for independence (complete with observed and expected cell counts, degrees of freedom, test statistic, and *p*-value) is shown in Figure 14-4. The *p*-value for this test is 0.157, which is greater than the typical $\alpha$ level (prespecified cutoff) of 0.05, so you can't reject Ho.

Because the Chi-square test for independence and the *Z*-test tests are equivalent when you have a two-by-two table, the *p*-value from the Chi-square test for independence is identical to the *p*-value from the *Z*-test for two proportions. If you compare the *p*-values from Figures 14-3 and 14-4, you can see that for yourself.

**Chi-Square Test: Gender, Cell Phone**

```
Expected counts are printed below observed counts
Chi-Square contributions are printed below expected counts

              Y       N   Total
      M      45      55     100
          50.00   50.00
          0.500   0.500

      F      55      45     100
          50.00   50.00
          0.500   0.500

Total       100     100     200

Chi-Sq =  2.000,  DF =  1, P-Value = 0.157
```

**Figure 14-4:** Minitab output testing independence of gender and cell-phone ownership.

Also, note that if you take the *Z*-test statistic for this example (from Figure 14-3), which is –1.41, and square it, you get 2.02, which is equal to the Chi-square test statistic for the same data (last line of Figure 14-4). It is also the case that when the square of the *Z*-test statistic (when testing for the equality of two proportions) is equal to the corresponding Chi-square test statistic for independence.

## The car accident–cell phone connection

Researchers are doing a great deal of study of the effects of cell-phone use while driving. One study published in the *New England Journal of Medicine* observed and recorded data in 1997 on 699 drivers who had cell phones and were involved in motor vehicle collisions resulting in substantial property damage but no personal injury. Each person's cell-phone calls on the day of the collision and during the previous week were analyzed through the use of detailed billing records. A total of 26,798 cell-phone calls were made during the 14-month study period.

One conclusion the researchers made was that ". . . the risk of a collision when using a cell phone is four times higher than the risk of a collision when a cell phone was not being used." They basically conducted a test to see whether cell-phone use and having a collision are independent, and when they found out they were not, they were able to examine the relationship further using appropriate ratios. In particular, they found that the risk of a collision is four times higher for those drivers using cell phones than for those who aren't.

Researchers also found out that the relative risk was similar for drivers who differed in personal characteristics, such as age and driving experience. (This finding means that they conducted similar tests to see whether the results were the same for drivers of different age groups and drivers of different levels of experience, and the results always came out about the same. Therefore, age and the experience of the driver were not related to the collision outcome.)

The research also shows that ". . . calls made close to the time of the collision were found to be particularly hazardous ($p < 0.001$). Hands-free cell phones offered no safety advantage over hand-held units ($p$-value not significant) . . ." ***Note:*** The items in parentheses show the typical way that researchers report their results — using $p$-values. The $p$ in both cases of parentheses represent the $p$-value of each test.

In the first case, the $p$-value is very tiny, less than 0.001, indicating strong evidence for a relationship between collisions and cell-phone use at the time. The second $p$-value in parentheses was stated to be insignificant, meaning that it was substantially more than 0.05, the usual $\alpha$ level people use. This second result indicates that whether or not the drivers used hands-free equipment didn't affect the chances of a collision happening. That is, the proportion of collisions using hands-free cell phones versus using regular cell phones were found to be statistically the same (they could've easily occurred by chance under independence). Whether you use a regular or hands-free cell phone, may this study be a lesson to everyone!

The Chi-square test and $Z$-test are equivalent only if the table is a two-by-two table (two rows and two columns) and if the $Z$-test is two tailed (the alternative hypothesis is that the two proportions aren't equal, instead of using Ha: one proportion is greater than or less than the other). If the $Z$-test is not two tailed, a Chi-square test isn't appropriate. If the two-way table has more than two rows or columns, use the Chi-square test for independence (because you no longer have only two proportions if you have many categories, so the $Z$-test isn't applicable).

## Chapter 15

# Using Chi-Square Tests for Goodness-of-Fit (Your Data, Not Your Jeans)

### *In This Chapter*

- Understanding what goodness-of-fit really means
- Using the Chi-square model to test for goodness-of-fit
- Looking at the conditions for goodness-of-fit tests

Many phenomena in life may appear to be random in the short term, but actually occur according to some preconceived, preselected, or predestined model over the long term. For example, while you don't know whether it will rain tomorrow, your local meteorologist can give you her model for the percentage of days that it rains, snows, is sunny, or cloudy, based on the last five years. Whether or not this model is still relevant this year is anyone's guess, but it's a model nonetheless. As another example, a biologist can produce a model for predicting the number of goslings raised by a pair of geese per year, even though you have no idea what the pair in your backyard will do. Is his model correct? Here's your chance to find out.

In this chapter, you build models for the proportion of outcomes that fall into each category for a categorical variable. You then test these models by collecting data and comparing what you observe in your data to what you expect from the model. You do this through a goodness-of-fit test that's based on the Chi-square distribution. In a way, a goodness-of-fit test is likened to a reality check of a model for categorical data.

## Finding the Goodness-of-Fit Statistic

The general idea of a *goodness-of-fit* procedure involves determining what you expect to find and comparing it to what you actually observe in your own sample through the use of a test statistic. This test statistic is called the *goodness-of-fit test statistic,* because it measures how well your model (what you expected) fits your actual data (what you observed).

In this section, you see how to figure out the numbers that you should expect in each category given your proposed model, and you also see how to put those expected values together with your observed values to form the goodness-of-fit test statistic.

## What's observed versus what's expected

For an example of something that can be observed versus what's expected, look no further than a bag of tasty M&M'S Milk Chocolate Candies. (A ton of different kinds of M&M'S are out there, and each kind has its own variation of colors and tastes. But for this study, any reference I give to M&M'S is to the original milk chocolate candy – my favorite.) The percentage of each color of M&M'S that appear in a bag is something Mars (the company that makes M&M'S) spends a lot of time thinking about. Mars does have specific percentages of each color that they want in their M&M'S bags, which it determines through comprehensive marketing research based on what people like and want to see. Mars then posts their current percentages for each color of M&M'S on their Web site. Table 15-1 shows the percentage of M&M'S of each color in 2006.

**Table 15-1 Expected Percentage of Each Color of M&M'S Milk Chocolate Candies (2006)**

| *Color* | *Percentage* |
|---|---|
| Brown | 13% |
| Yellow | 14% |
| Red | 13% |
| Blue | 24% |
| Orange | 20% |
| Green | 16% |

Now that you know what to expect from a bag of M&M'S, the next question is how does Mars deliver? If you open a bag of M&M'S right now, would you get the percentages of each color that you're supposed to get? You know from your previous studies in statistics that sample results vary (for a quick review of this idea, see Chapter 3). So you can't expect each bag of M&M'S to have exactly the correct number of each color of M&M'S as listed in Table 15-1. However, in order to keep customers happy, Mars should get close to the expectations. How can you determine how close they do get?

You now know what percentages are expected to fall into each category in the entire population of all M&M'S (that means every single M&M'S Milk Chocolate Candy that's currently being made), from Table 15-1. This set of percentages is called the *expected model* for the data. You want to see whether the percentages in the expected model are actually occurring in the packages you buy. To start this process, you can take a sample of M&M'S (after all, you can't check every single one in the population) and make a table showing what percentage of each color you observed. Then you can compare this table of observed percentages to the expected model.

TIP

The expected percentages are either given to you, as they are for the M&M'S, or you can figure them out by using math techniques. For example, if you're examining a single die to determine whether or not it's a fair die, you know that if the die is fair, you should expect $\frac{1}{6}$ of the outcomes to fall into each category of 1, 2, 3, 4, 5, and 6.

As an example, I examined one 1.69-ounce bag of plain, milk-chocolate M&M'S (tough job, but someone has to do it), and you can see my results in Table 15-2. (Think of this bag as a random sample of M&M'S, even though it's not technically the same as reaching into a silo filled with M&M'S and pulling out a true random sample of 1.69 ounces. For the sake of argument, one bag is okay.)

**Table 15-2 Percentage of M&M'S Observed in One Bag (1.69 oz.)**

| *Color* | *Number Observed* | *Percentage Observed* |
|---|---|---|
| Brown | 4 | 7.14% |
| Yellow | 10 | 17.86% |
| Red | 4 | 7.14% |
| Blue | 10 | 17.86% |
| Orange | 15 | 26.79% |
| Green | 13 | 23.21% |
| ***TOTAL*** | 56 | 100.00% |

Now you look at what I observed in my sample (Table 15-2) and compare it to what I expected to get (Table 15-1, last column). Notice that I observed a lower percentage of brown and red M&M'S than expected and a lower percentage of blues than expected. I also observed a higher percentage of yellow, orange, and green M&M'S than expected. You know that sample results vary by random chance, from sample to sample, and that the difference I observed may just be due to this chance variation. But could the differences indicate that the expected percentages, reported by Mars, aren't being followed?

It stands to reason that if the differences between what you observed and what you expected are small, you should attribute that difference to chance and let the expected model stand. On the other hand, if the differences between what you observed and what you expected are large enough, you may have enough evidence to indicate that the expected model has some problems. How do you know which conclusion to make? The operative phrase is "if the differences are large enough." You need to quantify this term *large enough*. Doing so takes a bit more machinery, so keep reading.

## Calculating the goodness-of-fit statistic

The goodness-of-fit statistic is one number that puts together the total amount of difference between what you expect in each cell compared to the number you observe. The term *cell* is used to express each individual category within a table format. For example, with the M&M'S example, the first column of Tables 15-1 and 15-2 contain six cells, one for each color of M&M'S. For any cell, the number of items you observe in that cell is called the *observed cell count*. The number of items you expect in that cell (under the given model) is called the *expected cell count* for that cell. You get the expected cell count by taking the expected cell percentage times the sample size.

The expected cell count is just a proportion of the total, so it doesn't have to be a whole number. For example, if you roll a fair die 200 times, you should expect to roll ones $\frac{1}{6}$, or 16.67 percent, of the time. In terms of the number of ones you expect, it should be $0.1667 * 200 = 33.33$. Use the 33.33 in your calculations for goodness-of-fit; don't round to a whole number. Your final answer is more accurate that way.

The reason the goodness-of-fit statistic is based on the *number* in each cell rather than the *percentage* in each cell is because percents are a bit deceiving. If you know that 8 out of 10 people support a certain view, that's 80 percent. But 80 out of 100 is also 80 percent. Which one would you feel is a more precise statistic? The 80 out of 100 percent, because it uses more information. Using percents alone disregards the sample size. Using the counts (the number in each group) keeps track of the amount of precision you have.

For example, if you roll a fair die, you expect the percentage of ones to be $\frac{1}{6}$. If you roll that fair die 600 times, the expected *number* of ones will be $\frac{1}{6} * 600 = 100$. That number (100) is the expected cell count for the cell that represents the outcome of one. If you roll this die 600 times and get 95 ones, then 95 is the observed cell count for that cell.

The formula for the goodness-of-fit statistic is given by the following: $\sum_{\text{all cells}} \frac{(O-E)^2}{E}$ where $E$ is the expected number in a cell and $O$ is the observed number in a cell. The steps for this calculation are as follows:

1. **For the first cell, find the expected number for that cell *(E)* by taking the percentage expected in that cell times the sample size.**
2. **Take the observed value in the first cell *(O)* minus the number of items that are expected in that cell *(E)*.**
3. **Square that difference.**
4. **Divide the answer by the number that's expected in that cell.**
5. **Repeat steps 1 through 4 for each cell.**
6. **Add up the results to get the goodness-of-fit statistic.**

REMEMBER

The reason you divide by the expected cell count in the goodness-of-fit statistic (step four) is to take into account the magnitude of any differences you find. For example, if you expect 100 items to fall in a certain cell and you get 95, the difference is 5. But in terms of a percentage, this difference is only $\frac{5}{100}$ = 5 percent. However, if you expected 10 items to fall into that cell and you observed 5 items, the difference is still 5, but in terms of a percentage, it's $\frac{5}{10}$ = 50 percent. This difference is much larger in terms of its impact. The goodness-of-fit statistic operates much like a percentage difference. The only added element is to square the difference to make it positive. (That's done because whether you expected 10 and got 15, or whether you expected 10 and got 5 makes no difference to others, you're still off by 50 percent.)

Table 15-3 shows the step-by-step calculation of the goodness-of-fit statistic for the M&M'S example, where $O$ indicates observed cell counts and $E$ indicates expected cell counts. To get the expected cell counts, you take the expected percentages shown in Table 15-1 and multiply by 56, because 56 is the number of M&M'S I had in my sample. The observed cell counts are the ones found in my sample, shown in Table 15-2.

**Table 15-3 Goodness-of-Fit Statistic for M&M'S Example**

| *Color* | *O* | *E* | *O – E* | $(O-E)^2$ | $\frac{(O-E)^2}{E}$ |
|---|---|---|---|---|---|
| Brown | 4 | 0.13 * 56 = 7.28 | 4 – 7.28 = –3.28 | 10.76 | 1.48 |
| Yellow | 10 | 0.14 * 56 = 7.84 | 10 – 7.84 = 2.16 | 4.67 | 0.60 |
| Red | 4 | 0.13 * 56 = 7.28 | 4 – 7.28 = –3.28 | 10.76 | 1.48 |

*(continued)*

**Table 15-3 *(continued)***

| *Color* | *O* | *E* | *O – E* | $(O-E)^2$ | $\frac{(O-E)^2}{E}$ |
|---|---|---|---|---|---|
| Blue | 10 | 0.24 * 56 = 13.44 | 10 – 13.44 = –3.44 | 11.83 | 0.88 |
| Orange | 15 | 0.20 * 56 = 11.20 | 15 – 11.20 = 3.80 | 14.44 | 1.29 |
| Green | 13 | 0.16 * 56 = 8.96 | 13 – 8.96 = 4.04 | 16.32 | 1.82 |
| ***TOTAL*** | 56 | 56 | | | **7.55** |

The goodness-of-fit statistic for the M&M'S example turns out to be 7.55, the bolded number in the lower-right corner of Table 15-3. This number represents the total squared difference between what I expected and what I observed, adjusted for the magnitude of each expected cell count. The next question is how to interpret this value of 7.55. Is it large enough to indicate that colors of M&M'S in the bag aren't following the percentages posted by Mars? The next section addresses how to make sense of these results.

# Interpreting the Goodness-of-Fit Statistic By Using Chi-Square

After you get your goodness-of-fit statistic, your next job is to interpret it. To do this, you need to figure out the possible values you could have gotten and where your statistic fits in among them. You can accomplish this task with a Chi-square goodness-of-fit test.

The values of a goodness-of-fit statistic actually follow a Chi-square distribution with $k - 1$ degrees of freedom, where $k$ is the number of categories in your particular population (see Chapter 14 for the full details on Chi-square). You can use the Chi-square table (Table A-3 in the Appendix) to determine how far out your particular goodness-of-fit statistic is, compared to all the others that were possible to get. If your Chi-square statistic is large compared to other values on the Chi-square distribution, the model doesn't fit; there's too much of a difference between what you observed and what you expected under the model. However, if your goodness-of-fit statistic is small, you can't reject the model. (What constitutes a high or low value of a Chi-square test statistic varies for each problem.) This section provides the details on using the Chi-square distribution to test for goodness-of-fit.

TIP

The goodness-of-fit statistic follows the main characteristics of the Chi-square distribution. The smallest possible value of the goodness-of-fit statistic is zero. If the M&M's found in my sample (continuing the example from the previous section) followed the exact percentages found in Table 15-1, the goodness-of-fit statistic would be zero. That's because the observed counts and the expected counts would be the same, so the values of the observed cell count minus the expected cell count would all be zero, so calculating the goodness-of-fit statistic here would result in zero.

The largest possible value of Chi-square isn't specified, although some values are more likely to occur than others. Each Chi-square distribution has its own set of likely values, as you can see in Figure 15-1. (Figure 15-1 shows a simulated Chi-square distribution with 6 – 1 = 5 degrees of freedom (relevant to the M&M's example). This figure basically gives a breakdown of all the possible values you could have for the goodness-of-fit statistic in this situation and how often they occur. You can see on Figure 15-1 that a Chi-square test statistic of 7.55 isn't unusually high, indicating that the model for M&M's colors probably can't be rejected. However, more particulars are needed before you can formally make that conclusion.

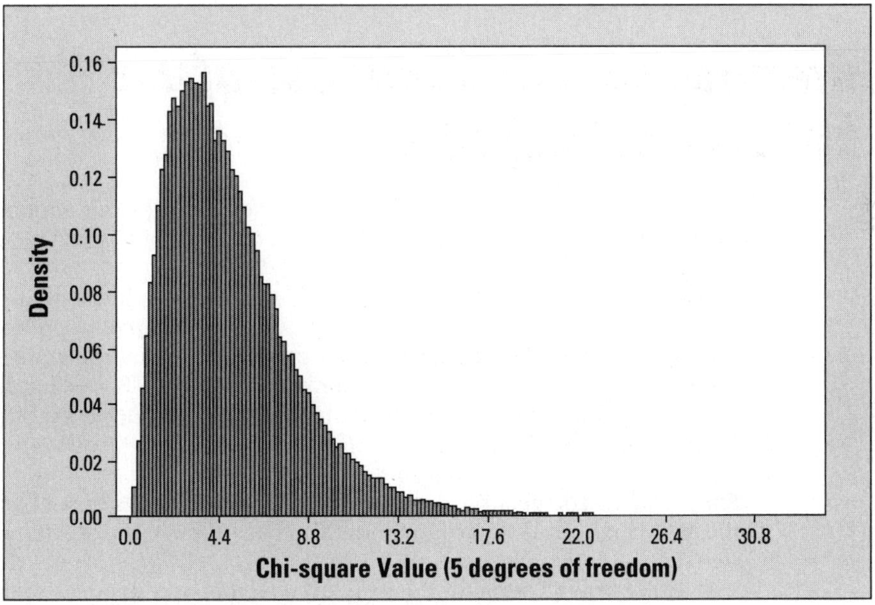


**Figure 15-1:** Chi-square distribution with 5 degrees of freedom.

## Checking the conditions before you start

Every statistical technique seems to have a catch, and this case is no exception. In order to use the Chi-square distribution to interpret your goodness-of-fit statistic, you have to be sure you have enough information to work with in each cell. The stat gurus usually recommend that the expected count for each cell turns out to be greater than or equal to five. If it doesn't, one option is to combine categories together to increase the numbers.

In the M&M'S example, the expected cell counts are all above seven (see Table 15-3), so the conditions are met. If this weren't the case, you could use a larger sample size, because you calculate the expected cell counts by taking the expected percentage in that cell times the sample size. If you increase the sample size, you increase the expected cell count. A higher sample size also increases your chances of detecting a real deviation from the model. This idea is related to the power of the test (see Chapter 3 for information on power).

WARNING!

After you collect your data, it's not really right to go back and take a new and larger sample. It's best to set up your sample size ahead of time, and you can do this by determining what sample size you need to get the expected cell counts to be at least five. For example, if you roll a fair die, you expect $\frac{1}{6}$ of the outcomes to be ones. If you only take a sample of six rolls, you have an expected cell count of $\frac{1}{6} * 6 = 1$, which isn't enough. However, if you roll the die 30 times, your expected cell count is $\frac{1}{6} * 30 = 5$, which is just enough to meet the condition.

## The steps of the Chi-square goodness-of-fit test

Assuming the necessary condition is met (see the previous section), you can get down to actually conducting a formal goodness-of-fit test.

The general version of the null hypothesis for the goodness-of-fit test is Ho: The model holds for all categories, versus the alternative hypothesis Ha: The model doesn't hold for at least one category. Each situation will dictate what proportions should be listed in Ho for each category. (For example, if you're rolling a fair die, you have Ho: proportion of 1s = $\frac{1}{6}$; proportion of 2s = $\frac{1}{6}$; . . . ; proportion of 6s = $\frac{1}{6}$.)

Following are the general steps for the Chi-square goodness-of-fit test, with the M&M'S example illustrating how you can carry out each step:

1. **Write down Ho using the percentages that you expect in your model for each category.**

   Using a subscript to indicate the proportion *(p)* of M&M's you expect to fall into each category (see Table 15-1), your null hypothesis is Ho: $p_{brown} = 0.13$, $p_{yellow} = 0.14$, $p_{red} = 0.13$, $p_{blue} = 0.24$, $p_{orange} = 0.20$, and $p_{green} = 0.16$. All these proportions must hold in order for the model to be upheld.

2. **Write your Ha: This model doesn't hold for at least one of the percentages.**

   Your alternative hypothesis, Ha, in this case, would be: One (or more) of the probabilities given in Ho isn't correct. In other words you know that at least one of the colors of M&M'S has a different proportion of colors than what is stated in the model.

3. **Calculate the goodness-of-fit statistic using the steps in the previous section.**

   The goodness-of-fit statistic for M&M'S, from the previous section, is 7.55. As a reminder, you take the observed number in each cell minus the expected number in that cell, square it, and divide by the expected number in that cell. Do that for every cell in the table and add up the results. For the M&M'S example that total is equal to 7.55, the goodness-of-fit statistic.

4. **Look up the Chi-square distribution with $k - 1$ degrees of freedom, where $k$ is the number of categories you have (use Table A-3 in the Appendix).**

   You compare this statistic (7.55) to the Chi-square distribution with $6 - 1 = 5$ degrees of freedom (because you have $k = 6$ possible colors of M&M'S).

   Looking at Figure 15-1 you can see that the value of 7.55 is nowhere near the high end of this distribution, so you likely don't have enough evidence to reject the model provided by Mars for M&M'S colors.

5. **Find the *p*-value of your goodness-of-fit statistic.**

   You can use Table A-3 in the Appendix to find the *p*-value (the probability of being beyond your test statistic; see Chapter 3) of your test statistic using the Chi-square distribution. (For more info on the Chi-square distribution, see Chapter 14.)

   REMEMBER: Because the Chi-square table (Table A-3 in the Appendix) can only list a certain number of results for each of the degrees of freedom, the exact *p*-value for your test statistic may fall between two *p*-values listed on the table.

   To find the *p*-value for the test statistic in the M&M'S example (7.55), you go to Table A-3 (Appendix) and find the row for 5 degrees of freedom and look at the numbers (the degrees of freedom is $k - 1 = 6 - 1 = 5$, where $k$ is the number of categories). You see that the number 7.55 is less than the first value in the row (9.24), which has a *p*-value of 0.10. (Find the *p*-value by looking at the column heading above the number.) So the *p*-value for 7.55, which is the area to the right of 7.55 on Figure 15-1, must be greater than 0.10, because 7.55 is to the left of 9.24 on that Chi-square distribution.

TIP

Many computer programs exist (online or via a graphing calculator) that will find exact *p*-values for a Chi-square test, saving time and headaches when you have access to them (the technology, not the headaches). Using one such online "*p*-value calculator" I found that the exact *p*-value for the goodness-of-fit test for the M&M'S example (test statistic 7.55, 5 degrees of freedom for Chi-square) is 0.1828 = 0.18. To find online *p*-value calculators, simply type in the name of the distribution and the word **p-value** in an Internet search engine. For this example, type in **Chi-square p-value.**

6. **If your *p*-value is less than your predetermined cutoff ($\alpha$), reject Ho. The model doesn't hold. If your *p*-value is greater than $\alpha$, you can't reject the model.**

REMEMBER

A typical value of $\alpha$ is 0.05. Some data analysts might use a higher value (up to 0.10) and others might go lower (for example 0.010.) See Chapter 3 for more information on choosing $\alpha$ and comparing your *p*-value to it.

Going again to the M&M'S example, the *p*-value, 0.18, is greater than 0.05, so you fail to reject Ho. You can't say the model is wrong. So, Mars does appear to deliver on the percentages of M&M'S of each color, as advertised. At least you can't say they don't. (I'm sure Mars already knew that.)

REMEMBER

While some hypothesis tests are two-sided tests, the goodness-of-fit test is always a right-tailed test, meaning that you have a greater than sign (>) in the alternative hypothesis, Ha (see Chapter 3 for the skinny on hypothesis testing). You're only looking at the right tail of the Chi-square distribution when you're doing a goodness-of-fit test. That's because a small value of the goodness-of-fit statistic means that the observed data and the expected model don't differ much, so you stick with the model. If the value of the goodness-of-fit statistic is way out on the right tail of the Chi-square distribution, however, that's a different story. That situation means the difference between what you observed and what you expected is larger than what you should get by chance, and, therefore, you have enough evidence to say the expected model is wrong.

# Part V
# Rebels without a Distribution

The 5th Wave By Rich Tennant

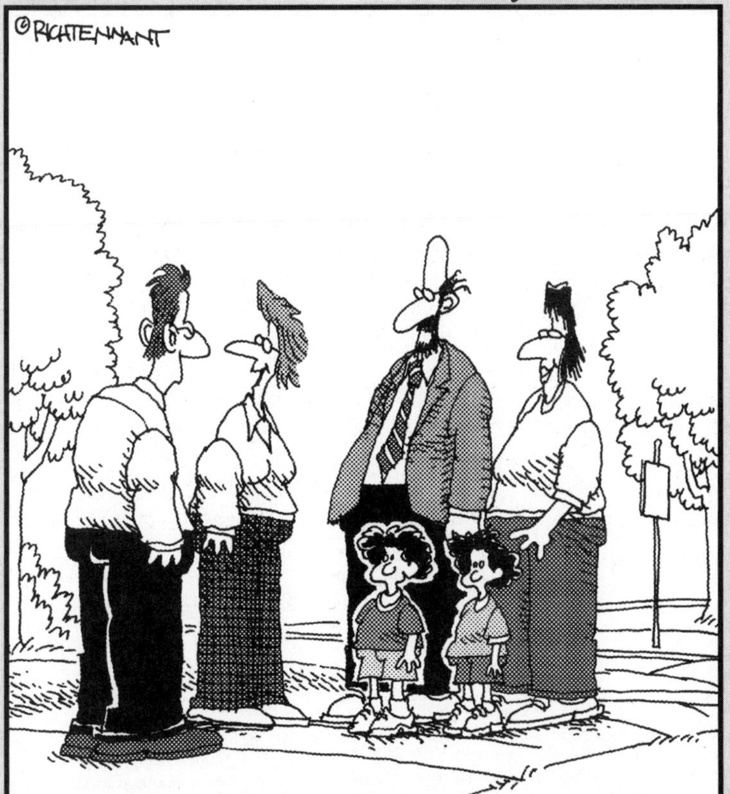


"This is my old Intermediate Stats professor, his wife Doris, and their two children, Wilcox and Kruskal."

## *In this part . . .*

Suppose you're driving home and one of the streets is blocked. What do you do? You back up and find another way to get home. Nonparametric statistics is that alternative route you take if the regular parametric statistical methods aren't allowed. Beyond that, this alternate route actually turns out to be better when the regular route isn't available. In this part, you see how.

# Chapter 16

# Going Nonparametric

*In This Chapter*

- Seeing the need for nonparametric techniques
- Distinguishing regular methods from nonparametric methods
- Laying the groundwork: The basics of nonparametric statistics

Many researchers do analyses involving hypothesis tests, confidence intervals, Chi-square tests, regression, and ANOVA. But nonparametric statistics doesn't seem to gain the same popularity as the other methods. It's more in the background — an unsung hero, if you will. However, nonparametric statistics is, in fact, a very important and very useful area of statistics because it gives you accurate results when other, more common methods fail.

In this chapter, you see the importance of nonparametric techniques and why they should have a prominent place in your data-analysis toolbox. You also discover some of the basic terms and techniques involved with nonparametric statistics.

## Arguing for Nonparametric Statistics

Nonparametric statistics plays an important role in the world of data analysis. Nonparametric techniques can save the day when you can't use other methods. The problem is that researchers often disregard, or don't even know about, nonparametric techniques and don't use them when they should. In that case, you never know what kind of results you get; what you do know is they could very well be wrong.

In the following sections, you see the advantages and the flexibility of using a nonparametric procedure. You also find out the downside is minimal, which makes it a win-win situation most of the time.

## No need to fret if conditions aren't met

Many of the techniques that you typically use to analyze data, including many shown in this book, have one very strong condition on the data that must be met in order to use them. That is the population(s) from which your data are collected must follow a typically required normal distribution. These methods are called *parametric* methods.

There are a couple of ways to help you decide whether a population has a normal distribution, based on your sample:

- You can graph the data, using a histogram, and see whether it appears to have a bell shape (a mound of data in the middle, trailing down on each side).
- You can make a normal probability plot, which compares your data to that of a normal distribution, using an *x-y* graph (similar to the ones used when you graph a straight line). If the data do follow a normal distribution, your normal probability plot will show a straight line. If the data do not follow a normal distribution, the normal probability plot will not show a straight line; it may show a curve off to one side or the other, for example.

To make a histogram in Minitab, enter your data into a column. Go to Graph> Histogram, and click OK. Click on your variable in the left-hand box, and it appears in the Graph Variables box. Click OK, and you get a histogram.

To make a normal probability plot in Minitab, enter your data in a column. Go to Graph>Probability Plot and click OK. Click on your variable in the left-hand column, and it appears in the Graph Variables column. Click OK, and you see your normal probability plot.

When you find that the normal distribution condition is clearly not met, that's where nonparametric methods come in. *Nonparametric methods* are those data-analysis techniques that don't require the data to have a specific distribution. Nonparametric procedures may require one of the following two conditions (and these are only in certain situations):

- The data come from a symmetric distribution (which looks the same on each side when you cut it down the middle).
- The data from two populations come from the same type of distribution (they have the same general shape).

Note also that the normal distribution centers solely on the mean as its main statistic (for example, the *Z*-value for the hypothesis test for one population mean is calculated by taking the data value, subtracting the mean, and dividing by the standard deviation). So the condition that the population has a normal distribution automatically says you are working with the mean. However, many nonparametric procedures work with the *median,* which is a much more flexible statistic because the median isn't affected by outliers or skewness as the mean is.

## The median's in the spotlight for a change

Many times, any particular statistics question at hand revolves around the center of a population —that is, the number that represents a typical value, or a central value, in the population. One of those measures of center is the *mean.* The *population mean* is the average value over the entire population, which is something that is typically not known (that's why you take a sample). Many data analysts focus heavily on the population mean; they want to estimate it, test it, compare the means of two or more populations, or predict the mean value of a $y$ variable given an $x$ variable. However, the mean isn't the only measure of the center of a population; you also have the good ol' median.

You may recall that the *median* of a data set is the value that represents the exact middle, when you order the data from smallest to largest. For example, in the data set 1, 5, 4, 2, 3, you order the data to get 1, 2, 3, 4, 5 and find that the number in the middle is 3, the median. If the data set has an even number of values, for example, 2, 4, 6, 8, then you average the two middle numbers to get your median (5 in this case).

As you may recall from your introductory statistics course, you can find the mean and the median and compare them to each other. You organize your data into a histogram, and you look at its shape. If the data set is symmetric, meaning it looks the same on either side when you draw a line down the middle, the mean and median are the same. Figure 16-1a shows an example of this situation. In this case, the mean and median are both 5.

If the histogram is skewed to the right, meaning that you have a lot of smaller values and a few larger values, the mean increases due to those few large values, but the median isn't affected. In this case, the mean is larger than the median. Figure 16-1b shows an example of this situation. In this case, the mean is 4.5 and the median is 4.0.

When a data set is skewed left, you have many larger values that pile up, but only a few smaller values. In this case, the mean goes down because of the few small values, but the median still isn't affected. In this case, the mean is lower than the median. Figure 16-1c pictures this case, with a 6.5 mean and a 7.0 median.

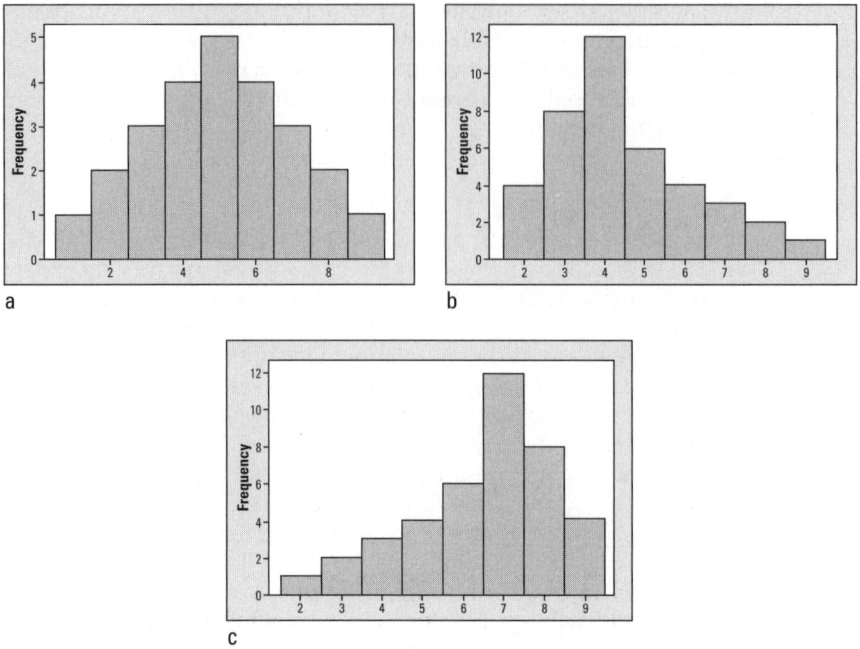


**Figure 16-1:** Symmetric and skewed histograms.

My point is that the median is important! It's a measure of the center of a population, or a sample data set. The median competes with the mean and often wins. Researchers use nonparametric procedures when they want to estimate, test, or compare the median(s) of one or more populations. They also want to use the median in cases where their data are symmetric but don't necessarily follow a normal distribution, or when they want to focus on a measure of center that's not influenced by *outliers* (extreme values either above or below the mean) or skewedness.

For example, if you look at house prices in your neighborhood, you may find a large number of houses within a certain relatively small price range, and then you have a few homes that cost a great deal more. If a real estate agent wants to sell a house and intends to justify a high price for it, she may report the mean price of homes in your neighborhood because the mean is affected by outliers. The mean is higher than the median in this case. But if the agent wants to help someone buy a house, she wants to look at the median of the house prices in the neighborhood, because the median isn't affected by those few higher-priced homes and is lower than the mean.

Now suppose you want to come up with a number that describes the typical house price in your entire county. Should you use the mean or the median? You gathered techniques in your introductory statistics class for estimating the mean of a population (see Chapter 3 for a quick review), but you probably didn't hear about how to come up with a confidence interval for the median of a population. Oh sure, you can take a random sample and calculate the median of that sample. But you need a margin of error to go with it. And I'll tell you something — the formula for the margin of error for the mean doesn't work for the margin of error associated with the median. (See Chapter 17 for the margin of error for the median.)

## So, what's the catch?

You may be wondering, what's the catch if I use a nonparametric technique? A downside must be around here somewhere. Well, many researchers believe that nonparametric techniques water down statistical results; for example, say you find an actual difference between two population means, and the populations really do have a normal distribution. A parametric technique, the hypothesis test for two means, would likely detect this difference (if the sample size was large enough).

The question is, if you use a nonparametric technique (which doesn't need the populations to be normal), do you risk the chance of not finding the difference? The answer is maybe, but the risk isn't as big as you think. More often than not, nonparametric procedures are only slightly less efficient than parametric procedures (meaning they don't work quite as well at detecting a significant result, or at estimating a value as parametric procedures are when the normality condition is met, but this difference in efficiency is small). But the big payoff occurs when the normal distribution conditions aren't met. Parametric techniques can make the wrong conclusion, and corresponding nonparametric techniques can lead to a correct answer. Many researchers don't know this, so spread the word!

REMEMBER

The bottom line: Always check for normality first. If you're very confident that the normality condition is met, go ahead and use parametric procedures because they are more precise. If you have any doubt about the normality condition, use nonparametric procedures. Even if the normality condition is met, nonparametric procedures are only a little less precise than parametric procedures. If the normality condition isn't met, nonparametrics provide appropriate and justifiable results where parametric procedures may not.

# Getting the Basics of Nonparametric Statistics

Because you may not have run into nonparametric statistics during your intro to stats class, figuring out some of the basics needs to be your first step toward using nonparametric techniques. In this section, you get used to some of the terminology and major concepts involved in nonparametric statistics. These terms and concepts are commonly used in Chapters 17 through 20 of this book (and hopefully in your intermediate stats course).

# Sign

The *sign* is a value of 0 or 1 that's assigned to each number in the data set. The sign for a value in the data set represents whether that data value is larger or smaller than some specified number. The value of +1 is given if the data value is greater than the specified number, and the value of 0 is given if the data value is less than or equal to the specified number. For example, suppose your data set is 10, 12, 13, 15, 20, and your specified number for comparison is 16. Because 10, 12, 13, and 15 are all less than 16, they each receive a sign of 0. Because 20 is greater than 16, it receives a sign of +1.

Several uses of the sign statistic appear in nonparametric statistics. You can use signs to test to see if the median of a population equals some specified value. Or you can use signs to analyze data from a matched-pairs experiment (where subjects are matched up according to some variable and a treatment is applied and compared). You can also use signs in combination with other nonparametric statistics. For example, you can combine signs with ranks to develop statistics for comparing the median of two populations. (Ranks are discussed in the next section and are used in a hypothesis test for two population medians in Chapter 18.)

In the following sections, you see exactly how the sign statistic is used to test the median of a population and analyze data in a matched pairs experiment.

## Testing the median

You can use signs to test whether the median of a population is equal to some value *m*. You do this by conducting a hypothesis test based on signs. You have Ho: Median = *m* versus Ha: Median ≠ *m* (or, you can use a > or < sign in Ha also). Your test statistic is the sum of the signs for all the data. If this sum is significantly greater or significantly smaller than what is expected if Ho were true, you reject Ho. Exactly how large or how small the sum of the signs must be to reject Ho is given by the sign test (Chapter 17).

Suppose you're testing whether the median of a population is equal to 5. That is, you're testing Ho: Median = 5 versus Ha: Median ≠ 5. You collect the following data: 4, 4, 3, 3, 2, 6, 4, 3, 3, 5, 7, 5. Ordering the data, you get 2, 3, 3, 3, 3, 4, 4, 4, 5, 5, 6, 7. Now you find the sign for each value in the data set, determined by whether the value is greater than 5. The sign of the first data value, 2, is 0, because it's below 5. Each of the 3s receives a sign of 0, as do the three 4s, and the 5s, for the same reason. Only the numbers 6 and 7 receive a sign of +1, being the only values in the data set that are greater than 5 (the number of interest for the median).

By summing the signs, you're in essence counting the number of values in the data set that are greater than the given quantity in Ho. For example, the total of all the signs of the ordered data values is 0 + 0 + 0 + 0 + 0 + 0 + 0 + 0 + 0+ 0 +

1 + 1 = 2, and you can see that the total number of data values above 5 (the number of interest for the median) is 2. The fact that the total of the signs (2) is much less than half the sample size gives you some evidence that the median is probably not 5 here, because the median represents the middle of the population. If the median were truly 5 in the population, your sample should yield about 6 values below it and 6 values above.

### Doing a matched-pairs experiment

You can use signs in a matched-pairs experiment (where you use the same subject twice or pair them on some important variables). For example, you can use signs to test whether or not a certain treatment resulted in an improvement in patients, compared to a control. In the cases where the sign statistic is used, improvement is measured not by the mean of the differences in the responses for treatment versus control (as in a paired *t*-test), but by the median of the differences in the responses.

Suppose you're testing a new antihistamine for allergy patients. You take a sample of 100 patients and have each patient assess the severity of his allergy symptoms before and after taking the medication on a scale from 1 (best) to 10 (worst). (Of course, you do a controlled experiment where some of the patients get a placebo to adjust for the fact that some people may perceive their symptoms going away just because they took something, anything.)

In this study, you're not interested in what level their symptoms are at, but in how many patients had a lower level of symptoms after taking the medicine. So you take the symptom level before the experiment minus the symptom level after the experiment. If that difference is positive, the medicine appears to have helped, and you give that person a sign of +1 (in other words, count them as a success). If the difference is zero, the medicine had no effect, and you give that person a sign of 0. Remember, though, that the difference could be negative, indicating that the symptoms before were lower than the symptoms after; in other words, the medicine made their symptoms worse. This scenario results in a sign of 0 as well.

After you've found the sign for each value or pair in the data set, you're ready to analyze it by using the sign test or the signed rank test (see Chapter 17).

## Rank

*Ranks* are a nice way to use important information from a data set without using the actual values of the data themselves. Rank comes into play in nonparametric statistics when you're not interested in what the values of the data are, but where they stand, compared to some supposed value for the median or to the ranks of values in another data set from another population. (You can see ranks in action in Chapter 18.)

The *rank of a value* in a data set is the number that represents its place in the ordering, from smallest to largest, within the data set. For example, if your data set is 1, 10, 4, 2, 1,000, you can assign the ranks in the following way: 1 gets the rank of one (because it's the smallest), 2 gets the rank of two, 4 gets the rank of three (being the third smallest number in the ordered data set), 10 gets the rank of four, and 1,000 gets the rank of five (being the largest).

Now suppose your data set is 1, 2, 20, 20, 1,000. How would the ranks be assigned? You know that 1 would get the rank of one (being the smallest), 2 would get the rank of two, and 1,000 would get the rank of five (being the largest). But what about the two 20s in this data set? Should the first 20 get a rank of three and the second 20 get the rank of four? That order doesn't seem to make sense, because you can't distinguish between the two 20s.

When two values in a data set are the same, you take the average of the two ranks the values need to fill and assign each tied value that average rank. If you have a tie between three numbers, you have three ranks, so take the sum of the ranks divided by three.

In this case, because both 20s are vying for the ranks of three and four, assign each of them the rank of 3.5, the average of the two ranks they must share. I show the final ranking for the data set 1, 2, 20, 20, 1,000 in Table 16-1.

**Table 16-1 Ranks of the Values in the Data Set 1, 2, 20, 20, 1,000**

| *Data Value* | *Rank Assigned* |
|---|---|
| 1 | 1 |
| 2 | 2 |
| 20 | 3.5 |
| 20 | 3.5 |
| 1,000 | 5 |

The lowest a rank can be is one, and the highest a rank can be is $n$, where $n$ is the number of values in the data set. If you have a negative value in a data set, for example, if your data set is –1, –2, –3, you still assign the ranks one through three to those data values. Never assign negative ranks to negative data. (By the way, when you order the data set –1, –2, –3, you get –3, –2, –1, so –3 gets the rank of one, –2 gets the rank of two, and –1 gets the rank of three.)

## Signed rank

A *signed rank* combines the idea of the sign and the rank of a value in a data set, with a small twist. The sign indicates whether that number is greater than, less than, or equal to a specified value. The rank indicates where that number falls in the ordering of the data set from smallest to largest.

To calculate the signed rank for each value in the data set, follow these steps:

1. **Assign a sign of +1 or 0 to each value in the data set, according to whether it's greater than some value specified in the problem.**

   If it's greater than the specified value, give it a sign of +1; if it's less than or equal to the specified value, give it a sign of 0.

2. **Rank the original data from smallest to largest, according to their absolute values.**

   Statisticians call these values the *absolute ranks.*

3. **Multiply the sign times the absolute rank to get the signed rank for each value in the data set.**

The absolute value of any number is the positive version of that number. The notation for absolute value is $| \ |$, where the number goes between those lines. For example, $|-2| = 2$ and $|+2| = 2$. Remember that $|0| = 0$.

One scenario in which you can use signed ranks is an experiment where a response variable is compared for a treatment group versus a control group. You can test for difference due to a treatment by collecting the data in pairs, either both from the same person (pretest versus post-test) or from two individuals that are matched up to be as similar as possible.

For example, suppose you compare four patients regarding their weight loss on a diet program. You're really wondering whether the overall change in weight is less than zero for the population. Two factors are important:

- Whether or not the person lost weight
- How the person's weight change measures up, compared to everyone else in the data set

You measure the person's weight before the program (the pretest) as well as his weight after the program (the post-test). The change is the important facet of the data you're interested in, so you apply the signs to the changes in weight. You give the change a sign of +1 if the person lost weight (constituting a success for the program) and a sign of 0 if the person stayed the same or gained weight (thus not contributing to the success of the program). You convert all the changes in weight loss to their absolute values, and then you rank the absolute values (in other words, you've found the absolute ranks of the changes in weight). The signed rank is the product of the sign and the absolute rank. After determining the signed rank, you can really compare the effectiveness of the program. Large signed ranks indicate a big weight loss; small signed ranks don't.

For example, weight changes of –20, –10, +1, and +5 have signs of +1, +1, 0, 0. The absolute values of the weight changes are 20, 10, 1, and 5. Their absolute ranks, respectively, are 4, 3, 1, and 2. The signed ranks are $4 * 1 = 4$, $3 * 1 = 3$, $1 * 0 = 0$, and $2 * 0 = 0$.

## Rank sum

A *rank sum* is just what it sounds like: The sum of all the ranks. You typically use rank sums in situations when you're comparing two or more populations to see whether one has a central location that's higher than the other. (In other words, if you looked at the populations in terms of their histograms, one would be shifted to the right of the other on the number line.)

Here's a way in which researchers use rank sums: Suppose you're looking at quiz scores for two classes, and they don't have a normal distribution, hence you want to use nonparametric techniques to compare them. The total possible points on this quiz is 30. You collect random samples of five quiz scores from each of the classes. Suppose the sample data from class number one is: 22, 23, 20, 25, 26, and the sample data from class number two is: 23, 30, 27, 28, 25. The twist here is to combine all the data into one big data set, rank all the values, and sum the ranks for the first sample and then the second sample. Then compare the two rank sums. If one rank sum is higher, this outcome may indicate that a particular class did better on the quiz.

In the quiz example, the ordered data for the combined classes is 20, 22, 23, 23, 25, 25, 26, 27, 28, 30. Their ranks, respectively, are 1, 2, 3.5, 3.5, 5.5, 5.5, 7, 8, 9, and 10. The ranks from the first class are 1 (associated with the score 20); 2 (22); 3.5 (23); 5.5 (25); and 7 (26). The rank sum for the first class is $1 + 2 + 3.5 + 5.5 + 7 = 19$, which is quite a bit lower than the rank sum for the second class ($3.5 + 5.5 + 8 + 9 + 10 = 36$). This result tells you that the second class did better on the quiz than the first class, for this sample.

In Chapter 18, you can see how to use a rank sum test to see whether the shapes of two population distributions are the same, meaning the values they take on and how often those values occur in each population. In Chapter 19, you can find even more on rank sums and also discover how to conduct Kruskal-Wallis tests.

WARNING!

Note that taking the mean of each data set and comparing them by using a two-sample *t*-test would be wrong in the quiz example because the quiz scores admittedly don't have a normal distribution. Indeed if the quiz were easy, you'd get many high scores and few low ones, and the population would be skewed left. On the other hand, if the quiz were hard, you'd get many low scores and few high ones, and the population would be skewed right (don't think too much about that scenario). In either case, you need a nonparametric procedure. See Chapter 18 for more on the nonparametric equivalent of the *t*-test.

# Chapter 17

# The Sign Test and Signed Rank Test

### *In This Chapter:*

- Testing and estimating the median: The sign test
- Figuring out when and how to use the signed rank test

Situations often arise where your data doesn't meet the conditions to test or estimate the mean, or you just don't have enough data (the biggest hurdle is whether the data come from a population with a normal distribution), or, your data is just of a different type than quantitative data, such as ranks (where you don't collect numerical data, but instead just order the data from low to high or vice versa).

In these situations, your best bet is a *nonparametric procedure* (see Chapter 16 for background info). These procedures have very few assumptions tied to them. Moreover, you can find that nonparametric procedures are easy to carry out and that their formulas make sense. Most importantly, they give accurate results compared to the use of parametric procedures when the conditions of parametric procedures aren't met or aren't appropriate.

In this chapter, you use the sign test and the Wilcoxon signed rank test to test or estimate the median of one population. These nonparametric procedures are the counterparts to the one-sample and matched pairs *t*-tests, which require data from a normal population.

## Reading the Signs: The Sign Test

The *sign test* is a nonparametric alternative for the one sample *t*-test. What makes the sign test so nice is that it's based on a very basic distribution, the binomial (for full info on the binomial, see your intro stats text).

The only condition of the sign test is that the data are ordinal or quantitative — not categorical. However, this is no big deal; if you are interested in the median, you wouldn't collect categorical data anyway.

Here are the steps for conducting the sign test. Note that Minitab can do Steps 4–7 for you; however, understanding what Minitab does behind the scenes is important, as always.

1. **Set up your null hypothesis: Ho: $m = m_o$.**

   The true value of the median is *m,* and $m_o$ is the claimed value of the median (the value you're testing).

2. **Set up your alternative hypothesis. Your choices are Ha: $m \neq m_o$, or Ha: $m > m_o$, or Ha: $m < m_o$.**

   Which Ha you choose depends on what conclusion you want to make in the case that Ho is rejected. For example if you only want to know when the median is greater than m, use Ho: $m > m_o$. See Chapter 3 for more on setting up alternative hypotheses.

3. **Collect a random sample of (ordinal or quantitative) data from the population.**

4. **Assign a plus or minus sign to each value in the data set.**

   If an observation is less than $m_o$, assign it a minus (–) sign. If the observation is greater than $m_o$, give it a plus (+) sign. If the observation equals $m_o$, disregard it and let the sample size decrease by one.

5. **Count up all the plus signs — this sum is your test statistic.**

6. **Find the *p*-value for your test statistic.**

   Look up your test statistic on Table A-2 (binomial distribution) corresponding to your sample size *n,* the value of $p = 0.50$, and *k* equal to the test statistic from step five. If Ha has a < sign, add up all the probabilities for $x \leq k$. If Ha has a > sign, added up all the probabilities for $x \geq k$. If Ha has a ≠ sign, add up the probabilities of being greater than or equal to *k* and double this value. This gives you the *p*-value of the test.

7. **Make your conclusion.**

   If the *p*-value from step six is less than the prespecified value of $\alpha$ (typically 0.05), reject Ho and say the median is greater than, less than, or ≠ $m_0$, depending on Ha. Otherwise, you can't reject Ho.

To run a sign test in Minitab, enter your data in a single column. Go to Stat> Nonparametric>One-sample Sign. Click on your variable in the left-hand box, and click Select. The variable will appear in the Variables box. Then click OK, and you get the results of the sign test.

In the sections that follow, I show you two different ways in which you can use the sign test:

- To test or estimate the median of one population
- To test or estimate the median difference of data where the observations come in pairs, either from the same individual (pretest versus post-test) or individuals paired up according to relevant characteristics

Now that you know what you're getting into, take a deep breath and jump in!

## Testing the median

Situations arise when you aren't interested in the mean, but rather the median of a population (see Chapter 16) — for example, when the data doesn't have a normal, or even a symmetric, distribution. When you want to estimate or test the median of a population (call it *m*), the sign test is a great option.

Suppose you're a real estate agent selling homes in a particular neighborhood and you hear from other agents that the median house price in that neighborhood is $110,000. You think the median is actually higher. Because you're interested in the median price of a home rather than the mean price, you decide to test this claim by using a sign test.

Following the steps of the sign test, you first set up your null hypothesis. Because the original claim is that the median price of a home is $110,000, you have Ho: $m = \$110{,}000$. Next, you set up the alternative hypothesis. Because you believe the median is higher than $110,000, your alternative hypothesis is Ha: $m > \$110{,}000$.

In step three of the sign test, you take a random sample of ten homes in your neighborhood. You can see the data in Table 17-1. Its histogram is shown in Figure 17-1. Now the question is, is the median selling price of all homes in the neighborhood equal to $110,000, or is it more than that (as you suspect)?

**Table 17-1 Sample of House Prices in a Neighborhood**

| *House* | *Price* | *Sign (Compared to $110,000)* |
|---|---|---|
| 1 | $132,000 | + |
| 2 | $107,000 | – |
| 3 | $111,000 | + |
| 4 | $105,000 | – |

*(continued)*

**Table 17-1 *(continued)***

| *House* | *Price* | *Sign (Compared to $110,000)* |
|---|---|---|
| 5 | $100,000 | – |
| 6 | $113,000 | + |
| 7 | $135,000 | + |
| 8 | $120,000 | + |
| 9 | $125,000 | + |
| 10 | $126,000 | + |

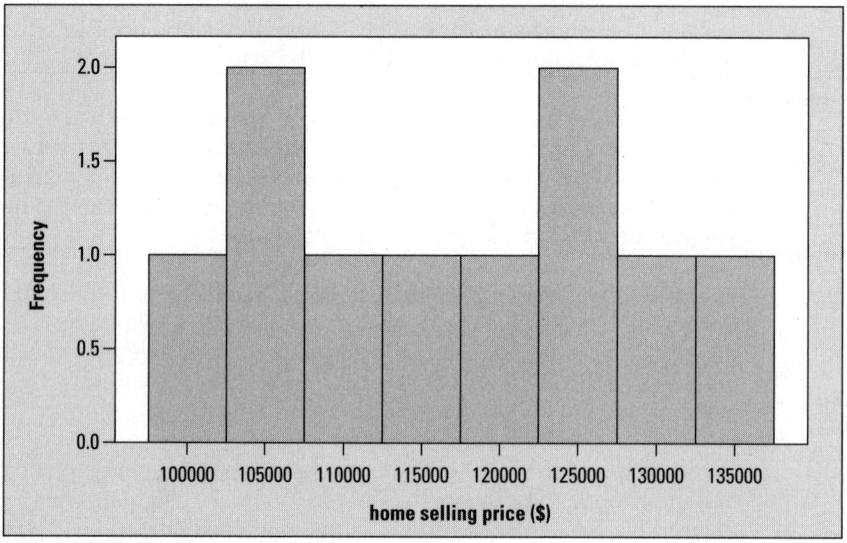


**Figure 17-1:** Histogram of ten house selling prices.

In step four, you assign a plus sign to any house price that is more than $110,000, and you assign a minus sign to any house that is less than $110,000. (See column three of Table 17-1.)

Step five of the sign test involves finding your test statistic. Your test statistic is 7, the number of "+" signs in your data set (Table 17-1), representing the number of houses in your sample whose prices were above $110,000.

In step six, you compare your test statistic to the binomial distribution (see Table A-2 in the Appendix) to find the *p*-value. While examining Table A-2, you look at the row where $n = 10$ (the sample size) and $k = 7$ (the test statistic) and the column where $p = 0.50$ (because if the population median equals $m_0$, 50 percent of the values in the population should be above it, and 50 percent below it). According to the table, you find the probability that $x$ equals 7 is 0.117. Because you have a right-tailed test (meaning Ha has a > sign in it), you add up the probabilities of being at or beyond 7 to get the *p*-value. The *p*-value in this case is 0.117 + 0.044 + 0.010 + 0.001 = 0.172. Note that all remaining probabilities are too small to include Table A-2, so they don't appear in this sum.

Step seven is the conclusion step. You compare the *p*-value (0.172) to the prespecified $\alpha$ (I always use 0.05). Because the *p*-value is greater than 0.05, you can't reject Ho. There is not enough evidence to say the median house selling price is more than $110,000.

Figure 17-2 shows these results as calculated by Minitab. These numbers confirm what was just shown.

**Sign Test for Median: Selling Price**

```
Sign test of median = 110000 versus > 110000

                    N   Below  Equal  Above        P  Median
Selling  Price     10       3      0      7   0.1719  116500
```

**Figure 17-2:** Sign test conducted by Minitab.

If your data are close to normal and the mean is the more appropriate measure of center for your situation, don't use the sign test; use the one sample *t*-test (or *Z*-test). The sign test isn't quite as powerful (able to reject Ho when it should) as the *t*-test in situations where the conditions for the *t*-test are met. More importantly, though, don't run to the *t*-test to reanalyze your data if the sign test doesn't reject Ho. That would be improper and unethical. In general, the idea of following a nonparametric procedure by a parametric procedure in hopes of getting more significant results is considered by statisticians to be *data fishing;* in other words, analyzing data in different ways until a statistically significant result appears.

## Estimating the median

You can also use the sign test to find a confidence interval for one population median. This comes in handy when you're interested in estimating what the median value of a population is. For example, what is the median income of a household in the United States? Or what is the median salary of people just coming out of an MBA program?

Following are the steps for conducting a confidence interval for the median by using the test statistic for the sign test, assuming your random sample of data has already been collected. Note that Minitab can calculate the confidence interval for you (steps two to five), but knowing how Minitab does the steps is important:

1. **Determine your level of confidence, $1 - \alpha$ (that is, how confident you want to be that this process will correctly estimate $m$ over the long term).**

   The typical confidence level data analysts use is 95 percent. See Chapter 3 for more information.

2. **Go to Table A-2 in the Appendix (binomial distribution), and find the section for $n$ equal to your sample size, and the column where $p = 0.50$ (because the median is the point where 50 percent of the data lies below and 50 percent lies above).**

   You will find probabilities for values of $x$ from 0 to $n$ in that section.

3. **Starting at each end ($x = 0$ and $x = n$) and moving one step at a time toward the middle of the $x$ values, add up the probabilities for those values of $x$ until you pass the total of $\alpha$ (which is one minus your confidence level).**

4. **Record the number of steps that you had to make just before you passed the value of $1 - \alpha$. Call this number $c$.**

5. **Take your data set and order it from smallest to largest. Starting at each end, work your way to the middle until you reach the $c^{th}$ number from the bottom and the $c^{th}$ number from the top.**

   This result is your confidence interval for the median.

You can use these steps to find a confidence interval for the median in the house-price example from the preceding section. As the first step, let your confidence level be set at $1 - \alpha = 0.95$. In step two, go to Table A-2 (Appendix) and look at the section where $n = 10$ (the sample size) and $p = 0.50$. These values are listed in Table 17-2.

**Table 17-2 Binomial Probabilities to Help Calculate a Confidence Interval for the Median ($n = 10$, $p = 0.50$)**

| $x$ | $p(x)$ |
|---|---|
| 0 | 0.001 |
| 1 | 0.010 |

| *x* | *p(x)* |
|---|---|
| 2 | 0.044 |
| 3 | 0.117 |
| 4 | 0.205 |
| 5 | 0.246 |
| 6 | 0.205 |
| 7 | 0.117 |
| 8 | 0.044 |
| 9 | 0.010 |
| 10 | 0.001 |

In step three, you start with the outermost values of $x$ ($x = 0$ and $x = 10$) and sum those probabilities to get 0.001 + 0.001 = 0.002. Because you haven't yet passed 0.05 (the value of $\alpha$), you go in to the second-innermost values of $x$ ($x = 1$ and $x = 9$). Add their probabilities to what you have so far to get 0.002 (old total) + 0.010 + 0.010 = 0.022. You are still not past 0.05 ($\alpha$), so continue one more step. Add the third-innermost probabilities for $x = 2$ and $x = 8$ to the grand total to get 0.022 (old total) + 0.044 + 0.044 = 0.110. You've now passed the value of $\alpha = 0.05$. That means the value of $c$ equals 2 because at the third-innermost values of $x$, you passed 0.05, and you back off one step from there to get your value of $c$.

Step four says to order your data (Table 17-1) from smallest to largest. This gives you (in dollars): 100,000, 105,000, 107,000, 111,000, 113,000, 120,000, 125,000, 126,000, 132,000, and 135,000. For step five, work your way in from each end of the data set to take the second-innermost values (because $c = 2$). This gives you the numbers \$105,000 and \$132,000. Put these two numbers together to form an interval, and you conclude that a 95-percent confidence interval for the median selling price for a home in this neighborhood is between \$105,000 and \$132,000.

To find a $1 - \alpha$ percent confidence interval for the median using Minitab based on the sign test, enter your data into a single column. Go to Stat>Nonparametrics>One-sample Sign. Click on the variable in the left-hand column for which you want the confidence interval, and it appears in the Variables column. Click the circle that says Confidence Interval, and type in the value of $1 - \alpha$ you want for your confidence level. (The default is 95 percent, written as 95.) Click OK to get the confidence interval.

## Testing matched pairs

The most useful application of the sign test is in testing matched pairs of data, that is, data that come in pairs and represent two observations from the same person (pretests versus post-tests, for instance) or one set of data from each pair of people who are matched according to relevant characteristics. In this section, you see how you can compare data from a matched-pairs study to look for a treatment effect, using a sign test for the median.

The idea of using a sign test for the median difference with matched-pairs data is similar to using a *t*-test for the mean differences with matched-pairs data. A test of the median (rather than the mean) is used when the data don't necessarily have a normal distribution, or if you're only interested in the median difference rather than the mean difference.

First, you set up your hypotheses, Ho: The median is zero (indicating no difference between the pairs). Your alternative hypothesis is Ha: The median is $\neq 0$, $> 0$, or $< 0$, depending on whether you want to know if the treatment made any difference, made a positive difference, or made a negative difference compared to the control. Then you collect your data (two observations per person or a pair of observations from each pair of people you have matched up). After that, you use Minitab to conduct steps four to seven of the sign test.

For example, suppose you wonder whether taking a test while chewing gum decreases test anxiety. You pair 20 students according to relevant factors such as GPA, score on previous midterms, and so on. One member of each pair is randomly selected to chew gum during the exam, and the other member of the pair doesn't. You measure test anxiety on each person via a very short survey right after they turn in their exams. You measure the results on a scale of 1 (lowest anxiety level) to 10 (highest anxiety level). The data based on a sample of ten pairs is shown in Table 17-3.

**Table 17-3 Testing the Effectiveness of Chewing Gum in Lowering Test Anxiety**

| *Pair* | *Gum* | *No Gum* | *Difference (Gum/No Gum)* | *Sign* |
|---|---|---|---|---|
| 1 | 9 | 10 | –1 | – |
| 2 | 6 | 8 | –2 | – |
| 3 | 3 | 1 | +2 | + |
| 4 | 3 | 5 | –2 | – |
| 5 | 4 | 4 | 0 | none |
| 6 | 2 | 7 | –5 | – |
| 7 | 2 | 6 | –4 | – |

| *Pair* | *Gum* | *No Gum* | *Difference (Gum/No Gum)* | *Sign* |
|---|---|---|---|---|
| 8 | 8 | 10 | –2 | – |
| 9 | 6 | 8 | –2 | – |
| 10 | 1 | 3 | –2 | – |

REMEMBER

The actual levels of test anxiety aren't important here; what matters is the difference between anxiety levels within each pair. So, instead of looking at all the individual anxiety levels, you can look at the difference in anxiety levels for each pair. This method gives you one data set, not two. (In this case, to calculate the differences in each pair, you can use the formula test anxiety without gum minus text anxiety with gum, and look for an overall difference that's positive.) Typically, in the case of matched-pairs data, you're testing whether the median difference equals zero. In other words, Ho: $m = 0$; the same holds in the test anxiety example.

The differences in anxiety levels for each pair in your data set now become a single data set (see column four of Table 17-3). You can now use the regular sign test methods to analyze this data, using Ho: $m = 0$ (no median difference in test anxiety of gum versus no gum) versus Ha: $m < 0$ (chewing gum reduces test anxiety).

Assign each difference a plus or minus sign, depending on whether it's greater than zero (plus sign) or less than zero (minus sign.) Your test statistic is the total number of plus signs, 1, and the relevant sample size is $10 - 1 = 9$. (You don't count the data that hit the median of zero right on the head.)

Now compare this test statistic to the binomial distribution with $p = 0.50$ and $n = 9$, using Table A-2 in the Appendix. You have a test statistic of $k = 1$, and you want to find the probability that $x \leq 1$ (because you have a left-tailed test, see step six of the sign test from a previous section). Under the column for $p = 0.50$ in the section for $n = 9$, you get the probability of 0.018 for $x = 1$ and 0.002 for $x = 0$. Add these values to get 0.020, your $p$-value. This result means that you reject Ho at the prespecified $\alpha$ level of 0.05. You conclude that, based on this data, chewing gum on an exam is related to lower test anxiety.

# Going a Step Further with the Signed Rank Test

The signed rank test is more powerful at detecting real differences in the median than the sign test is. The most common use of the signed rank test is with matched-pairs data, to test for a median difference due to some treatment (like chewing-gum use during an exam and its affect on test anxiety). In this section, you see what the signed rank test is, how it is carried out, and an application involving the test of a weight-loss program.

## A limitation of the sign test

The sign test has the advantage of being very simple and easy to do by hand. However, because it only looks at whether a value is above or below the median, it doesn't take the magnitude of the difference into account.

Looking at Tables 17-1 and 17-3 from the previous section on the sign test, you see that for each data value, the test statistic for the sign test only counts whether or not each data value is greater than or equal to the median in the null hypothesis, $m_0$. It doesn't count how great those differences are. For example, in Table 17-3, you can see that the sixth pair had a huge reduction in test anxiety when chewing gum (from 7 down to 2), but the first pair had a very small reduction in test anxiety (from 10 down to 9). Yet both of these differences received the same outcome (a minus sign) in the test statistic.

This shows a bit of a limitation in the sign test in that it doesn't take into account how much the values in the data differ from the median. The sign test is less powerful (less able to detect when Ho is false) than it could be. So if you want to test the median and you want to take the magnitude of the differences into account (and you're willing to go through some math hoops to get there), you can conduct the *signed rank test,* (also known as the *Wilcoxon signed rank test*). The next section walks you through it.

## Stepping through the signed rank test

Just like the sign test, the only condition of the signed rank test is that the data are ordinal or quantitative.

Following are the steps in carrying out the signed rank test on paired data:

1. **Set up your hypotheses.**

   The null hypothesis is Ho: $m = 0$. Your choices for an alternative hypothesis are Ha: $m \neq 0$, or Ha: $m > 0$, or Ha: $m < 0$, depending on whether you want to detect any difference, a positive difference or a negative difference in the pairs.

2. **Collect a random sample of paired data.**

3. **For each observation, calculate the difference for each pair of observations.**

4. **Calculate the absolute value of each of the differences.**
5. **Rank the absolute values from smallest to largest.**

   If two of the absolute values are tied, give each one the average rank of the two values. For example if the fourth and fifth numbers, in order, are tied, give each one the rank of 4.5.
6. **Add up the ranks that correspond to those original differences from step three that are positive.**

   The sum of the positive differences is your signed rank test statistic, denoted by SR.
7. **Find the *p*-value.**

   Look at all possible ways that the absolute differences could've appeared in a sample, with either plus or minus signs, assuming that Ho is true. Find all their test statistics (SR values) from all these possible arrangements by using steps four to six and compare your SR value to those. The percentage of SR values that are at or beyond your test statistic is your *p*-value.
8. **Make your conclusion.**

   If the *p*-value is less than the pre-specified $\alpha$ (typically 0.05), reject Ho and conclude the median difference is not zero. Otherwise, you can't reject Ho.

Before you go crazy looking at step seven, don't worry; Minitab can do steps four to eight for you.

To conduct the Wilcoxon signed rank test using Minitab, enter the differences from step three in a single column. Go to Stat>Nonparametrics>1-Sample Wilcoxon. Click on the name of the variable for your differences in the left-hand box, and it appears in the right-hand Variables box. Click on the circle that says Test Median, and indicate which Ha you want (>0, <0, or ≠). Click OK, and your test is done. (Note that Minitab calculates the test statistic for the signed rank test a little differently than what you will get by hand, although the results will be close. The reason for the slight calculation difference is beyond the scope of this book.)

How do you handle situations where a piece of data is exactly equal to the median? Most of the time (including all data sets you will encounter) this occurrence is rare, and can be handled by ignoring those data values and reducing the sample size by one for each time this occurs.

You can see this test in action in the following example of looking to see whether a weight-loss program actually works. I first show you each step as if you were doing the process by hand. Then you see the results in Minitab.

## Losing weight with signed ranks

Suppose you want to test whether or not a weight-loss plan is effective. You want to look at the median weight loss for people on the plan by using a matched-pairs experiment. You want the magnitude of weight loss to factor into the analysis. That means you use a signed rank test to analyze the data.

Following the steps from the preceding section, you first set up your hypotheses. Test Ho: $m = 0$, where $m$ represents the median weight loss (before the program versus after the program). Your alternative hypothesis is Ha: $m > 0$, indicating the median difference in weight loss is positive.

For step two, you take a random sample of, say, three people and measure them before and after an 8-week weight loss program. Step three says that for each person, you calculate the difference in weight (weight before the program minus weight after the program); a positive difference means the person lost weight, and a negative difference means they gained weight. (Note the small value of $n = 3$ here is used for illustrative purposes only.)

The data and relevant statistics for the weight-loss signed rank test are shown in Table 17-4. You can see the differences in weight (before – after) in column four of Table 17-4.

**Table 17-4 Data on Weight Loss Before and After Program**

| *Person* | *Before* | *After* | *Difference* | *\|Difference\|* | *Rank* |
|---|---|---|---|---|---|
| 1 | 200 | 205 | –5 | 5 | 1 |
| 2 | 180 | 160 | +20 | 20 | 2* |
| 3 | 134 | 110 | +24 | 24 | 3* |

** Represents ranks associated with a positive difference in weight loss*

In step four, take the absolute values of the differences. You can see those in column five of Table 17-4. Step 5 says to rank those absolute differences; column six reflects the ranks of those absolute values, from 1 to 3.

In step six, you find your test statistic, which is the sum of the ranks corresponding to positive differences. (In other words, you only count ranks of people who lost weight.) For this data set, those ranks you can count are indicated by * in Table 17-4. The sum turns out to be $2 + 3 = 5$. This number, 5, is your test statistic; you can call it SR to designate the signed rank test statistic.

Step seven says to calculate the *p*-value. Now you need to compare that test statistic to some distribution to see where it stands. To do this, you determine all the possible ways that the three absolute differences (column five of Table 17-4) — 5, 20, and 24 — could have appeared in a sample, with their actual differences taking on plus signs or minus signs. (Assume Ho is true, that is, the actual differences have a 50 percent chance of being positive or negative, like the flip of a coin.)

Then you find all their test statistics (SR values) from all of these possible arrangements and compare your SR value, 5, to those. The percentage of the other SR values that are at or beyond your test statistic is your *p*-value.

Here's how step seven looks for the weight-loss example. First, you have eight possible ways that you can have absolute differences of 5, 20, and 24 by including either plus or minus signs on each difference (two possible signs for each equals 2 * 2 * 2 = 8). Those eight possibilities are listed in separate columns of Table 17-5. SR denotes the sum of the positive ranks in each case (these are the test statistics for each possible arrangement).

**Table 17-5 Possible Samples with Absolute Differences of 5, 20, and 24**

| *1* | *2* | *3* | *4* | *5* | *6* | *7* | *8* | *Rank of \|Diff.\|* |
|---|---|---|---|---|---|---|---|---|
| 5 | –5* | 5 | 5 | –5* | –5* | 5 | –5* | 1 |
| 20 | 20 | –20* | 20 | –20* | 20 | –20* | –20* | 2 |
| 24 | 24 | 24 | –24* | 24 | –24* | –24* | –24* | 3 |
| SR = 6 | SR = 5 | SR = 4 | SR = 3 | SR = 3 | SR = 2 | SR = 1 | SR = 0 | — |

** denotes negative differences*

To make sense of Table 17-5, consider the following: The three absolute differences you have in your data set are 5, 20, and 24, which have ranks 1, 2, and 3, respectively (which you can see in Table 17-4). You can find the eight different combinations of 5, 20, and 24 that exist, where you can put either a minus or plus sign on any of those values. For each scenario, I found the signed rank statistic by summing the ranks for only those differences that are positive (the person lost weight). Those ranks are the column 9 values in Table 17-5 for data values without an asterisk (*).

For example, column seven has two negative differences, –20 and –24, and one positive difference of 5 (whose rank among the absolute differences is 1; see column 9). Summing the positive ranks you get a signed rank statistic (SR) of one because 5 is the only positive number. (You can see in column two the data that you actually observed in the sample.)

Now compare the test statistic, 5, to all the values of SR in the last row of Table 17-5. Because you're using Ha: $m > 0$, you can find the percentage of signed ranks (SR) that are at or above the value of 5. You have two of them out of eight, so your *p*-value (the percentage of possible test statistics beyond or the same as yours if Ho were true), is $\frac{2}{8} = 0.25$ or 25 percent.

Finally, you arrive at step eight! Because the *p*-value (0.25) is greater than the pre-specified value of $\alpha$ (typically 0.05), you can't reject Ho, and you can't say there's a positive weight loss via this program. (With a sample size of only 3, it's difficult to find any real difference, so the weight-loss program may actually be working and this small data set just couldn't determine that.)

Figure 17-3 shows the Minitab output for this test, using the data from Table 17-4. The *p*-value turns out to be 0.211; this is due to a slight difference in the way that Minitab calculates the test statistic. Note the estimated median found in Figure 17-3 refers to a calculation made over all possible samples, and the medians you would get from them.

**Wilcoxon Signed Rank Test: Wt loss**

```
Test of median = 0.000000 versus median > 0.000000

                N
                for    Wilcoxon                 Estimated
           N   Test   Statistic        P           Median
Wt loss    3                5.0    0.211            14.75
```

**Figure 17-3:** Computer output for signed rank test of weight-loss data.

You can also use the SR statistic to estimate the median of one population (or the median of the difference in a matched-pairs situation). To find a $1 - \alpha$ percent confidence interval for the median using Minitab based on the signed rank test, enter your data into a single column. (If your data represents differences from a matched-pairs data set, enter those differences as one column.) Go to Stat>Nonparametrics>1-Sample Wilcoxon. Click on the name of the variable in the left-hand column, and it appears in the Variables column on the right-hand side. Click the circle that says Confidence Interval, and type in the value of $1 - \alpha$, your confidence level. Click OK.

# Chapter 18

# Pulling Rank with the Rank Sum Test

### *In This Chapter*

- Comparing two populations by using medians not means
- Conducting the rank sum test

In introductory statistics when you want to compare two populations, you conduct a hypothesis test for two population means. You may remember that, in order to conduct a test for two means, one of the following two conditions must be met:

- The populations have normal distributions (with no restrictions on the sample sizes).
- The populations don't have normal distributions, but the sample sizes are large enough (the larger, the better).

If either of these conditions are met, you go ahead and use the *Z*-distribution to analyze your data (because in the second case the Central Limit Theorem says it's okay; for more on this theory, see *Statistics For Dummies* [Wiley] or your introductory statistics text). If neither of these conditions are met, you can't use the *Z*-distribution to conduct the test. However, this result doesn't mean you can't do anything and have to throw in the towel. When conditions for parametric procedures (ones involving normal distributions) aren't met, a nonparametric alternative is always there to save the day.

In this chapter, you see a nonparametric test that compares the centers of two populations — the *rank sum test.* This test focuses on the median, the measure of center that's most appropriate in situations where the data isn't symmetric. Two other names you may also see used for this test are the *two-sample Wilcoxon rank sum test* and the *Mann-Whitney test.* (The two different names acknowledge two sets of independent inventors of the same test at around the same time; one of which [Whitney] is a professor emeritus in my department at Ohio State, and I know the guy — how cool is that?). However, for this book (and because I don't want to play favorites), I stick with just calling it the rank sum test.

# Conducting the Rank Sum Test

In this section you see the conditions for the rank sum test and steps for conducting the test. An example is provided in the next section, "Which real estate agent sells homes faster?"

## Checking the conditions

Before you can think about conducting the rank sum test to compare the medians of two populations, you have to make sure your data sets meet the conditions for the test. The conditions for the rank sum test are the following:

- **The two random samples, one taken from each population, are independent of each other.**

  The first condition is taken care of in the way you collect your data. Just make sure you aren't using matched pairs, for example, using data from the same person in a pretest and post-test manner. Then the two sets of data would be dependent.

- **The two populations have the same distribution. (That is, their histograms have the same shape.)**

  You can check the second condition by making histograms to compare the shapes of the sample data from the two populations.

- **The two populations have the same variance. (In other words, the amount of spread in the values is the same.)**

  You can check the third condition by finding the variances or standard deviations of the two samples. They should be close, meaning that they shouldn't be different enough for you to want to write home about it. (A hypothesis test for two variances actually exists, but that's outside the scope of this book.)

Notice that the centers of the two populations need not be equal; that's what the test is going to decide.

TECHNICAL STUFF

More sophisticated methods for checking conditions two and three fall outside the scope of this book. However, checking the conditions as I describe above allows you to find and stay clear of any major problems.

## Stepping through the test

The rank sum test is a test for the equality of the two population medians — call them $\eta_1$ and $\eta_2$. After you've checked the conditions for using the rank sum test, you conduct the test by following these steps. (Note: Minitab can run this test for you, but knowing what it is doing behind the scenes is important.)

1. **Set up Ho: $\eta_1 = \eta_2$ versus Ha: $\eta_1$ >, <, (one-sided test) or $\neq \eta_2$ (two-sided test) depending on whether you're looking for a positive difference, a negative difference, or any difference between the two population medians.**

2. **Think of the data as one combined group and assign overall ranks to the values from lowest (rank = 1) to highest.**

   In the case of ties, give both values the average of the ranks they would have normally been given. For example, suppose the third and fourth numbers (in order) both have the same rank; assign each of them a rank of 3.5 (the average of 3 and 4.)

3. **Sum the ranks assigned to the sample that has the smallest sample size; call this statistic *T*.**

   The reason the smallest sample is used is convention — statisticians like to be consistent. If the sample sizes are equal, sum the ranks for the first sample to get *T*. If the value of *T* is small (relative to the total sum of all the ranks from both data sets), that means the numbers from the first sample tend to be smaller than the second sample, hence the median of the first population may be smaller than the median of the second one.

4. **Look at Table A-4, the rank sum table (in the Appendix), and find the column and row for the sample sizes of group one and two, respectively.**

   You see two critical values, $T_L$ (the lower critical value) and $T_U$ (the upper critical value). These critical values are the boundaries between rejecting Ho and not rejecting Ho.

5. **Compare your test statistic, *T*, to the critical values on Table A-4 to conclude whether you can reject Ho — that the population medians are different.**

   The method you use to compare these values depends on the type of test you're conducting:

   - **One-sided test (Ha has a > or < sign in it):** Table A-4 shows the critical values for $\alpha$ level 0.025. For a right-sided test (that means you have Ha: $\eta_1 > \eta_2$) reject Ho if $T \geq T_U$. For a left-sided test (that means where Ha: $\eta_1 < \eta_2$) reject Ho if $T \leq T_L$. If you reject Ho, conclude that the population medians are different and that one of them is greater than the other depending on Ha. (Otherwise you can't conclude that there's a difference in their medians.)

- **Two-sided test:** Table A-4 shows the critical values for $\alpha$ level 0.05. Reject Ho if $T$ falls outside of the interval $(T_L, T_U)$; that is, reject Ho if $T \leq T_L$ or $\geq T_U$. Conclude that the population medians are not equal. (Otherwise you can't conclude there is a difference in their medians.)

To conduct a rank sum test in Minitab, enter your data from the first sample in Column 1, and your data from the second sample in Column 2. Go to Stat>Nonparametrics>Mann-Whitney. Click on the name of your Column 1 variable; it appears in the First Sample box. Click on the name of your Column 2 variable; it appears in the Second Sample box. Under Alternative, there is a pull-down menu to select whether your Ha is not equal, greater than, or less than (as indicated by your particular problem). Click OK, and the test is done.

## Stepping up the sample size

After the sample sizes reach a certain point, the table values run out. Table A-4 (which shows the critical values for rejecting Ho in the rank sum test) only shows the critical values for sample sizes between three and ten. If both sample sizes are larger than ten, you use a two-sample $Z$-test to get an approximation for your answer. That's because for large sample sizes the test statistic $T$ for the rank sum test resembles a normal distribution. (So why not use it? It's a lot easier!) The larger the two sample sizes are, the better the approximation will be.

So if both sample sizes are more than ten, you conduct steps one through three of the rank sum test as before. Then, instead of looking up the value of $T$ on Table A-4 in step four of the rank sum test, you change it to a $Z$-value (a value on the standard normal distribution) by subtracting its mean and dividing by its standard error.

The formula you use to get this $Z$-value for the test statistic is

$$Z = \frac{T - \frac{n_1(n_1+n_2+1)}{2}}{\sqrt{\frac{n_1 n_2(n_1+n_2+1)}{12}}}$$

, where $T$ is given by step three in the previous section, $n_1$ is the sample size for the first data set (taken from the first population) and $n_2$ is the sample size for the second data set (taken from the second population). After you have the $Z$-value, follow the same procedures that you do for any test involving a $Z$-value, such as the test for two population means.

That is, find the *p*-value by looking up the *Z*-value on the bottom row of the *t*-table, which you can find in Table A-1 (in the Appendix), and finding the area beyond it. (If the test is
a two-sided test, double the *p*-value.) If your *p*-value is less than $\alpha$, reject Ho. Otherwise fail to reject Ho.

In the case where *n* is large and you use a *Z*-value for the test statistic, you can still use Minitab (in fact, that is recommended to save the tedium of working through a big example by hand). The Minitab directions are shown just after the steps earlier in this section.

# Performing a Rank Sum Test: Which Real Estate Agent Sells Homes Faster?

Suppose you want to choose a real estate agent to sell your house, and two agents are in your area. Your most important criteria is to get the house sold fast, so you decide to find out whether one agent sells homes faster. You choose a random sample of eight homes each agent sold in the last year, and for each home, you record the number of days it was on the market before being sold. You can see the data in Table 18-1.

**Table 18-1 Days on the Market for Homes Sold by Two Real Estate Agents**

| *Suzy Sellfast* | *Tommy Nowait* |
|---|---|
| 48 | 109 |
| 97 | 145 |
| 103 | 160 |
| 117 | 165 |
| 145 | 185 |
| 151 | 250 |
| 220 | 251 |
| 300 | 350 |

Check out the data summarized in *boxplots* (a graph summarizing the data by showing its minimum, first quartile, median, third quartile, and maximum values) in Figure 18-1a and the descriptive statistics in Figure 18-1b. In the following sections, you use this data to see the rank sum test in action. Be prepared to be amazed.

To make two boxplots side by side in Minitab, go to Graph>Boxplots>Simple Multiple Y's. Click on each of your two variables in the left-hand box; they will appear in the right-hand Variables box. Click OK.

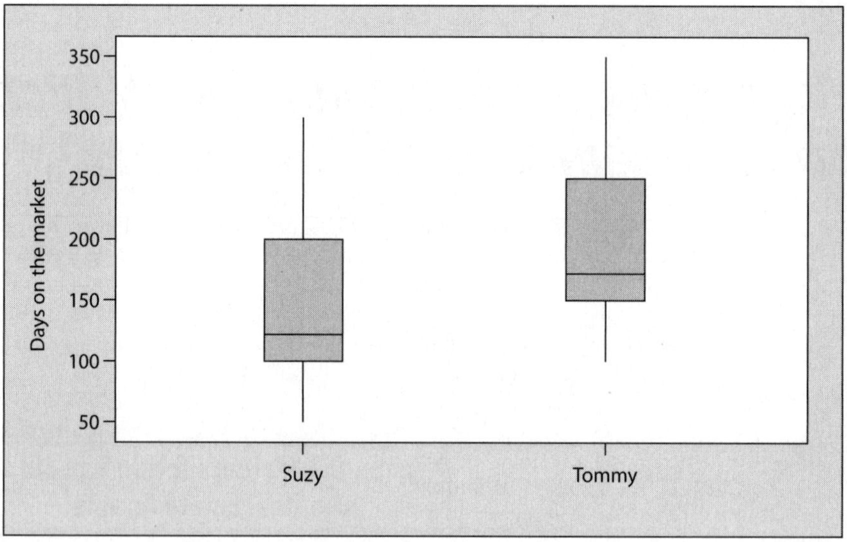


a

Descriptive Statistics: Suzy, Tommy

| Variable | Total Count | Mean | StDev | Minimum | Median | Maximum |
|---|---|---|---|---|---|---|
| Suzy | 8 | 147.6 | 79.2 | 48.0 | 131.0 | 300.0 |
| Tommy | 8 | 201.9 | 77.4 | 109.0 | 175.0 | 350.0 |

b

**Figure 18-1:** Boxplots and descriptive statistics for real estate agent data.

## Checking the conditions for this test

Checking the conditions, you know that the data from the two samples are independent, assuming that Suzy and Tommy are competitors. Next, the boxplots in Figure 18-1a show the same basic shape and amount of variability for each data set. (You don't have enough data to make histograms to check this further.) So based on this data, it isn't unreasonable to assume that the two population distributions of days on the market are the same for the two agents. In Figure 18-1b, the sample standard deviations are close: 79.2 days for Suzy and 77.4 days for Tommy. Because the data meets the conditions for the rank sum test, you can have the confidence to go ahead and apply it to analyze your data.

To find descriptive statistics (such as the standard deviation) in Minitab, go to Stat>Basic Statistics>Display Descriptive Statistics. Click on Options. Click on the box for each statistic you want to calculate. If a box is checked for a statistic you don't want, click on it again and the check mark disappears.

Figure 18-1b shows that the median for Suzy (131 days on the market) is less than the median for Tommy (175 days). It may appear Suzy sells homes faster than Tommy. However, the results aren't exactly clear-cut. A portion of the two boxplots (Figure 18-1a) overlap with each other. You may not be able to declare Suzy the clear winner as being the fastest real estate agent. You need a hypothesis test to make that final determination.

## Testing the hypotheses

The null hypothesis for the real estate agent test (from previous sections) is Ho: $\eta_1 = \eta_2$, where $\eta_1$ = median days on the market for the population of all Suzy's homes sold in the last year, and $\eta_2$ = median days on the market for the population of all Tommy's homes sold in the last year. The alternative hypothesis is Ha: $\eta_1 \neq \eta_2$.

After you looked at the data, you developed a hunch that if one of the agents sold homes faster, it was Suzy. However, before you saw the data, you had no preconceived notion as to whom was faster. You must base your Ho and Ha on what your thoughts were *before* you looked at the data, not after. Setting up your hypotheses after you collect the data is unfair and unethical.

After you determine your Ho and Ha, the time has come to test your data. So, keep reading to figure out what this test looks like in a real-life example.

### Combining and ranking

The first step in the data analysis is to combine all the data together and rank the days on the market from lowest (rank = 1) to highest. You can see the overall ranks for the combined data in Table 18-2.

In the case of ties, you give both of the values the average of the ranks they normally would have received. You can see in Table 18-2 that two values of 145 are in the data set. Because they represent the sixth and seventh numbers in the ordered data set, you give each of them the same rank of $(6+7)/2 = 6.5$.

**Table 18-2 Ranks of Combined Data from the Real Estate Example**

| *Suzy Sellfast* | *Overall Rank* | *Tommy Nowait* | *Overall Rank* |
|---|---|---|---|
| 48 | 1 | 109 | 4 |
| 97 | 2 | 145 | 6.5 |
| 103 | 3 | 160 | 9 |
| 117 | 5 | 165 | 10 |
| 145 | 6.5 | 185 | 11 |
| 151 | 8 | 250 | 13 |
| 220 | 12 | 251 | 14 |
| 300 | 15 | 350 | 16 |

### *Finding the test statistic*

After you've ranked your data, you can determine which group is group one, so you can find your test statistic, $T$. Because the sample sizes are equal, let group one be Suzy, because her data is given first. Now sum the ranks from Suzy's data set. The sum of Suzy's ranks is $1 + 2 + 3 + 5 + 6.5 + 8 + 12 + 15 = 52.5$; this value of $T$ is your rank sum test statistic.

### *Determining whether you can reject $H_o$*

Suppose you want to use $\alpha = 0.05$ for this test; using this cutoff means that you use Table A-4 (see Appendix), because you have a two-sided test at level $\alpha = 0.05$. Looking at Table A-4, you go to the column for $n_1 = 8$ and the row for $n_2 = 8$. You see $T_L = 49$ and $T_U = 87$. You reject Ho if $T$ is outside this range; in other words, reject Ho if $T \leq T_L = 49$ or if $T \geq T_U = 87$. Your statistic $T = 52.5$ doesn't fall outside this range; you don't have enough evidence to reject Ho at the $\alpha = 0.05$ level. So you can't say that you see a difference in the median days on the market for Suzy and Tommy.

These results may seem very strange given the fact that the medians for the two data sets were so different: 131 days on the market for Suzy compared to 175 days on the market for Tommy. However you have two strikes against you in terms of being able to find a real difference here:

- **The sample sizes are quite small (only eight in each group).** A small sample size makes it very hard to get enough evidence to reject Ho.
- **The standard deviations are both in the high 70s, which is quite large compared to the medians.**

Both of these problems make it hard for the test to actually find anything through all the variability the data shows.

To conduct the rank sum test by using Minitab, click on Stat>Nonparametric> Mann-Whitney. Select your two samples and choose your alternate Ha as >, <, or $\neq$. The Confidence Level is equal to one minus your value of $\alpha$. After you make all of these settings, click on OK.

Figure 18-2 shows the Minitab output when you conduct the rank sum test on the real estate data. To interpret the results in Figure 18-2, you must note that the Mann-Whitney test is just another word for the rank sum test. Also, Minitab writes ETA rather than $\eta$ for the medians. The results at the bottom of the output say that the test for equal (versus nonequal) medians is significant at the level 0.1149, when adjusting for ties. This is your *p*-value adjusted for ties. (Note that if no ties are present in your data, you use the results just above that line. That gives you the *p*-value not adjusted for ties.)

**Mann-Whitney Test and CI: Suzy, Tommy**

```
        N   Median
Suzy    8    131.0
Tommy   8    175.0


Point estimate for ETA1-ETA2 is -49.0
95.9 Percent CI for ETA1-ETA2 is (-137.0, 36.0)
W = 52.5
Test of ETA1 = ETA2 vs ETA1 not = ETA2 is significant at 0.1152
The test is significant at 0.1149 (adjusted for ties)
```

**Figure 18-2:** Using the rank sum test to figure out who sells homes faster.

To make your final conclusion, compare your *p*-value to your pre-specified level of $\alpha$ (typically 0.05.) If your $\alpha$ level is 0.1149 (or larger), you reject Ho; otherwise you can't. In this case, because 0.1149 is greater then 0.05, you can't reject Ho. That means you don't have enough evidence to say the population medians for days on the market for Suzy's versus Tommy's houses are different based on this data. These results confirm your conclusions from the previous section.

The Minitab output in Figure 18-2 also provides a confidence interval for the difference in the medians between the two populations, based on the data from these two samples. The difference in the sample medians (Suzy – Tommy) is 131.0 – 175.0 = –44.0. Adding and subtracting the margin of error (these calculations are beyond the scope of this book), Minitab finds the confidence interval for the difference in medians (Suzy – Tommy) is –137.0, +36.0. The difference in the population medians could be anywhere from –137.0 to 36.0. Because 0, the value in Ho, is in this interval, you can't reject Ho in this case. So again, you can't say that the medians are different, based on this (limited) data set.

## Using a rank sum test to compare Olympic judges

Rank sum tests can be used to compare two groups of judges of a competition, to see whether there is a difference in their scores. For example, in the Olympic ice-skating events, the gender of the judges is sometimes suspected to play a role in the scores they give to certain skaters. Suppose you have a men's ice-skating competition and you have ten judges: five males and five females. You want to know whether male and female judges score the competitors in the same way, so you do a rank sum test to compare their median scores. Your hypotheses are Ho: male and female judges have the same median score versus Ha: they have different median scores. For your sample, you let each judge score the same individual. You rank their scores in order from lowest to highest and label M for a male judge and F for a female judge. Your results are the following: F M M M M F F F F M. The value of the test statistic $T$ is the sum of the ranks for group one (say the males), which gives you $T = 2 + 3 + 4 + 5 + 10 = 24$. Now compare that to the critical values in Table A-4 (Appendix), where both sample sizes equal five, and you get $T_L = 18$ and $T_U = 37$. Because your test statistic, $T = 24$, is inside this interval, you fail to reject Ho: judging is the same for male and female judges. You just don't have enough evidence to say that they differ.

# Chapter 19

# Do the Kruskal-Wallis and Rank the Sums with Wilcox

### *In This Chapter*

- Comparing more than two population medians with the Kruskal-Wallis test
- Determining which populations are different by using the Wilcoxon rank sum test

Statisticians who are in the nonparametrics business make it their jobs to always find a nonparametric equivalent to a parametric procedure (one that doesn't depend on the normal distribution). And in the case of comparing more than two populations, these stats superheroes didn't let us down. In this chapter, you see how the Kruskal-Wallis test works to compare more than two populations as a nonparametric procedure. If Kruskal-Wallis tells you at least two populations differ, you also figure out how to use the Wilcoxon rank sum test to determine which population is different.

## *Doing the Kruskal-Wallis Test to Compare More than Two Populations*

The Kruskal-Wallis test compares the medians of several (more than two) populations to see whether or not they are different. The basic idea of Kruskal-Wallis is to collect a sample from each population, rank all the combined data from smallest to largest, and then look for a pattern in how those ranks are distributed among the various samples. For example, if one sample gets all the low ranks and another sample gets all the high ranks, perhaps their population medians are different. Or if all the samples have an equal mix of all the ranks, perhaps the medians of the populations are all deemed to be the same. In this section, you see exactly how the Kruskal-Wallis test is conducted using ranks and sums and all that good stuff, and you see it applied to an example comparing airline ratings.

Suppose your boss flies a lot, and she wants you to determine which of three airlines gets the best ratings from customers. You know that ratings involve data that is just not normal (pun intended), so you opt to use the Kruskal-Wallis test. You take three random samples of nine people each from three different airlines. You ask each person to rate his satisfaction with the one airline for which you chose that person to rate. Each person uses a scale from 1 (the worst) to 4 (the best). You can see the data from your samples in Table 19-1.

**Table 19-1 Customer Ratings of Three Airlines**

| *Airline A Rating* | *Airline B Rating* | *Airline C Rating* |
|---|---|---|
| 4 | 2 | 2 |
| 3 | 3 | 3 |
| 4 | 3 | 3 |
| 4 | 3 | 2 |
| 3 | 4 | 2 |
| 3 | 4 | 1 |
| 2 | 3 | 3 |
| 3 | 4 | 2 |
| 4 | 3 | 2 |

In looking at the data in Table 19-1, it appears that airlines A and B have better ratings than airline C. However, the data has a lot of variability in it, so you have to conduct a hypothesis test before you can make any general conclusions beyond this data set.

You may be thinking of using ANOVA to analyze this data (the test that compares the means of several populations and is found in Chapter 9). But the data from each airline is ratings from 1 to 4, and this blows the strongest condition of ANOVA — the data from each population must follow a normal distribution. (A *normal distribution* is continuous, meaning it takes on all real numbers in a certain range. Data that are whole numbers like 1, 2, 3, and 4 don't fall under this category.)

But don't sweat; a nonparametric alternative fits the bill. The Kruskal-Wallis test compares the medians of several (more than two) populations to see whether they are all the same or not. In other words, it's like ANOVA, except it's done with medians not means.

In this section, you discover how to check the conditions of the Kruskal-Wallis test, set it up, and carry it out step by step.

## Checking the conditions

Following are all of the conditions of the Kruskal-Wallis test that must be met:

- The random samples taken from each population are independent. (This means matched-pairs data like in Chapter 17 are out of this picture.)
- All the populations have the same distribution. (That is, their shapes are the same as seen on a histogram.)
- The variances of the populations are the same. That means the amount of spread in the population values is the same from one population to the next.

Note that these conditions mention shape and spread, but they don't mention the center of the distributions. That's what the test is trying to determine, whether the populations are centered at the same place.

TECHNICAL STUFF

In nonparametrics, you often see the word *location* in reference to a population distribution rather than the *center,* although the two words mean about the same thing. Location indicates where the distribution is sitting on the number line. If you have two bell-shaped curves with the same variance, and one has mean 10 and the other has mean 15, the second distribution is located five units to the right of the first. In other words, it's location is a five-unit shift to the right of the first distribution. In nonparametrics, where you don't have bell-shaped distributions, you typically use the median as a measure of location (center) of a distribution. So throughout this discussion, you could use the word *median* instead of location (although location leaves it a bit more open).

Regarding the airline survey, you know that the samples are independent, because you didn't use the same person to rate more than one airline. The other two conditions have to do with the distributions the samples came from; each population must have the same shape and the same spread. You can examine both conditions by looking at boxplots of the data (see Figure 19-1) and descriptive statistics, such as the median, standard deviation, and the rest of the summary statistics making up the boxplots (see Figure 19-2).

The boxplots in Figure 19-1 all have the same shape, and their standard deviations, shown in Figure 19-2, are very close. All of this evidence taken together allows you to go ahead with the Kruskal-Wallis test. (Now looking at the overlap in the boxplots for airlines A and B, in Figure 19-1, you can also make an early prediction that airlines A and B have similar ratings; whether C is different enough from A and B is impossible to say without running the hypothesis test.)

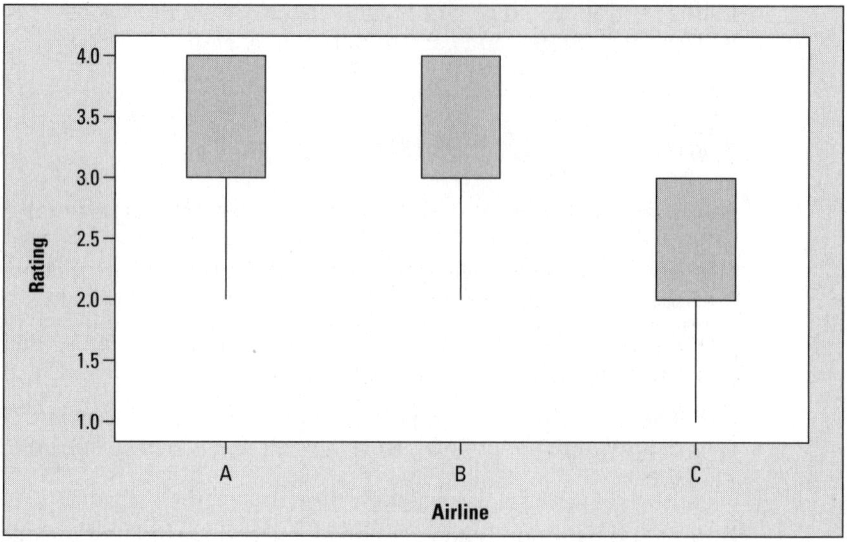


**Figure 19-1:** Boxplots comparing the ratings of three airlines.

**Figure 19-2:** Descriptive statistics comparing the ratings of three airlines.

**Descriptive Statistics: Rating**

| Variable | Airline | StDev | Minimum | Q1 | Median | Q3 | Maximum |
|---|---|---|---|---|---|---|---|
| Rating | A | 0.707 | 2.000 | 3.000 | 3.000 | 4.000 | 4.000 |
| | B | 0.667 | 2.000 | 3.000 | 3.000 | 4.000 | 4.000 |
| | C | 0.667 | 1.000 | 2.000 | 2.000 | 3.000 | 3.000 |

TIP

Either a boxplot or a histogram can tell you about the shape and spread of a distribution (as well as the center). The *boxplot* is a common type of graph to use for nonparametric procedures because it displays the median (the nonparametric statistic of choice) rather than the mean. A *histogram* is at its best showing the shape of the data; it doesn't directly tell where the center is — you just have to eyeball it. Go ahead with the boxplot versus the histogram for the airline data.

COMPUTER OUTPUT

To make boxplots of each sample of data show up side by side on one graph (called side-by-side boxplots, cleverly) in Minitab, click on Graph>Box Plots and select the Multiple Y's Simple version. In the left-hand box, click on each of the column names for your data sets. They each appear in the Graph Variables window on the right. Click OK and you get a set of boxplots that are side by side, all on the same graph using the same scale (slick, huh?).

## Setting up the test

The Kruskal-Wallis test assesses Ho: All *k* populations have the same location versus Ha: The location of at least two of the *k* populations are different. (Here, *k* is the number of populations you're comparing.)

In Ho, you see that all the populations have the same location (which means they all sit on top of each other on the number line and are in essence the same population). Ha is looking for the opposite situation in this case. However, the opposite of "the locations are all equal" isn't "the locations are all different." The opposite is that at least two of them are different. Failure to recognize this difference will lead you to believe all the populations differ when, in reality, there may only be two that differ, and the rest are all the same. That's why you see Ha stated the way it is in the Kruskal-Wallis test. (The same idea holds for comparing means using ANOVA; see Chapter 9.)

For the airline satisfaction example (see Table 19-1), your setup looks like this: Ho: The satisfaction ratings of all three airlines have the same median versus Ha: The median satisfaction ratings of at least two airlines are different.

## Conducting the test step by step

After you've determined your hypotheses, and checked the conditions, you must carry out the test. Here are the steps for conducting the Kruskal-Wallis test, using the airline example to show how each step works:

1. **Rank all the numbers in the entire data set from smallest to largest (using all samples combined); in the case of ties, use the average of the ranks that the values would have normally been given.**

   For an example of a tie, say that on a scale from 1 to 4, the observations 1, 1, 1 would normally have gotten ranks 1, 2, 3 if they were different, but because they're equal, give each one the average of 1, 2, 3, which is $\frac{(1+2+3)}{3} = 2$. Figure 19-3 shows the results for ranking and summing the data in the airline example.

   In Figure 19-3, you can see how to rank the ties. For example, you have only one 1, which is given rank 1. Then you have seven 2s, which normally would have gotten ranks 2, 3, 4, 5, 6, 7, and 8. Because the 2s are all equal, you give each of them the average of all these ranks, which is $\frac{(2+3+4+5+6+7+8)}{7} = 5$. Similarly, you see twelve 3s, whose ranks would be 9 through 20. Because they're all equal, give them each a rank equal to $\frac{(9+10+\ldots+20)}{12} = 14.5$. Finally, you see seven 4s, each with rank 24, which is the average of their would-be ranks, ranging from 21 to 27.

| Airline A | | Airline B | | Airline C | |
|---|---|---|---|---|---|
| Rating | Rank | Rating | Rank | Rating | Rank |
| 4 | 24 | 2 | 5 | 2 | 5 |
| 3 | 14.5 | 3 | 14.5 | 3 | 14.5 |
| 4 | 24 | 3 | 14.5 | 3 | 14.5 |
| 4 | 24 | 3 | 14.5 | 2 | 5 |
| 3 | 14.5 | 4 | 24 | 2 | 5 |
| 3 | 14.5 | 4 | 24 | 1 | 1 |
| 2 | 5 | 3 | 14.5 | 3 | 14.5 |
| 3 | 14.5 | 4 | 24 | 2 | 5 |
| 4 | 24 | 3 | 14.5 | 2 | 5 |
| | $T_1 = 159$ | | $T_2 = 149.5$ | | $T_3 = 69.5$ |

**Figure 19-3:** Rankings and rank sum for the airline example.

2. **Total the ranks for each of the samples; call those totals $T_1, T_2, \ldots, T_k$, where $k$ is the number of populations.**

   The totals of the ranks in each column of Figure 19.3 for the airline data are $T_1 = 159$, $T_2 = 149.5$, and $T_3 = 69.5$. In the steps that follow, you use these rank totals in the Kruskal-Wallis test statistic (denoted KW). (Note $T_1$ and $T_2$ are close to equal, but $T_3$ is much lower, giving the idea that airline C may be the odd man out.)

3. **Calculate the Kruskal-Wallis test statistic, $KW = \frac{12}{n(n+1)} \sum \frac{T_j^2}{n_j} - 3(n+1)$, where $n$ is the total number of observations (all sample sizes combined).**

   Continuing with the airline example, the Kruskal-Wallis test statistic is $KW = \frac{12}{27(27+1)}\left(\frac{159^2}{9} + \frac{149.5^2}{9} + \frac{69.5^2}{9}\right) - 3(27+1)$, which equals $0.0159 * 5{,}829.056 - 3(28) = 8.52$.

4. **Find the $p$-value.**

   You find the $p$-value for your KW test statistic by comparing it to the Chi-square distribution with $k - 1$ degrees of freedom (Table A-3 in the Appendix). For the airline example, you look at the Chi-square table (Table A-3 in Appendix) and find the row for with $3 - 1 = 2$ degrees of freedom. Then look at where your test statistic (8.52) falls in that row. Because 8.52 lies between 7.38 and 9.21 (shown on the table in row two) that means the $p$-value for 8.52 lies between 0.025 and 0.010 (shown in their respective column headings.)

5. **Make your conclusion about whether you can reject Ho by examining the *p*-value.**

   You can reject Ho: All populations have the same location, in favor of Ha: At least two populations have differing locations, if the *p*-value associated with KW is $< \alpha$, where $\alpha$ is 0.05 (or your prespecified $\alpha$ level). Otherwise, you must fail to reject Ho.

   Following the airline example, because the *p*-value is between 0.010 and 0.025, which are both less than $\alpha = 0.05$, you can reject Ho. You conclude that the ratings of at least two of the three airlines are different.

To conduct the Kruskal-Wallis test by using Minitab, enter your data in two columns, the first column represents the actual data values and the second column represents which population the data came from (for example, 1, 2, 3). Then click on Stat>Nonparametrics>Kruskal-Wallis. In the left-hand box, click on column one; it appears on the right side as your *response variable*. Then click on column two in the left-hand box. This column appears on the right side as the *factor variable*. Click OK, and the KW test is done. The main results of the KW test are shown in the last two lines of the Minitab output.

The results of the Minitab data analysis of the airline data are shown in Figure 19-4. On the second-to-last line of Figure 19-4, you can see the KW test statistic for the airline example is 8.52, which matches the one you found by hand (whew!). The exact *p*-value from Minitab is 0.014.

**Kruskal-Wallis Test: Rating versus Airline**

```
Kruskal-Wallis Test on Rating

Airline           N    Median    Ave Rank       Z
A                 9     3.000        17.7    1.70
B                 9     3.000        16.6    1.21
C                 9     2.000         7.7   -2.91
Overall          27                  14.0

H = 8.52    DF = 2    P = 0.014
H = 9.70    DF = 2    P = 0.008 (adjusted for ties)
```

**Figure 19-4:** Comparing ratings of three airlines by using the Kruskal-Wallis test.

However, quite a few ties are in this data set, and the formulas adjust a bit for that (in ways that go outside the scope of this book). Taking those ties into account, the computer gives you KW = 9.70 with a *p*-value of 0.008. The total evidence here says the same result loud and clear — reject Ho: The ratings for the three airlines have the same location. You conclude that the ratings of at least two of the airlines are different. (But which ones? The answer comes in the next section.)

## Leveling the playing field

Most people want life — from football to food portions — to be fair. And nothing appears to be more unfair than car insurance rates, right? You've heard the ads; one company claims to offer the lowest possible rates one day and a competing company makes the same claim the very next day. Who can you believe? You decide to grab the wheel and run your own test. You take a random sample of 20 different car and driver combinations (for example, a 40-year-old female with a Ford pickup, or a 78-year-old lady driving a Caddy) and you get the corresponding car insurance estimates from each company for each car and driver combo based on a six-month premium. Knowing that the distribution of prices for each company has no real reason to be normal (as in distribution) you go for the Kruskal-Wallis test of their medians. You rank all the premiums from smallest to largest, you sum the ranks that correspond to estimates from each company, and you compare them using the KW statistic. In the end, you might very well find that the companies' prices don't look that different after all, because the prices they talk about in their advertisements represent a selective sample of the population of all their prices, and your sample gets more at the heart of the pricing that is actually going on overall. The moral of the story is don't listen to everything you hear about car insurance rates. Get a cross section of prices and do the Kruskal-Wallis. Your pocketbook will thank you for it.

# Pinpointing the Differences: The Wilcoxon Rank Sum Test

Suppose you reject Ho in the Kruskal-Wallis test. That means you have enough evidence to conclude that at least two of the populations have different medians. But you don't know which ones are different. When someone finds that a set of populations don't all share the same median, the next question is very likely to be, "Well then, which ones are different?" To find out which populations are different after the Kruskal-Wallis test has rejected Ho, you can use the Wilcoxon rank sum test (also known as the Mann-Whitney test; refer to Chapter 18).

You can't go looking for differences in specific pairs of populations until you've first established that the populations aren't all the same (that is, Ho is rejected in the Kruskal-Wallis test). If you don't make this check first, you can encounter a ton of problems, not the least of which being much-increased chance of making the wrong decision.

In the following sections, you see how pairwise comparisons are conducted and interpreted in order to find out where the differences lie among the *k* population medians you're studying.

## Pairing off with pairwise comparisons

The rank sum test is a nonparametric test that compares two population locations (for example, their medians). When you have more than two populations, you conduct the rank sum test on every pair of populations in order to see whether differences exist. This procedure is called conducting *pairwise comparisons* or *multiple comparisons.* (See Chapter 10 for info on the parametric version of multiple comparisons.) For example, because you're comparing three airlines in the airline satisfaction example (see Table 19-1), you have to run the rank sum test three times to compare airlines A and B, A and C, and B and C, respectively. So you need three pairwise comparisons to figure out which populations are different.

REMEMBER

To determine how many pairs of comparisons you need if you're given $k$ populations, you use the formula $\frac{k(k-1)}{2}$. You have $k$ populations to choose from first, and then $k - 1$ populations left to compare them with. Finally, you don't care what the order is among the populations (as long as you keep track of them); so you divide by two because you have two ways to order any pair (for example, comparing A and B gives you the same results as comparing B and A). In the airlines example, you have $k = 3$ populations, so you should have $\frac{k(k-1)}{2} = \frac{3(3-1)}{2} = 3$ pairs of populations to compare, which matches what was determined previously. (For more information and examples on how to count the number of ways to choose or order a group of items by using permutations and combinations, see another book I authored, *Probability For Dummies* [Wiley].)

## Carrying out comparison tests to see who's different

The Wilcoxon rank sum test assesses Ho: The two populations have the same location versus Ha: The two populations have different locations. Here are the general steps for using the Wilcoxon rank sum test for making comparisons (for detailed step-by-step instructions for the Wilcoxon rank sum test see Chapter 18):

1. **Check the conditions for the test by using descriptive statistics and histograms for the last two and proper sampling procedures for the first one:**
    - The two samples must be from independent populations
    - The populations must have the same distribution (shape)
    - The populations must have the same variance

2. **Set up your Ho: Medians are equal versus Ha: Medians aren't equal.**
3. **Combine all the data and rank the values from smallest to largest.**
4. **Add up all the ranks from the first sample (or the smallest sample if the sample sizes are not equal).**

   This result is your test statistic, *T*.
5. **Compare *T* to the critical values in Table A-4 (Appendix) in the row and column corresponding to the two sample sizes.**

   If *T* is at or beyond the critical values (less than or equal to the lower one or greater than or equal to the upper one), reject Ho and conclude the two population medians are different. Otherwise, you can't reject Ho.
6. **Repeat Steps 1–5 on every pair of samples in the data set and draw conclusions.**

   Sort through all the results to see the overall picture of which pairs of populations have the same median and which ones don't.

To conduct the Wilcoxon rank sum test for pairwise comparisons in Minitab, refer to Chapter 18. Note that Minitab calls this test by its other name, the Mann-Whitney test.

You can see the Minitab results of the three Wilcoxon rank sum tests comparing airlines A and B, A and C, and B and C, respectively, in Figures 19-5a, 19-5b, and 19-5c.

Before you make any judgments about your hypotheses, you must analyze your data. Figure 19-5a compares the ratings of airlines A and B. The *p*-value (adjusted for ties) is 0.7325, which is much higher than the 0.05 you need to reject Ho. So you can't conclude that airlines A and B have satisfaction ratings with different medians. Figure 19-5b shows that the *p*-value for comparing airlines A and C is 0.0078. Because this *p*-value is a lot smaller than the typical $\alpha$ level of 0.05, this is very convincing evidence that airlines A and C don't have the same median ratings. Figure 19-5c also has a small *p*-value (0.0107), which gives evidence that airlines B and C have significantly different ratings.

## Examining the medians to see how they're different

Now that you know two or more populations have different medians, the next question to answer is how they are different; which one has the higher median, which one has the lower median. In this section, you see how to take the results of your pairwise comparisons combined with some descriptive statistics to get your answers.

**Mann-Whitney Test and CI: Airline A, Airline B**

```
    N   Median
A   9    3.000
B   9    3.000

Point estimate for ETA1-ETA2 is -0.000
95.8 Percent CI for ETA1-ETA2 is (-1.000,1.000)
W = 89.5
Test  of ETA1 = ETA2 vs ETA1 not =  ETA2 is significant at 0.7573
The test is significant at 0.7325   (adjusted for ties)
```

a

**Mann-Whitney Test and CI: Airline A, Airline C**

```
    N   Median
A   9    3.000
C   9    2.000

Point estimate for ETA1-ETA2 is 1.000
95.8 Percent CI for ETA1-ETA2 is (0.000,2.000)
W = 114.5
Test of ETA1 = ETA2 vs ETA1 not = ETA2 is significant at 0.0118
The test is significant at 0.0078 (adjusted for ties)
```

b

**Mann-Whitney Test and CI: Airline B, Airline C**

```
    N   Median
B   9    3.000
C   9    2.000

Point estimate for ETA1-ETA2 is 1.000
95.8 Percent CI for ETA1-ETA2 is (0.000,2.000)
W = 113.0
Test of ETA1 = ETA2 vs ETA1 not = ETA2 is significant at 0.0171
The test is significant at 0.0107 (adjusted for ties)
```

c

**Figure 19-5:** Wilcoxon rank sum tests comparing ratings of two airlines at a time.

After you've rejected Ho for a multiple comparison, that means the two populations you examined have different medians. There are two ways to proceed from here to see how the medians differ:

- You can look at side-by-side boxplots of all the samples and compare their medians (located at the line in the middle of each box).
- You can calculate the median of each sample and see which ones are higher and which ones are lower (from the populations you have concluded are statistically different).

From the previous section, you see that the pairwise comparisons for the airline data conducted by Wilcoxon rank sum tests conclude that the ratings of airlines A and B aren't found to be different, but both of them are found to be different from airline C.

But you can say even more; you can say how the differing airline compares to the others. Going back to Figure 19-2, you see the medians of both airlines A and B are 3.0, while the median of airline C is only 2.0. That difference means airlines A and B have similar ratings, but airline C has lower ratings than A and B.

The boxplots in Figure 19-1 confirm these results. By looking at these boxplots first, you may have had an idea that A and B were the same, but you didn't know whether airline C was statistically significantly different from airlines A and B. And now you know it is.

# Chapter 20

# Pointing Out Correlations with Spearman's Rank

### In This Chapter

- Understanding correlation from a nonparametric point of view
- Finding and interpreting Spearmen's rank correlation

Data analysts commonly look for and try to quantify relationships between two variables, *x* and *y*. Depending on the type of data you're dealing with in *x* and *y*, there are different procedures to use for quantifying their relationship.

When *x* and *y* variables are *quantitative* (that is, their possible outcomes are measurements or counts), the correlation coefficient (also known as the *Pearson's correlation coefficient*) measures the strength and direction of their linear relationship. (See Chapter 4 for all the info on Pearson's correlation coefficient, denoted by *r*.) If *x* and *y* are both *categorical* variables (their possible outcomes are categories that have no numerical meaning; for example male and female), you use Chi-square procedures and conditional probabilities to look for and describe their relationship. All of that machinery is laid out in Chapters 13 and 14.

Then there is a third type of variable, called *ordinal* variables (their values fall into categories, but the possible values can be placed into an order and given a numerical value that has some meaning, for example, grades on a scale of A = 4, B = 3, C = 2, D = 1, and E = 0 or a student's evaluation of a teacher on a scale from best [5] to worst [1]). To look for a relationship between two ordinal variables like these, use Spearman's rank correlation; it's the nonparametric counterpart to Pearson's correlation coefficient (Chapter 4). In this chapter, you see why ordinal variables don't meet Pearson's conditions, and you see how to use and interpret Spearman's rank correlation to correctly quantify and interpret the relationship between two ordinal variables.

# Pickin' On Pearson and His Precious Conditions

Pearson's correlation coefficient is the most common correlation measure out there, and many data analysts think it's the only one out there. Trouble is, Pearson's correlation has certain conditions that must be met before using it. If those conditions are not met, Spearman's correlation is waiting in the wings. In this section, you see the conditions for Pearson's correlation and how they are easy pickin's for Spearman's rank correlation.

The Pearson correlation coefficient *r* (the correlation) is a number that measures the direction and strength of the linear relationships between two variables *x* and *y*. (For more info on the correlation, see Chapter 4.)

Several conditions have to be met for ol' Pearson:

- **The variables *x* and *y* must have a linear relationship (as shown on a scatterplot; see Chapter 4).**
- **Both variables *x* and *y* must be numerical (or quantitative).** That is, they must represent measurements with no restriction on their level of precision. For example, numbers with many places after the decimal point (such as 12.322 or 0.219) must be possible.
- **The *y* values must have a normal distribution for each *x*, with the same variance at each *x*.**

One of the most common instances where Pearson's conditions aren't met is when the two variables are ordinal. *Ordinal data* comes in categories that can be assigned numerical values that make sense. However, typically with ordinal variables, you won't see many different categories offered or compared for simplicity reasons. This means there won't be enough numerical values to try to build a linear regression model for two ordinal variables like you can with two quantitative variables. (Because there are typically not enough categories offered with an ordinal variable, Pearson's conditions aren't met.) That also makes condition three impossible.

As well, if you have a gender variable with categories male and female, you can assign the numbers 1 and 2 to each gender, but those numbers have no numerical meaning. Gender isn't an ordinal variable; rather it is a *categorical variable* (a variable that places individuals into categories only). Categorical variables, such as gender, also don't lend themselves to linear relationships, so they don't meet Pearson's conditions either. (To explore relationships between categorical variables, see Chapter 14.)

### Who are these guys? A look at the people behind the statistics

Some people are lucky enough to have a statistic actually named after them. Typically, the person who came up with the statistic in the first place, recognizing a need for it and coming up with a solution, gets the honor. If the new statistic gets picked up and used by others, it eventually takes on the name of its inventor.

Spearman's rank correlation is named after its inventor, Charles Edward Spearman, who lived from 1863 to 1945. He was an English psychologist who studied experimental psychology and worked in the area of human intelligence. He was a professor for many years at the University College London. Spearman followed closely the work of Francis Galton, who *originally* developed the concept of correlation. Spearman developed his rank correlation in 1904.

Pearson's correlation coefficient was developed several years prior, in 1893 by Karl Pearson, one of Spearman's fellow colleagues at University College London and another follower of Galton. Pearson and Spearman didn't get along. Pearson had an especially strong and volatile personality, and had problems getting along with quite a few people in fact. Such is the way of some of the more brilliant people of the 19th century.

# Scoring with Spearman's Rank Correlation

Spearman's rank correlation doesn't require the relationship between the variables $x$ and $y$ to be linear, nor does it require the variables to be numerical. You use Spearman's rank when the variables are ordinal and/or quantitative. Rather than examining a linear relationship between $x$ and $y$, Spearman's rank correlation tests whether two ordinal and/or quantitative variables are dependent (in other words, related to each other).

***Note:*** Spearman's rank applies to ordinal data only. To test to see if two categorical (and non-ordinal) variables are independent, you use a Chi-square test; see Chapter 14.

REMEMBER

Spearman's rank correlation is the same as Pearson's correlation except that it's calculated based on the ranks of the $x$ variable and the ranks of the $y$ variable rather than their actual values. You interpret the value of Spearman's rank correlation, $r_s$ the same way you interpret Pearson's correlation, $r$ (see Chapter 4). The values of $r_s$ can go between –1 and +1. The higher the magnitude of $r_s$ (in the positive or negative directions), the stronger the relationship between $x$ and $y$. If $r_s$ is zero, this indicates that $x$ and $y$ are independent. However, if the correlation between $x$ and $y$ is not zero, you can't say whether or not they're independent.

In this section, you see how to calculate and interpret Spearman's rank correlation and apply it to an example.

## Figuring Spearman's rank correlation

The notation for Spearman's rank correlation is $r_s$, where $s$ stands for Spearman. To find $r_s$, you do the steps listed in this section. Minitab does the work for you in steps two through six, although some professors may ask you to do the work by hand (not me of course).

1. **Collect the data in the form of pairs of values $x$ and $y$.**

2. **Rank the data from the $x$ variable where $1$ = lowest to $n$ = highest, where $n$ is the number of pairs of data in the data set. (This gives you a new set of data for the $x$ variable called the *ranks* of the $x$ values.)**

   If any of the values appear more than once, Minitab assigns each tied value the average of the ranks they would normally be given if they were not tied.

3. **Complete step two with the data from the $y$ variable. (This gives you a new data set called the *ranks* of the $y$-values.)**

4. **Find the standard deviation of the ranks of the $x$-values, using the usual formula for standard deviation, $s_x = \sqrt{\frac{\Sigma(x-\overline{x})^2}{n-1}}$; call it $s_x$. In a similar manner find the standard deviation of the ranks of the $y$-values using $s_y = \sqrt{\frac{\Sigma(y-\overline{y})^2}{n-1}}$; call it $s_y$.**

   (Note that n is the sample size, $\overline{x}$ is the mean of the ranks of the $x$ values, and $\overline{y}$ is the mean of the ranks of the $y$ values.

5. **Find the *covariance* of the $x$-$y$ values, using the formula $\text{Cov}(x,y) = \frac{\Sigma\Sigma(x-\overline{x})(y-\overline{y})}{n-1}$; call it $s_{xy}$.**

   The covariance of x and y is a measure of the total deviation of the $x$ and $y$ values from the point $(\overline{x}, \overline{y})$.

6. **Calculate the value of Spearman's rank correlation by using the formula $r_s = \frac{s_{xy}}{s_x s_y}$.**

Notice that the formula for Spearman's rank correlation is just the same as the formula for Pearson's correlation coefficient, except the data Spearman uses for his correlation formula is the ranks of *x* and the ranks of *y*, rather than the original *x*- and *y*-values as used by Pearson. So Spearman just cares about the order of the values of the *x*'s and the *y*'s, not their actual values.

To calculate Spearman's rank correlation straightaway by using Minitab, rank the *x*-values, rank the *y*-values, and then find the correlation of the ranks. That is, go to Data>Rank and click on the *x* variable to get *x* ranks. Then do the same thing to get the *y* ranks. Now go to Stat>Basic Statistics>Correlation, click on the two columns representing ranks, and click OK.

## Watching Spearman at work: Relating aptitude to performance

Knowing the process of how to calculate Spearman's rank correlation is one thing, but if you can apply it to real-world situations, you'll be the golden child of the statistics world (or at least your intermediate stats class). So, try to put yourself in this section's scenario to get the full effect of Spearman's rank correlation.

You're a statistics professor, and you give exams every now and then (it's a dirty job, but someone's got to do it). After looking at students' final grades over the years (yes, you're an old professor, or at least in your mid-forties), you notice that students who do well in your class tend to have a better aptitude (background ability) for math and statistics. You want to check out this theory, so you give students a math and statistics aptitude test on the first day of the course; you want to compare students' aptitude test scores with their final grades at the end of the course.

Now for the specifics. Your variables are $x$ = aptitude test score (using a 100-point pretest on the first day of the course) and $y$ = final grade, on a scale from 1 to 5 where 1 = F (failed the course); 2 = D (passed); 3 = C (average); 4 = B (above average); and 5 = A (excellent). The *y* variable, final grade, is an ordinal variable, and the *x* variable, aptitude, is a numerical variable. You want to find out whether there's a relationship between *x* and *y*. You collect data on a random sample of 20 students; the data are shown in Table 20-1. This is step one of the process of calculating Spearman's rank correlation (from the steps listed in the previous section).

**Table 20-1 Aptitude Test Scores and Final Grades in Statistics**

| *Student* | *Aptitude* | *Final Grade* |
|---|---|---|
| 1 | 59 | 3 |
| 2 | 47 | 2 |
| 3 | 58 | 4 |
| 4 | 66 | 3 |
| 5 | 77 | 2 |
| 6 | 57 | 4 |
| 7 | 62 | 3 |
| 8 | 68 | 3 |
| 9 | 69 | 5 |
| 10 | 36 | 1 |
| 11 | 48 | 3 |
| 12 | 65 | 3 |
| 13 | 51 | 2 |
| 14 | 61 | 3 |
| 15 | 40 | 3 |
| 16 | 67 | 4 |
| 17 | 60 | 2 |
| 18 | 56 | 3 |
| 19 | 76 | 3 |
| 20 | 71 | 5 |

Using Minitab for the aptitudes and final grades example, you get a correlation of 0.379. The following discussion walks you through steps two through six as you do this correlation yourself. This is likely what you may be asked to do on an exam.

Steps two and three of finding Spearman's rank correlation are to rank the aptitude test scores *(x)* from lowest (1) to highest; then rank the final grades *(y)* from lowest (1) to highest. Note that the final exam grades have several ties, so you use average ranks. For example, in column three of Table 20-1 you see a single 1, which gets rank 1. Then you see four 2s. Their ranks, had they not been tied, would be 2, 3, 4, and 5. The average of these four ranks is $r_s = \frac{2+3+4+5}{4} = \frac{14}{4} = 3.5$. Each of the 2s in column three, therefore, receive rank 3.5.

Table 20-2 shows the original data, the ranks of the aptitude scores ($x$), and the ranks of the final grades ($y$) as calculated by Minitab.

**Table 20-2 Aptitude Test Scores, Final Exam Grades, and Rank**

| *Student* | *Aptitude* | *Rank of Aptitude* | *Final Grade* | *Rank of Final Grade* |
|---|---|---|---|---|
| 1 | 59 | 9 | 3 | 10.5 |
| 2 | 47 | 3 | 2 | 3.5 |
| 3 | 58 | 8 | 4 | 17.0 |
| 4 | 66 | 14 | 3 | 10.5 |
| 5 | 77 | 20 | 2 | 3.5 |
| 6 | 57 | 7 | 4 | 17.0 |
| 7 | 62 | 12 | 3 | 10.5 |
| 8 | 68 | 16 | 3 | 10.5 |
| 9 | 69 | 17 | 5 | 19.5 |
| 10 | 36 | 1 | 1 | 1.0 |
| 11 | 48 | 4 | 3 | 10.5 |
| 12 | 65 | 13 | 3 | 10.5 |
| 13 | 51 | 5 | 2 | 3.5 |
| 14 | 61 | 11 | 3 | 10.5 |
| 15 | 40 | 2 | 3 | 10.5 |
| 16 | 67 | 15 | 4 | 17.0 |
| 17 | 60 | 10 | 2 | 3.5 |
| 18 | 56 | 6 | 3 | 10.5 |
| 19 | 76 | 19 | 3 | 10.5 |
| 20 | 71 | 18 | 5 | 19.5 |

For step four of the process of finding Spearman's rank correlation, you have Minitab calculate the standard deviation of the aptitude test score ranks (located in column two of Table 20-2) and the standard deviation of the final grades (located in column four of Table 20-2). In step five, you have Minitab calculate the covariance of the ranks of aptitude test scores and final grade ranks. These statistics are shown in Figure 20-1.

**Descriptive Statistics: Ranks of X, Ranks of Y**

| Variable | StDev |
|---|---|
| Ranks of X | 5.92 |
| Ranks of Y | 5.50 |

**Covariances: Ranks of X, Ranks of Y**

| | Ranks of X | Ranks of Y |
|---|---|---|
| Ranks of X | 35.0000 | |
| Ranks of Y | 12.3421 | 30.2632 |

**Figure 20-1:** Standard deviations and covariance of ranks of aptitude *(x)* and final grade *(y)*.

For the sixth and final step of finding Spearman's rank correlation, calculate $r_s$ by taking the covariance of the ranks of $x$ and $y$, divided by the standard deviation of the ranks of $x$ ($s_x$) times the standard deviation of the ranks of $y$ ($s_y$). You get $\frac{12.34}{5.92 * 5.50} = 0.379$. This matches the value for Spearman's correlation that was found by Minitab straightaway.

This correlation of 0.379 is fairly low, indicating a weak relationship between aptitude scores before the course and final grades at the end of the course. The moral of the story? If you aren't the sharpest tack in the bunch, you can still hope, and if you come in on top, you may not go out the same way. Although, there is still something to be said about working hard during the course (buying *Intermediate Statistics For Dummies* certainly doesn't hurt!).

# Part VI

# The Part of Tens

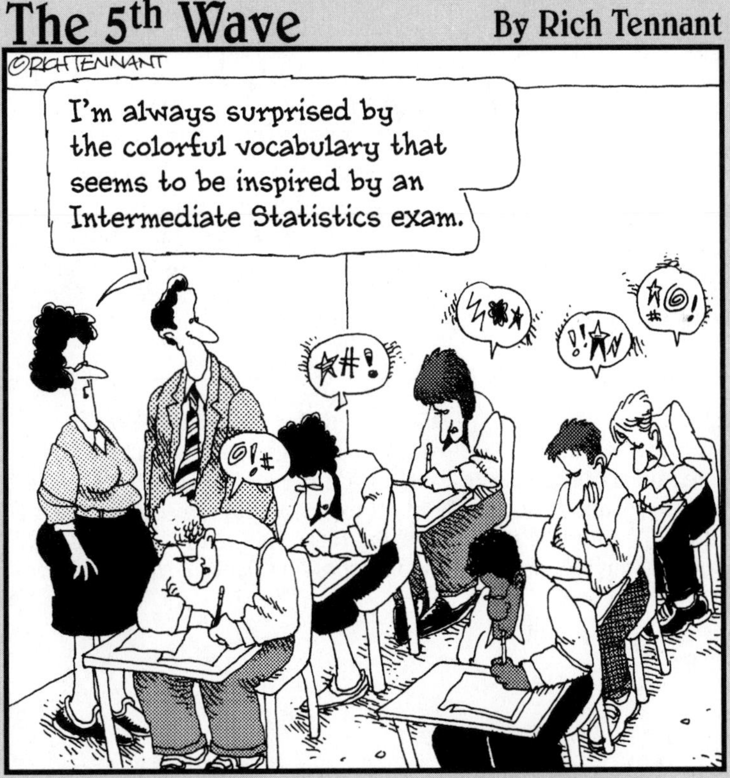

The 5th Wave
By Rich Tennant
©RICHTENNANT
I'm always surprised by the colorful vocabulary that seems to be inspired by an Intermediate Statistics exam.

## In this part . . .

You get a quick, concise reference that you can use to brush up on your problem-solving strategies. This part also gives you a reminder of some of the most common misconceptions that can occur and how to avoid them. In short, this part helps you start and end each problem right.

## Chapter 21

# Ten Errors in Statistical Conclusions

### *In This Chapter*

- Recognizing and avoiding mistakes when interpreting statistical results
- Knowing how to decide whether or not someone's conclusions are credible

Intermediate statistics is all about building models and doing data analysis. It focuses on looking at data and figuring out the story behind it. It's about making sure that the story is told correctly, fairly, and comprehensively. In this chapter, I discuss some of the most common errors I've seen as a teacher and statistical consultant for many moons. You can use this list to pull ideas together for homework and reports or as a quick review before a quiz or exam. Trust me — your professor will love you for it!

## These Statistics Prove . . .

Be skeptical of anyone who uses the words *these statistics* and *prove* in the same sentence. The word *prove* is a definitive, end-all-be-all, case-closed, lead-pipe-lock sort of concept, and statistics by nature isn't definitive. Instead, statistics gives you evidence for or against your theory, model, or claim, based on the data you collected; then it leaves you to your own conclusions. Because the evidence is based on data, and data changes from sample to sample, the results can change as well — that's the challenge, the beauty, and sometimes the frustration of statistics. The best you can say is that your statistics suggest, lead you to believe, or give you sufficient evidence to conclude — but never go as far as to say that your statistics prove anything.

# It's Not Technically Statistically Significant, But . . .

After you set up your model and test it with your data, you have to stand by the conclusions no matter how much you believe they're wrong. Statistics must lend objectivity to every process.

Suppose Barb, a researcher, has just collected and analyzed the heck out of her data, and she still can't find anything. However, she knows in her heart that her theory holds true, even if her data can't confirm it. Barb's theory is that dogs have ESP — in other words, a "sixth sense." She bases this theory on the fact that her dog seems to know when she's leaving the house, when he's going to the vet, and when a bath is imminent, because he gets sad and finds a corner to hide in.

Barb tests her ESP theory by studying ten dogs, placing a piece of dog food under one of two bowls and asking each dog to find the food by pushing on a bowl. (Assume the bowl is thick enough that the dogs can't cheat by smelling the food.) She repeats this process ten times with each dog and records the number of correct answers. If the dogs don't have ESP, you would expect that they would be right 50 percent of the time, because each dog has two bowls to choose from and each bowl has an equal chance of being selected.

As it turns out, the dogs were right 55 percent of the time. Now this percentage is technically higher than the long-term expected value of 50 percent, but it's not enough (especially with so few dogs and so few trials) to warrant statistical significance. In other words, Barb doesn't have enough evidence for the ESP theory. But when Barb presents her results at the next conference she attends, she puts a spin on her results by saying "The dogs were correct 55 percent of the time, which is more than 50 percent. These results are *technically* not enough to be statistically significant, but I believe they do show some evidence that dogs have ESP."

Some statistically incorrect researchers use this kind of conclusion all the time — skating around the statistics when they don't go their way. This game is very dangerous, because the next time someone tries to replicate Barb's results (and believe me, someone always does), they find out what you knew from the beginning (through ESP?): When Barb starts packing for a trip, her dog senses trouble coming and hides. That's all.

# This Means X Causes Y

Do you see the word that makes statisticians nervous? Because the words *this* and *means* seem pretty tame, and $x$ and $y$ are just letters of the alphabet, it's got to be that word *cause.* Of all the words on a final exam that aren't supposed to be there, *cause* probably tops the list.

Here's an example of what I mean. For your final report in stats class, you study which factors are related to your final exam score. You collect data on 500 statistics students, asking each one a variety of questions, such as "What was your grade on the midterm?"; "How much sleep did you get the night before the final?"; and "What is your GPA?" You conduct a multiple linear regression analysis (using techniques from Chapter 5), and you conclude that study time and the amount of sleep the night before are the most-important factors in determining exam scores. You write up all your analyses in a paper, and at the very end you say, "These results demonstrate that more study time and a good night of sleep the night before causes your exam grade to be higher."

I was with you until you said the word *cause.* You can't say that more sleep or more study time causes an increase in exam score. The data you collected shows that people who get a lot of sleep and study a lot do get good grades, and those who don't don't get the good grades. But that result doesn't mean you can take a flunky and just have him sleep and study more, and all will be okay. This theory is like saying that because an increase in height is related to an increase in weight, you can get taller by gaining weight.

The problem is that you didn't take an individual person, change his sleep time and study habits, and see what happened in terms of exam performance (using two different exams of the same difficulty). That study requires a *designed experiment.* When you conduct a *survey,* you have no way of controlling other related factors going on, which can muddy the waters.

The only way to control for other factors is to do a randomized experiment (complete with a treatment group, a control group, and controls for other factors that may ordinarily affect the outcome). Claiming causation without conducting a randomized experiment is a very common error some researchers make when they draw conclusions.

# I Assumed the Data Was Normal . . .

The operative word here is *assumed.* To break it down simply, an assumption is something you believe without checking. Assumptions can lead to wrong analyses and incorrect results — all without the person doing the assuming even knowing it.

Many analyses have certain requirements. For example, data should come from a normal distribution (the classic distribution that has a bell shape to it). If someone says "I assumed the data was normal," she just assumed that the data came from a normal distribution. But is having a normal distribution an assumption you just make and then move on, or is more work involved? You guessed it — more work.

For example, in order to conduct a one-sample *t*-test (see Chapter 3), your data must come from a normal distribution unless your sample size is large, in which you get an approximate normal distribution anyway by the Central Limit Theorem (remember those three words from intro stats?). Here, you aren't making an assumption, but examining a *condition* (something you check before proceeding). You plot the data, see if it meets the condition, and if it does, you proceed. If not, you can use nonparametric methods instead (Chapter 16).

Nearly every statistical technique for analyzing data has at least some condition(s) on the data in order for you to use it. Always find out what those conditions are, and check to see whether your data meets them. Be aware that many statistics textbooks wrongly use the word *assumption* when they actually mean *condition.* It's a subtle, but very important, difference.

## I'm Only Reporting "Important" Results

As a data analyst, you must not only avoid the pitfall of reporting only the significant, exciting, and meaningful results, but you also have to be able to detect when someone else is doing so. Some number crunchers examine every possible option and look at their data in every possible way before settling on the analysis that got them the desired result.

You can probably see the problem here. Every technique has a chance for error along with it. If you're doing a *t*-test, for example, and the $\alpha$ level is 0.05, over the long term 5 out every 100 *t*-tests you conduct will result in a false alarm just by chance (you declare a statistically significant result when it wasn't really there). So, if an eager researcher conducts 20 hypothesis tests on the same data set, odds are that at least one of those tests could result in a false alarm just by chance, on average. As this researcher conducts more and more tests, he's unfairly increasing his odds of "finding something" and running the risk of a wrong conclusion in the process.

It's not all the eager researcher's fault. He's pressured by a result-driven system. It's a sad state of affairs when the only results that get broadcasted on the news and appear in journal articles are the ones that show a statistically significant result (when Ho is rejected). Perhaps it was a bad choice when statisticians came up with the term *significance* to denote rejecting Ho — as if to say that rejecting Ho is the only important conclusion you can come to. What about all the times when Ho couldn't be rejected? For example, when doctors failed to conclude that drinking diet cola causes weight gain, or when pollsters didn't find that people were unhappy with the president? The public would be better served if researchers and the media were encouraged to spend at least some time reporting the statistically insignificant but still important results, along with the statistically significant ones.

TIP

The bottom line is this: In order to find out whether a statistical conclusion is correct, you can't just look at the analysis the researcher is showing you. You also have to find out about the analyses and results they're not showing you and ask questions. Avoid the urge to rush to reject Ho.

# A Bigger Sample Is Always Better

Bigger is better in some things, but not always with sample sizes. On one hand, the bigger your sample is, the more precise the results are (if no bias is present). A bigger sample also increases the ability of your data analysis to detect differences from a model or to deny some claim about a population (in other words, to reject Ho when you're supposed to). This ability to detect true differences from Ho is called the *power* of a test (see Chapter 3). However, some researchers can (and often do) take the idea of power too far. They increase the sample size to the point where even the tiniest difference from Ho sends them screaming to press that all-important reject Ho button.

Suppose research claims that the typical in-house dog watches an average of ten hours of TV per week. Bob thinks the true average is more, based on the fact that his dog Fido watches at least ten hours of cooking shows alone each week. Bob sets up the following hypothesis test: Ho: $\mu = 10$ versus Ha: $\mu > 10$. He takes a random sample of 100 dogs and has their owners record how much TV their dogs watch per week. The result turns out that the sample mean is 10.1 hours, and the sample standard deviation is 0.8 hours. This result isn't what Bob hoped for because 10.1 is so close to 10. He calculates the test statistic for this test using the formula $t = \frac{(\bar{x} - \mu)}{\frac{s}{\sqrt{n}}}$ and comes up with a value of $t = \frac{(10.1 - 10.0)}{\frac{0.8}{\sqrt{100}}} = \frac{0.1}{0.08}$, which equals 1.25 for $t$. Because the test is a right-tailed test (> in Ha), he can reject Ho at $\alpha$ if $t$ is beyond 1.645, and his $t$-value of 1.25 is far short of that value. Note that because $n = 100$ here, you find the value of 1.645 by looking at the very last row of the $t$-distribution table (Table A-1 in the Appendix). The row is marked with the infinity sign to indicate a large sample. So Bob can't reject Ho.

To add insult to injury, Bob's friend Joe conducts the same study and gets the same sample mean and standard deviation as Bob did, but Joe uses a random sample of 500 dogs rather than 100. Consequently, Joe's $t$-value is $t = \frac{(10.1 - 10.0)}{\frac{0.8}{\sqrt{500}}} = \frac{0.1}{0.036}$, which equals 2.78. Because 2.78 is greater than 1.645, Joe gets to reject Ho (to Bob's dismay).

Why did Joe's test find a result that Bob's didn't? The only difference was the sample size. Joe's sample was bigger, and a bigger sample size always makes the standard error smaller (see Chapter 3). The standard error sits in the denominator of the *t*-formula (as you just saw), so as it gets smaller, the *t*-value gets larger. A larger *t*-value makes it easier to reject Ho. (See Chapter 3 for more on precisions and margin of error.)

Now, Joe could technically give a big press conference or write an article on his results (his mom would be so proud), but you know better. You know that Joe's results are technically *statistically* significant, but not *practically* significant — they don't mean squat to any person or dog. After all, who cares that he was able to show evidence that dogs watch just a tiny bit more than ten hours of TV per week? This news isn't exactly earth-shattering.

TIP

Sample sizes should be large enough to provide precision and repeatability of your results, but there is such a thing as being too large, believe it or not. You can always take sample sizes big enough to reject any null hypothesis, even when the actual deviation from it is embarrassingly small. What can you do about this? When you read or hear that a result was deemed statistically significant, ask what the sample mean actually was (before it was put into the *t*-formula) and see how significant it is to you from a practical standpoint. Beware of someone who says, "These results are statistically significant, and the large sample size of 100,000 gives even stronger evidence for that."

# It's Not Technically Random, But . . .

When you take a sample on which to build statistical results, the operative word is *random.* You want the sample to be randomly selected from the population. The problem is that people oftentimes collect a sample that they think is *mostly* random or *sort of* random or random *enough* — and that doesn't cut it. The plan for taking a sample is either random or it isn't.

One day I gave each student in my class of 50 a number from 1 to 50, and I drew two numbers randomly from a hat. The two students I picked sat in the first row, and not only that, they sat right next to each other. Students immediately cried foul!

After these seemingly odd results appeared, I took the opportunity to talk to my class about truly random samples. A *random sample* is chosen in such a way that every member of the original population has an equal chance of being selected. Sometimes people who sit next to each other are chosen. In fact, if these seemingly strange results never happen, you may worry about the process; in a truly random process, you're going to get results that may seem odd, weird, or even fixed. That's part of the game.

In my consulting experiences, I always ask how my clients chose or plan to choose their samples. They always say they'll make sure it's random. But when I ask them how they'll do this, I sometimes get less-than-stellar answers. For example, someone needed to get a random sample from a population of 500 free-range chickens in a farmyard. He needed five chickens and said that he'd select them randomly by choosing the five that came up to him first. The problem is, animals that come up to you may be friendlier, more docile, older, or perhaps more tame. These characteristics aren't present in every chicken in the yard, so choosing a sample this way isn't random. The results are likely biased in this case.

Always ask the researcher how she selected a sample, and when you select your own samples, stay true to the definition of random. And don't use your own judgment to choose a random sample; use a computer to do it for you!

# 1,000 Responses Is 1,000 Responses

A newspaper article on the latest survey says that 50 percent of the respondents said blah blah blah. The fine print says the results are based on a survey of 1,000 adults in the United States. But wait — is 1,000 the actual number of people selected for the sample, or is it the final number of respondents? You may need to take a second look; those two numbers hardly ever match.

For example, Jenny wants to know what percentage of people in the U.S. have ever knowingly cheated on their taxes. In her statistics class, she found out that if she gets a sample of 1,000 people, the margin of error for her survey is only plus or minus 3 percent, which she thinks is groovy. So she sets out to achieve the goal of 1,000 responses to her survey. She knows that in these days it's hard to get people to respond to a survey, and she's worried that she may lose a great deal of her sample that way, so she has an idea. Why not send out more surveys than she needs, so that she gets 1,000 surveys back?

Jenny looks at several survey results in the newspapers, magazines, and on the Internet, and she finds that the response rate (the percentage of people who actually responded to the survey) is typically around 25 percent. (In terms of the real world, I'm being generous with this number, believe it or not. But think about it: How many surveys have your thrown away lately? Don't worry, I'm guilty of it too.) So, Jenny does the math and figures that if she sends out 4,000 surveys and gets 25 percent of them back, she has the 1,000 surveys she needs to do her analysis, answer her question, and have that small margin of error of plus or minus 3 percent.

Jenny conducts her survey, and just like clockwork, out of the 4,000 surveys she sends out, 1,000 come back. She goes ahead with her analysis and finds that 400 of those people reported cheating on their taxes (40 percent). She adds her margin of error, and reports, "Based on my survey data, 40 percent of Americans cheat on their taxes, plus or minus 3 percentage points."

Now hold the phone, Jenny. She only knows what those 1,000 people who returned the survey said. She has no idea what the other 3,000 people said. And here's the kicker: Whether or not someone responds to a survey is often related to the reason the survey is being done. It's not a random thing. Those nonrespondents (people who don't respond to a survey) carry a lot of weight in terms of what they're not taking time to tell you.

For the sake of argument, suppose that 2,000 of the people who originally got the survey were uncomfortable with the question because they *do* cheat on their taxes, and they just didn't want anyone to know about it, so they threw the survey in the trash. Suppose that the other 1,000 people don't cheat on their taxes, so they didn't think it was an issue and didn't return the survey. If these two scenarios were true, the results would look like this:

Cheaters = 400 (surveyed) + 2,000 (nonrespondents) = 2,400

These results raise the total percentage of cheaters to 2,400 divided by 4,000 — 60 percent. That's a huge difference!

You could go completely the other way with the 3,000 nonrespondents. You can suppose that none of them cheat, but they just didn't take time to say so. If you knew this info, you would get 600 (surveyed) + 3,000 (nonrespondents) = 3,600 noncheaters. Out of 4,000 surveyed, this is 90 percent. The truth is likely to be somewhere between the two examples I just gave you, but nonrespondents make it too hard to tell.

And the worst part is that the formulas Jenny uses for margin of error don't know that the information she put into them is based on biased data, so her reported 3 percent margin of error is wrong. The formulas happily crank out results no matter what. It's up to you to make sure that what you put into the formulas is good, clean info.

TIP

Getting 1,000 results when you send out 4,000 surveys is nowhere near as good as getting 1,000 results when sending out 1,000 surveys (or even 100 results from 100 surveys). Plan your survey based on how much follow-up you can do with people to get the job done, and if it takes a smaller sample size, so be it. At least the results have a better chance of being statistically correct.

## Of Course These Results Apply to the General Population!

Making conclusions about a much broader population than your sample actually represents is one of the biggest no-no's in statistics. This kind of problem is called *generalization,* and it occurs more often than you may think. People want their results instantly; they don't want to wait for them, so well-planned surveys and experiments take a back seat to instant Web surveys and convenience samples.

For example, a researcher wants to know how cable news channels have influenced the way Americans get their news. He also happens to be a statistics professor at a large research institution and has 1,000 students in his class. He decides that instead of taking a random sample of Americans, which would be difficult, time-consuming, and expensive, he just puts a question on his final exam to get his students' answers. His data analysis shows him that only 5 percent of his students read the newspaper and/or watch network news programs anymore; the rest watch cable news. For his class, the ratio of students who exclusively watch cable news compared to those students who don't is 20 to 1. The professor reports this and sends out a press release about it. The cable news channels pick up on it and the next day are reporting, "Americans choose cable news channels over newspapers and network news by a 20 to 1 margin!"

Do you see what's wrong with this picture? The problem is that the professor's conclusions go way beyond his study, which is wrong. He used the students in his statistics class to obtain the data that serves as the basis for his entire report and the resulting headline. Yet the professor reports the results about all Americans. I think it's safe to say that a sample of 1,000 college students taking a statistics class at the same time at the same college doesn't represent a cross section of America.

If the professor wants to make conclusions in the end about America, he has to select a random sample of Americans to take his survey. If he uses 1,000 students from his class, then his conclusions can only be made about that class and no one else.

TIP

To avoid or detect generalization, identify the population that you're intending to make conclusions about and make sure the sample you selected represents that population. If the sample represents a smaller group within that population, then the conclusions have to be downsized in scope also.

# I Just Decided to Leave It Out

It seems easier sometimes to just leave information out. I see this all too often when I read articles and reports based on statistics. But, this error isn't the fault of only one person or group. The guilty parties can include

- **The producers:** The researchers out there leave items out for a variety of reasons, including time and space constraints. After all, you can't write about every element of the experiment from beginning to end. However, other items they leave out may be indicative of a bigger problem. For example, reports often say very little about how they collected the data or chose the sample. Or they may discuss the results of a survey but not show the actual questions they asked. Ten out of 100 people may have dropped out of their experiment, and they don't tell you why. All these items are important to know before making a decision about the credibility of someone's results.

Another way in which some data analysts leave information out is by removing data that doesn't fit the intended model (in other words, "fudging" the data). Suppose a researcher records the amount of time surfing the Internet and relates it to age. He fits a nice line to his data indicating that younger people surf the Internet much more than older people and that surf time decreases as age increases. All is good except for Claude the outlier, who is 80-years-old and surfs the Internet day and night, leading his own bingo chat rooms and everything. What to do with Claude? If not for him, the relationship looks beautiful on the graph; what harm would it do to remove him? After all, he's only one person, right?

No way. Everything is wrong with this idea. Removing undesired data points from a data set is not only very wrong but also very risky. The only time it's okay to remove an observation from a data set is if you're certain beyond doubt that the observation is just plain wrong. For example, someone writes on a survey that she spends 30 hours a day surfing the Internet or that her IQ is 2,200.

- **The communicators:** When reporting statistical results, the media leaves out important information all the time, which is often due to space limitations and fast deadlines. However, part of it is a result of the current, fast-paced society that feeds itself on sound bytes. The best example is survey results, where they often leave out the size of the sample. You can't calculate margin of error without it.
- **The consumers:** The general public also plays a role in the leave-things-out mindset. People hear a news story and instantly believe it's true, ignoring any chance for error or bias in the results. You need to make a decision about what car to buy, and you ask your neighbors and friends rather than examine the research and the meticulous, comprehensive ratings that have resulted. Everyone neglects to ask questions as much as he should, at one time or another, which indirectly feeds the entire problem.

In the chain of statistical information, the producers (researchers) need to be comprehensive and forthcoming about the process they conducted and the results they got. The communicators of that information (the media) need to critically evaluate the accuracy of the information they're getting and report it fairly. The consumers of statistical information (the rest of us) need to stop taking results for granted and to rely on credible sources of statistical studies and analyses to help make those important life decisions.

In the end, if a data set looks too good, it probably is. If the model fits too perfectly, be suspicious. If it fits exactly right, run and don't look back! Sometimes what is left out speaks much louder than what is put in.

# Chapter 22

# Ten Practice Problems

### In This Chapter

- Translating a word problem into a statistics problem
- Picking out the necessary info

Many students miss out on the fact that the most important and difficult skill they need to develop in statistics is the ability to attack a problem correctly, especially real-world problems. This skill requires identifying what the problem is really asking, figuring out what the underlying statistical question is, and determining which statistical technique will do the job. Because professors give course exams usually over small chunks of information in certain chapters of a textbook, the synthesis process is underdeveloped. Then comes the final exam, where you're supposed to magically be able to put it all together, which can spell disaster.

In this chapter, you discover the important skill of attacking problems correctly and with confidence — a skill that can no doubt help you to be successful not only in your statistics course, but also in the workplace and everyday life. I help you determine which statistical technique you need to solve each problem (I don't actually solve the problems in this chapter but refer you to the appropriate chapters to get those details). The focus of this chapter is how to start a problem.

## Comparing Means with One-Way ANOVA

The key to knowing you need to use ANOVA is that you have a group of populations that you want to compare according to some quantitative variable $y$. Suppose you have a population of consumers for each of the four brands of cereal, and you're interested in seeing whether ages differ across any of the four populations. The response variable is age and the variable on which ages are being compared is the brand of cereal the population buys. Cereal brand is the variable on which *y* is being compared; cereal brand is called a *factor* in this case. The fact that you have a response variable being compared according to different values of a factor tips you off that one-way ANOVA is to be used in this situation; you have a quantitative response variable whose means are compared on some categorical variable called a factor.

If your were to compare two populations, you'd use a hypothesis test for two population means. If you have more than two means to compare, you must go in the direction of ANOVA (see Chapter 9 for the big ideas of ANOVA). The only factor to include in the model is cereal brand, because that variable is the only one on which you're making comparisons. So, you use a one-way ANOVA versus a two-way ANOVA (see Chapter 9).

ANOVA first tests to see whether there's an overall difference between population means, using the *F*-test. If you reject Ho: The population means are equal, you can conclude Ha: At least two of the population means are different. The ANOVA table for comparing data from four brands of cereal is shown in Table 22-1. Because a condition of ANOVA is that the populations are independent, you take a random sample of ten boxes of each type of cereal, for a total of 40 observations. (Table 22-1 gives you the general setup of the ANOVA table; you can determine the sums of squares from the particular data in the problem.)

**Table 22-1 ANOVA Table Setup for the Cereal Example**

| *Source* | *DF* | *SS* | *MS* | *F* |
|---|---|---|---|---|
| Treatment | 4 – 1 = 3 | SST | SST/3 | MST/MSE |
| Error | 40 – 4 = 36 | SSE | SSE/36 | |
| Total | 40 – 1 = 39 | SSTO | | |

# Doing Multiple Comparisons

When comparing multiple population means on some factor (such as brand of cereal), you first conduct a one-way ANOVA (see Chapter 9) to determine whether any differences at all are in the means. If you determine that at least two population means are different (in other words, Ho is rejected in the ANOVA procedure), your next step is to find out which ones are different and how they compare to the others. This situation is where multiple comparison procedures enter the picture (see Chapter 10). (If you can't say the means are different, you have no reason to proceed further.)

While many multiple comparison procedures exist, in this book I discuss LSD (least significant differences) and Tukey's (not *turkeys*) procedure. LSD compares all the pairs of $n$ different means while keeping an overall eye on the chance of making an error due to chance. Tukey uses pairs of confidence intervals and looks for overlap and groups the means in order of magnitude. (Each procedure has its pluses and minuses, but most statisticians use Tukey or LSD. See Chapter 10 for full details on multiple comparison procedures.)

Sometimes, this process of answering questions is flipped around. Instead of asking you a question that you use computer output to answer, your professor may give you computer output and ask you to determine what question was answered by the analysis. To do this, you look for clues that tell you what type of analysis was done, and fill in the details using what you already know about that particular type of analysis.

For example, your prof gives you computer output comparing the ages of ten consumers of each of the four cereal brands, labeled C1–C4 (see Figure 22-1). On the analysis, you can see the mean consumer ages for the four cereals being compared to each other, and the confidence intervals for the averages are also shown and compared. Seeing confidence intervals being compared tells you that you're dealing with a multiple comparison procedure.

Remember you're looking to see whether the confidence intervals for each cereal group overlap; if they don't, those cereals have different average ages of consumers. If they do overlap, those cereals have mean ages that can't be declared different. From Figure 22-1, you can see that cereals one and two aren't significantly different, but for cereal three, consumers have a higher average age than cereals one and two, while cereal four has a significantly higher age then the three others. Now after the multiple comparison procedure, you know which cereals are different and how they compare to the others.

```
                                 Individual 95% CIs For Mean Based on
                                 Pooled StDev
Level  N     Mean    StDev    -------+---------+---------+---------+--
C1     10    8.800   1.687     (--*--)
C2     10   11.800   1.033       (--*--)
C3     10   36.500   7.735                     (--*--)
C4     10   55.400  10.309                                (--*--)
```

**Figure 22-1:** Multiple comparison results for cereal example.

# Looking at Two Factors with Two-Way ANOVA

You use two-way ANOVA when you want to compare the means of $n$ populations that are classified according to two different categorical variables (factors). For example, suppose you want to see how four brands of detergent (brands A, B, C, and D) and water temperature (cold, warm, hot) work together to affect the whiteness of clothes being washed. Product-testing groups can use this information as well as the detergent companies to investigate or advertise how it measures up according to its competitors. The only way I can think of to measure whiteness is on some sort of scale from least white (say 1) to most white (say 10).

It makes sense that you would want to test different combinations of detergents and water temperatures to see how they affect the mean whiteness of the clothes. Because this question involves two different factors and their affects on some numerical (quantitative) variable, you know that you need to do a two-way ANOVA. (For all the information on two-way ANOVA, see Chapter 11; I just discuss the overall setup here.)

You can't assume that water temperature affects whiteness of clothes in the same way for each brand, so you need to include an interaction effect of brand and temperature in the two-way ANOVA model. Because brand of detergent has four possible types (or levels) and water temperature has three possible values (or levels), you have $4 * 3 = 12$ different combinations to examine in terms of how brand and temperature interact. Those combinations are: brand A in cold water, brand A in warm water, brand A in hot water, brand B in cold water, brand B in warm water, brand B in hot water, and so on.

The resulting model looks like this: $y = b_i + w_j + bw_{ij} + \varepsilon$, where $b$ represents the brand of detergent, $w$ represents the water temperature, $y$ represents the whiteness of the clothes after washing, and $bw_{ij}$ represents the interaction of brand $i$ of detergent ($i = 1, 2, 3,$ and 4) and temperature $j$ of the water ($j = 1, 2, 3$). ($\varepsilon$ represents the amount of variation in the $y$ values [whiteness] that isn't explained by either brand or temperature.)

Suppose that your experiment involves four brands of detergents, three water temperatures, and ten data values for each combination (for a total of $4 * 3 * 10 = 120$ data values). You can see the setup of the ANOVA table to analyze this data in Table 21-2. Now you're off and running (hopefully not in two directions) with a two-way ANOVA!

**Table 22-2 Two-Way ANOVA Table for Clothes Example**

| *Source* | *DF* | *SS* | *MS* | *F* |
|---|---|---|---|---|
| Brand of Detergent | 4 – 1 = 3 | SSb | SSb/3 | MSb/MSE |
| Water Temp | 3 – 1 = 2 | SSw | SSw/2 | MSw/MSE |
| Brand * Water Temp | (4 – 1) * (3 – 1) = 6 | SSbw | SSbw/6 | MSbw/MSE |
| Error | 120 – 4 * 3 = 108 | SSE | SSE/108 | |
| Total | 120 – 1 = 119 | | SSTO | |

# Predicting a Quantitative Variable by Using Regression

You use regression when you have a response variable, *y,* that's quantitative, and you're using another quantitative variable, *x,* to predict it. For example, suppose you're trying to estimate how much a house is going to cost (on average). You may think of many different factors that could come into play when estimating house cost, such as house size, location, number of bedrooms, number of bathrooms, or the cost of other homes in the neighborhood.

Suppose you focus on only one variable: house size. Certainly house size is one of the factors that builders and realtors use to base house price. For example, suppose the typical price for a new house in Columbus, Ohio is approximately $100 per square foot. Then a home that has 2,000 square feet would cost approximately $200,000 on average.

By trying to put yourself into the shoes of the person who's asking this question, you can get a much better idea of what the problem is really asking. Here, you have one variable — house price — and you're trying to estimate that variable. That tells you that house price is a response (outcome) variable, because you're using one variable (house size) to estimate house price. That means you're treating house size as an explanatory (input) variable — the variable on the *x*-axis. Trying to use *x* (house size) to estimate *y* (house price) is what you do in simple linear regression. And that technique is exactly what you need in order to answer this question.

If indeed house cost is based on a model of $100 per square foot, your data would be fit, using the straight line shown in Figure 22-2. (This figure shows a hypothetical data set of 22 homes.)

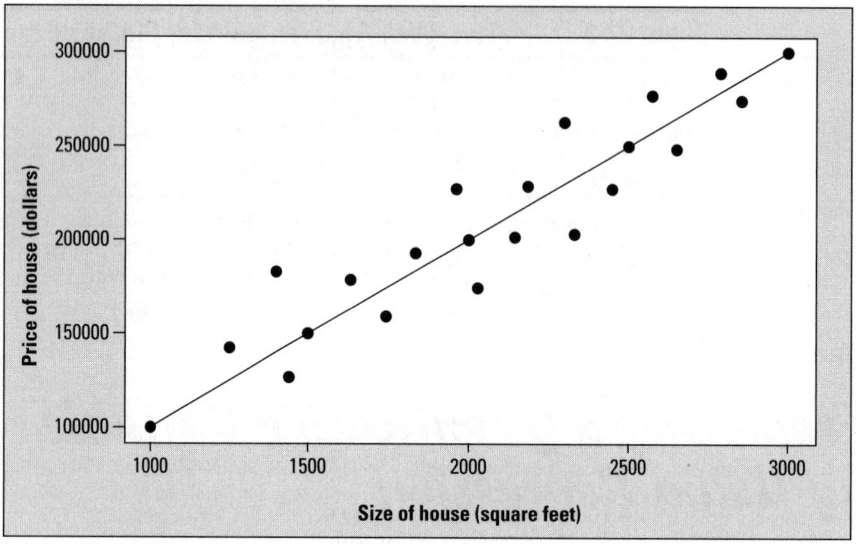


**Figure 22-2:** House price = 100 * size of home (in square feet).

You can add more *x* variables to the model to try to predict *y*. This procedure is called *multiple regression.* (For more information on simple linear regression, see Chapter 4; for multiple regression, see Chapter 5.)

# Predicting a Probability with Logistic Regression

You use logistic regression when you use a quantitative variable to predict or guess the outcome of some categorical variable with only two outcomes (for example, using barometric pressure to predict whether or not it will rain).

Because you're trying to use one variable *(x)* to make a prediction for another variable *(y)*, you may think about using regression — and you would be right. However, you have many types of regression to choose from, and you need to determine what kind is most appropriate here. You need the type of regression that uses a quantitative variable *(x)* to predict the outcome of some categorical variable *(y)* that has only two outcomes (yes or no).

So being the good intermediate statistics student that you are, you go to your trusty list of statistical techniques, and you look under regression. You see simple linear regression. . . no, you use that when you have one quantitative variable predicting another. Multiple regression? No. . . that method just expands simple linear regression to add more *x* variables. Nonlinear regression? Well no. . . that still works with two quantitative variables; it's just that the data forms a curve, not a line.

But then you come across logistic regression, and. . . eureka! You see that logistic regression handles situations where the $x$ variable is numerical and the $y$ variable is categorical with two possible categories. Just what you're looking for! Logistic regression, in essence, estimates the probability of $y$ being in one category or the other, based on the value of some quantitative variable, $x$. In the gender and height example, logistic regression predicts whether someone is a male (or female) based on his height. If a "1" indicates a male, then people who receive a probability of more than 0.5 of being male (based on their heights) are predicted to be male, and the people who receive a probability of less than 0.5 of being male (based on their heights) are predicted to be female. (For all the details on logistic regression, see Chapter 8.)

It may help at this point to sort out some situations that sound similar but have subtle differences that lead to very different analyses. You can use the following list to compare these subtle, but important, differences:

- If you want to compare three or more groups of numerical variables, use ANOVA (Chapter 10). (For only two groups use a *t*-test; see Chapters 3 and 9.)
- If you want to estimate one numerical variable from another, use simple linear regression (Chapter 4).
- If you want to estimate one numerical variable using many other numerical variables, use multiple regression (Chapter 5).
- If you want to estimate a categorical variable with two categories by using a numerical variable, you want to use logistic regression (Chapter 8.)
- If you want to compare two categorical variables to each other, head straight for a Chi-square test (Chapter 14).

# Using Nonlinear Regression for Curved Data

Nonlinear regression takes the stage when you want to predict some quantitative variable *(y)* by using another quantitative variable *(x),* but the pattern you see in the data collected resembles not a straight line, but rather a curve.

Suppose a manager is considering the purchase of a new office management software but is hesitating. She wants to know how long it typically takes someone to get up to speed using the software (that's what a learning curve shows — the decrease in time to do a task with more and more practice).

What is the statistical question here? She wants a model that shows what the learning curve looks like (on average). You have two variables — time to complete the task and trial number (for example, the first try is designated by 1, the second try by 2, and so on). Both of these variables are numerical, or quantitative, and you want to find a connection between two quantitative variables. At this point, you can start thinking regression.

A regression model produces a function (be it a line or otherwise) that describes a pattern or relationship. The relationship here is task time versus number of times the task is practiced. But what type of regression model do you use? After all, you can see four types in this book: simple linear regression, multiple regression, nonlinear regression, and logistic regression. You need more clues.

The word *curve* in learning curve is a clue that the relationship being modeled here may not be linear. That word sends the signal that you're talking about a nonlinear regression model (see Chapter 7). If you think about what a possible learning curve may look like, you can imagine task time on the *y*-axis, and the number of the trial on the *x*-axis. You may guess that, at first, the *y*-values will be high, because the first couple of times you try a new task, it takes longer. Then, as the task is repeated, the task time decreases, but at some point, more practice doesn't reduce task time much. So the relationship may be represented by some sort of curve, like the one I simulate in Figure 22-3 (which can be fit by using an exponential function).

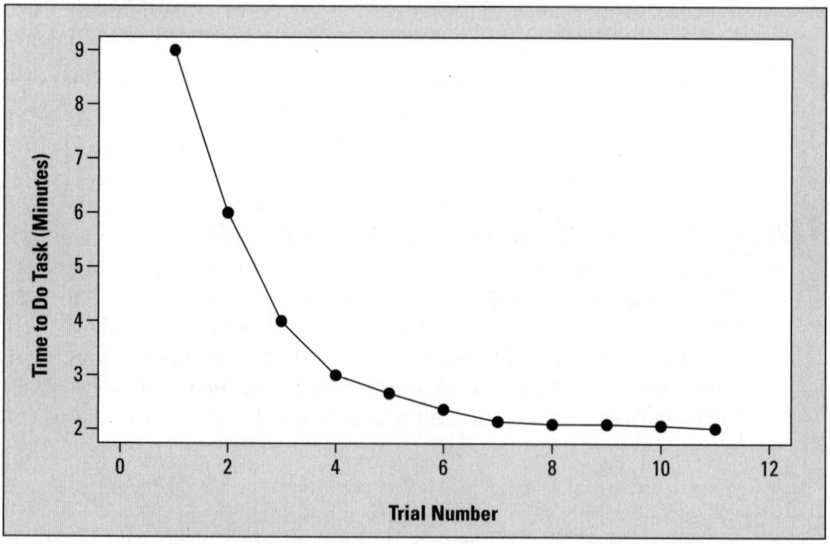


**Figure 22-3:** Learning curve for time performing a new task.

# Using Chi-Square to Test for Independence

When you read this section's heading, you may notice one thing right off the bat: the word *independence*. This word should remind you of the Chi-square test of independence. Don't forget to take a second look at how many and what type of variables you've got (and be sure you do this for every problem before you start it).

Suppose you want to study two variables — eating breakfast (yes or no data) and gender (male or female). Each of these variables is categorical, or qualitative. Whenever you have two variables, $x$ and $y$, that are both qualitative, use a Chi-square test to see whether or not those variables are independent. Use Ho: $x$ and $y$ are independent, versus Ha: $x$ and $y$ are dependent. (See Chapter 14 for more on the Chi-square test.)

Table 22-3 shows how you would set up the table for the Chi-square test for this particular question. You would enter the data in the cells marked by xx.

**Table 22-3 Table Setup for the Breakfast and Gender Question**

| | *Eat Breakfast* | *Don't Eat Breakfast* |
|---|---|---|
| Male | xx | xx |
| Female | xx | xx |

TIP

Note that the Chi-square test for independence is equivalent to testing whether two population proportions are equal, in other words Ho: $p_1 = p_2$ versus Ha: $p_1 \neq p_2$. That is, if you took this same data and analyzed it using a two-sample test for proportions, you'd test to see whether the proportion of breakfast eaters *(p)* is the same for males and females. If you reject Ho, that means breakfast eating is different for males and females. This result implies then that gender and eating breakfast are dependent. Similarly, if Ho isn't rejected, you conclude you can't find a difference in breakfast eating for males versus females, which tells you gender and breakfast eating may be independent.

# Checking Specific Models with the Goodness-of-Fit Test

You can use the Chi-square goodness-of-fit test to check to see whether a specified model fits. A *specified model* is a model in which each possible value of the variable $x$ is listed, along with its associated probability according to the model. For example, suppose you want to know whether the colors of Skittles candy are evenly mixed (that is, you have an equal percentage of each color). Think about circumstances where you want to know whether a situation is even, or fair. You may be flipping a coin. You can assess fairness in this instance by testing whether the probability of heads equals ½ (using a one-sample test for proportions; see *Statistics For Dummies* [Wiley] or your intro stats textbook for more information).

But in this case, instead of heads and tails, you have five possible outcomes representing each color of Skittles (purple, orange, red, green, and yellow). You want to know whether the proportion of each color of Skittles is the same. In other words, you want to test Ho: $p_1 = p_2 = p_3 = p_4 = p_5$, where each $p$ represents the proportion of a different color of Skittles. In this case, each proportion would have to equal 0.20 to spread the Skittle colors evenly. This model is very specific. Which statistical technique requires your model to be very specific? (Hold that thought for just a second. You're not quite done.)

You can find another clue by again looking at the number and type of variables you're working with. You have one variable — Skittles color — and that variable is categorical. So you're testing a model for one categorical variable and that model is very specific. You want to see whether that specific model fits. How do you do it? With a Chi-square goodness-of-fit test. (See Chapter 15 for all the information on the goodness-of-fit test.)

# Estimating the Median with the Signed Rank Test

Many times when you hear the word *median,* you think of the middle number in a data set. It's true that the median is the middle number, when you order all the values from smallest to largest. But you may be more accustomed to finding the median of a sample, rather than estimating the median of an entire population. Estimating the median of a population is quite a bit different from finding the median of a sample.

If you want to estimate the mean cost of tuition per year, you would right away think of a confidence interval for the population mean, based on $\bar{x}$, the sample mean, plus or minus a margin of error. You can't use the same formula to answer the given question about the median, but you can use the general idea. You can't know the median of a population any more than you can know the mean; populations are generally too large to measure every value, so you need to take a sample and calculate a confidence interval instead. That is, you need a sample statistic plus or minus a margin of error. (See Chapter 3 for the full breakdown on confidence intervals and the whole margin-of-error thing.) In this case, the sample statistic would be the sample median; it only makes sense. But what about the margin of error? Where do you turn for a formula for that?

Don't forget the unsung hero: nonparametric statistics. Anytime you're dealing with data that doesn't meet the conditions of the "normal" procedures, pull out your nonparametric tools. Anytime you're estimating or testing the median, the data at hand likely doesn't come from a normal distribution either. The reason the data doesn't come from a normal distribution is that in a normal distribution, the mean and the median are the same, and you can use the regular old (parametric) methods to estimate the median.

So far, you know that you need a confidence interval for the median that's based on nonparametric statistics. The signed rank test handles just that situation because its sole purpose is to rank data from smallest to largest and figure out where the middle lies (see Chapter 17 for all the details). The biggest challenge is to remember that nonparametric statistics are available, and you need to use them when you can't use parametric procedures. (Chapter 16 tells you what types of situations need nonparametric procedures.)

# Checking Model Fit by Using $R^2$

One of the most important ideas in intermediate statistics is using the right technique for the right data and to answer the right question. To know whether you have the right technique, you need to check the conditions for that technique, using your data, to make sure those conditions are being met in the population. (Each technique used in this book has a set of conditions presented along with it in its corresponding chapter and section.) Because most of these procedures are based on building a model from the data to make predictions, you also need to make sure that the final model you chose fits the data well, so you can sleep at night knowing you did the right thing.

Several different methods exist for checking the fit of models, and those methods differ according to the model you use, of course. However, one particular method is universal no matter what kind of model you fit. Always check the value of $R^2$, the *coefficient of determination* (also known as the *coefficient of extermination,* because it can kill off a model in a matter of seconds with a low number).

The coefficient of determination ($R^2$) measures the amount to which the model (which contains the $x$ variable or variables) explains or accounts for the amount of variability in the $y$ variable. The value of $R^2$ is a number between 0 and 1, and you can interpret it as a percentage. A high value of $R^2$ (at least 0.70, but the higher, the better) indicates that the model fits well; a value of $R^2$ below 0.70 indicates the model doesn't fit well (and the closer $R^2$ is to zero, the worse the model fits).

For example, say you want to conduct a regression analysis of exam score based on study time. Suppose you analyzed your data and got the computer output listed in Figure 22-4.

**Figure 22-4:** Regression analysis for exam data.

```
Predictor       Coef    SE Coef       T       P
Constant      51.410      1.290   39.84   0.000
C1            4.6227     0.2076   22.27   0.000


S = 2.36349   R-Sq = 98.2%   R-Sq(adj) = 98.0%
```

You can see in Figure 22-4 that the value of $R^2$ for this model is 98.2 percent. So study time in this case explains 98.2 percent of why those exam scores vary. Therefore, the model fits the data very well according to $R^2$.

The most important use of $R^2$ is in choosing the best model if given a variety of possibilities. Typically you choose the model with the highest value of $R^2$, adjusting for the number of variables in the model. This variation of $R^2$ is called $R^2$ adjusted (in Figure 22-4, $R^2$ adjusted is 98.0 percent, which is very high). (For the full scoop on model fit, see Chapter 6, where you can find stepwise procedures to choose the best multiple regression model given a choice of many variables.)

# Appendix

# Tables for Your Reference

This Appendix includes commonly used tables for five important distributions for intermediate statistics: the *t*-distribution, the binomial distribution, the Chi-square distribution, the distribution for the rank sum test statistic, and the *F*-distribution.

## t-Table

Table A-1 shows right-tail probabilities for the *t*-distribution (refer to Chapter 3). To use Table A-1, you need four pieces of information from the problem you're working on:

- The sample size *(n)*
- The mean of *x* (the given normal distribution)
- The standard deviation of your data *(s)*
- The value of *x* for which you want the right-tail probability

After you have this information, transform your value of *x* to a *t*-statistic (or *t*-value) by taking your value of *x*, subtracting the mean, and dividing by the standard error (see Chapter 3) by using the formula $t_{n-1} = \frac{\bar{x} - \mu}{\frac{s}{\sqrt{n}}}$.

Then look up this value of *t* on Table A-1 by finding the row corresponding to the degrees of freedom for the *t*-statistic ($n - 1$). Go across that row until you find two values between which your *t*-statistic falls. Then go to the top of those columns and find the probabilities there. The probability that *t* is beyond your value of *x* (the right-tail probability) is somewhere between these two probabilities. Note that the last line of the *t*-table shows df = ∞, which represents the values of the z-distribution because for large sample sizes *t* and *z* are close.

## Table A-1 *t*-Distribution

t-distribution showing area to the right

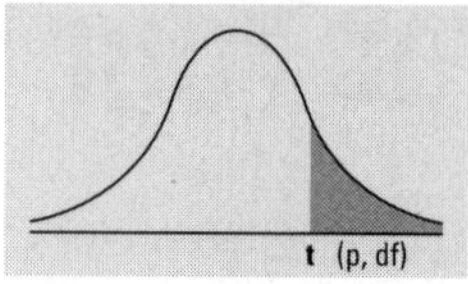


| df/p | 0.40 | 0.25 | 0.10 | 0.05 | 0.025 | 0.01 | 0.005 | 0.0005 |
|---|---|---|---|---|---|---|---|---|
| 1 | 0.324920 | 1.000000 | 3.077684 | 6.313752 | 12.70620 | 31.82052 | 63.65674 | 636.6192 |
| 2 | 0.288675 | 0.816497 | 1.885618 | 2.919986 | 4.30265 | 6.96456 | 9.92484 | 31.5991 |
| 3 | 0.276671 | 0.764892 | 1.637744 | 2.353363 | 3.18245 | 4.54070 | 5.84091 | 12.9240 |
| 4 | 0270722 | 0.740697 | 1.533206 | 2.131847 | 2.77645 | 3.74695 | 4.60409 | 8.6103 |
| 5 | 0.267181 | 0.726687 | 1.475884 | 2.015048 | 2.57058 | 3.36493 | 4.03214 | 6.8688 |
| 6 | 0.264835 | 0.717558 | 1.439756 | 1.943180 | 2.44691 | 3.14267 | 3.70743 | 5.9588 |
| 7 | 0.263167 | 0.711142 | 1.414924 | 1.894579 | 2.36462 | 2.99795 | 3.49948 | 5.4079 |
| 8 | 0.261921 | 0.706387 | 1.396815 | 1.859548 | 2.30600 | 2.89646 | 3.35539 | 5.0413 |
| 9 | 0.260955 | 0.702722 | 1.383029 | 1.833113 | 2.26216 | 2.82144 | 3.24984 | 4.7809 |
| 10 | 0260185 | 0.699812 | 1.372184 | 1.812461 | 2.22814 | 2.76377 | 3.16927 | 4.5869 |
| 11 | 0259556 | 0.697445 | 1.363430 | 1.795885 | 2.20099 | 2.71808 | 3.10581 | 4.4370 |
| 12 | 0259033 | 0.695483 | 1.356217 | 1.782288 | 2.17881 | 2.68100 | 3.05454 | 43178 |
| 13 | 0.258591 | 0.693829 | 1.350171 | 1.770933 | 2.16037 | 2.65031 | 3.01228 | 4.2208 |
| 14 | 0.258213 | 0.692417 | 1.345030 | 1.761310 | 2.14479 | 2.62449 | 2.97684 | 4.1405 |
| 15 | 0.257885 | 0.691197 | 1.340606 | 1.753050 | 2.13145 | 2.60248 | 2.94671 | 4.0728 |
| 16 | 0257599 | 0.690132 | 1.336757 | 1.745884 | 2.11991 | 2.58349 | 2.92078 | 4.0150 |
| 17 | 0.257347 | 0.689195 | 1.333379 | 1.739607 | 2.10982 | 2.56693 | 2.89823 | 3.9651 |
| 18 | 0.257123 | 0.688364 | 1.330391 | 1.734064 | 2.10092 | 2.55238 | 2.87844 | 3.9216 |
| 19 | 0.256923 | 0.687621 | 1.327728 | 1.729133 | 2.09302 | 2.53948 | 2.86093 | 3.8834 |
| 20 | 0.256743 | 0.686954 | 1.325341 | 1.724718 | 2.08596 | 2.52798 | 2.84534 | 3.8495 |
| 21 | 0.256580 | 0.686352 | 1.323188 | 1.720743 | 2.07961 | 2.51765 | 2.83136 | 3.8193 |
| 22 | 0256432 | 0.685805 | 1.321237 | 1.717144 | 2.07387 | 2.50832 | 2.81876 | 3.7921 |
| 23 | 0256297 | 0.685306 | 1.319460 | 1.713872 | 2.06866 | 2.49987 | 2.80734 | 3.7676 |
| 24 | 0.256173 | 0.684850 | 1.317836 | 1.710882 | 2.06390 | 2.49216 | 2.79694 | 3.7454 |
| 25 | 0.256060 | 0.684430 | 1.316345 | 1.708141 | 2.05954 | 2.48511 | 2.78744 | 3.7251 |
| 26 | 0.255955 | 0.684043 | 1.314972 | 1.705618 | 2.05553 | 2.47863 | 2.77871 | 3.7066 |
| 27 | 0.255858 | 0.683685 | 1.313703 | 1.703288 | 2.05183 | 2.47266 | 2.77068 | 3.6896 |
| 28 | 0.255768 | 0.683353 | 1.312527 | 1.701131 | 2.04841 | 2.46714 | 2.76326 | 3.6739 |
| 29 | 0.255684 | 0.683044 | 1.311434 | 1.699127 | 2.04523 | 2.46202 | 2.75639 | 3.6594 |
| 30 | 0.255605 | 0.682756 | 1.310415 | 1.697261 | 2.04227 | 2.45726 | 2.75000 | 3.6460 |
| ∞ | 0.253347 | 0.674490 | 1.281552 | 1.644854 | 1.95996 | 2.32635 | 2.57583 | 3.2905 |

# Binomial Table

Table A-2 shows probabilities for the binomial distribution (refer to Chapter 17). To use Table A-2, you need three pieces of information from the particular problem you're working on:

- The sample size, *n*
- The probability of success, *p*
- The value of *x* for which you want the cumulative probability

Find the portion of Table A-2 that's devoted to your *n*, and look at the row for your *x* and the column for your *p*. Intersect that row and column, and you can see the probability for *x*. To get the probability of being strictly less than, greater than, greater than or equal to, or between two values of *x*, you sum the appropriate values of Table A-2, using the steps found in Chapter 16.

## Table A-2 The Binomial Table

Numbers in the table represent the probabilities for values of $x$ from 0 to $n$.

Binomial probabilities:

$\binom{n}{x} p^x(1-p)^{n-x}$

| | | p | | | | | | | | | | |
|---|---|---|---|---|---|---|---|---|---|---|---|---|
| n | x | 0.1 | 0.2 | 0.25 | 0.3 | 0.4 | 0.5 | 0.6 | 0.7 | 0.75 | 0.8 | 0.9 |
| 1 | 0 | 0.900 | 0.800 | 0.750 | 0.700 | 0.600 | 0.500 | 0.400 | 0.300 | 0.250 | 0.200 | 0.100 |
| | 1 | 0.100 | 0.200 | 0.250 | 0.300 | 0.400 | 0.500 | 0.600 | 0.700 | 0.750 | 0.800 | 0.900 |
| 2 | 0 | 0.810 | 0.640 | 0.563 | 0.490 | 0.360 | 0.250 | 0.160 | 0.090 | 0.063 | 0.040 | 0.010 |
| | 1 | 0.180 | 0.320 | 0.375 | 0.420 | 0.480 | 0.500 | 0.480 | 0.420 | 0.375 | 0.320 | 0.180 |
| | 2 | 0.010 | 0.040 | 0.063 | 0.090 | 0.160 | 0.250 | 0.360 | 0.490 | 0.563 | 0.640 | 0.810 |
| 3 | 0 | 0.729 | 0.512 | 0.422 | 0.343 | 0.216 | 0.125 | 0.064 | 0.027 | 0.016 | 0.008 | 0.001 |
| | 1 | 0.243 | 0.384 | 0.422 | 0.441 | 0.432 | 0.375 | 0.288 | 0.189 | 0.141 | 0.096 | 0.027 |
| | 2 | 0.027 | 0.096 | 0.141 | 0.189 | 0.288 | 0.375 | 0.432 | 0.441 | 0.422 | 0.384 | 0.243 |
| | 3 | 0.001 | 0.008 | 0.016 | 0.027 | 0.064 | 0.125 | 0.216 | 0.343 | 0.422 | 0.512 | 0.729 |
| 4 | 0 | 0.656 | 0.410 | 0.316 | 0.240 | 0.130 | 0.063 | 0.026 | 0.008 | 0.004 | 0.002 | 0.000 |
| | 1 | 0.292 | 0.410 | 0.422 | 0.412 | 0.346 | 0.250 | 0.154 | 0.076 | 0.047 | 0.026 | 0.004 |
| | 2 | 0.049 | 0.154 | 0.211 | 0.265 | 0.346 | 0.375 | 0.346 | 0.265 | 0.211 | 0.154 | 0.049 |
| | 3 | 0.004 | 0.026 | 0.047 | 0.076 | 0.154 | 0.250 | 0.346 | 0.412 | 0.422 | 0.410 | 0.292 |
| | 4 | 0.000 | 0.002 | 0.004 | 0.008 | 0.026 | 0.063 | 0.130 | 0.240 | 0.316 | 0.410 | 0.656 |
| 5 | 0 | 0.590 | 0.328 | 0.237 | 0.168 | 0.078 | 0.031 | 0.010 | 0.002 | 0.001 | 0.000 | 0.000 |
| | 1 | 0.328 | 0.410 | 0.396 | 0.360 | 0.259 | 0.156 | 0.077 | 0.028 | 0.015 | 0.006 | 0.000 |
| | 2 | 0.073 | 0.205 | 0.264 | 0.309 | 0.346 | 0.312 | 0.230 | 0.132 | 0.088 | 0.051 | 0.008 |
| | 3 | 0.008 | 0.051 | 0.088 | 0.132 | 0.230 | 0.312 | 0.346 | 0.309 | 0.264 | 0.205 | 0.073 |
| | 4 | 0.000 | 0.006 | 0.015 | 0.028 | 0.077 | 0.156 | 0.259 | 0.360 | 0.396 | 0.410 | 0.328 |
| | 5 | 0.000 | 0.000 | 0.001 | 0.002 | 0.010 | 0.031 | 0.078 | 0.168 | 0.237 | 0.328 | 0.590 |
| 6 | 0 | 0.531 | 0.262 | 0.178 | 0.118 | 0.047 | 0.016 | 0.004 | 0.001 | 0.000 | 0.000 | 0.000 |
| | 1 | 0.354 | 0.393 | 0.356 | 0.303 | 0.187 | 0.094 | 0.037 | 0.010 | 0.004 | 0.002 | 0.000 |
| | 2 | 0.098 | 0.246 | 0.297 | 0.324 | 0.311 | 0.234 | 0.138 | 0.060 | 0.033 | 0.015 | 0.001 |
| | 3 | 0.015 | 0.082 | 0.132 | 0.185 | 0.276 | 0.313 | 0.276 | 0.185 | 0.132 | 0.082 | 0.015 |
| | 4 | 0.001 | 0.015 | 0.033 | 0.060 | 0.138 | 0.234 | 0.311 | 0.324 | 0.297 | 0.246 | 0.098 |
| | 5 | 0.000 | 0.002 | 0.004 | 0.010 | 0.037 | 0.094 | 0.187 | 0.303 | 0.356 | 0.393 | 0.354 |
| | 6 | 0.000 | 0.000 | 0.000 | 0.001 | 0.004 | 0.016 | 0.047 | 0.118 | 0.178 | 0.262 | 0.531 |
| 7 | 0 | 0.478 | 0.210 | 0.133 | 0.082 | 0.028 | 0.008 | 0.002 | 0.000 | 0.000 | 0.000 | 0.000 |
| | 1 | 0.372 | 0.367 | 0.311 | 0.247 | 0.131 | 0.055 | 0.017 | 0.004 | 0.001 | 0.000 | 0.000 |
| | 2 | 0.124 | 0.275 | 0.311 | 0.318 | 0.261 | 0.164 | 0.077 | 0.025 | 0.012 | 0.004 | 0.000 |
| | 3 | 0.023 | 0.115 | 0.173 | 0.227 | 0.290 | 0.273 | 0.194 | 0.097 | 0.058 | 0.029 | 0.003 |
| | 4 | 0.003 | 0.029 | 0.058 | 0.097 | 0.194 | 0.273 | 0.290 | 0.227 | 0.173 | 0.115 | 0.023 |
| | 5 | 0.000 | 0.004 | 0.012 | 0.025 | 0.077 | 0.164 | 0.261 | 0.318 | 0.311 | 0.275 | 0.124 |
| | 6 | 0.000 | 0.000 | 0.001 | 0.004 | 0.017 | 0.055 | 0.131 | 0.247 | 0.311 | 0.367 | 0.372 |
| | 7 | 0.000 | 0.000 | 0.000 | 0.000 | 0.002 | 0.008 | 0.028 | 0.082 | 0.133 | 0.210 | 0.478 |

*(continued)*

## Table A-2 *(continued)*

Binomial probabilities:

$\binom{n}{x} p^x (1-p)^{n-x}$

| n | x | p = 0.1 | 0.2 | 0.25 | 0.3 | 0.4 | 0.5 | 0.6 | 0.7 | 0.75 | 0.8 | 0.9 |
|---|---|---|---|---|---|---|---|---|---|---|---|---|
| 8 | 0 | 0.430 | 0.168 | 0.100 | 0.058 | 0.017 | 0.004 | 0.001 | 0.000 | 0.000 | 0.000 | 0.000 |
| | 1 | 0.383 | 0.336 | 0.267 | 0.198 | 0.090 | 0.031 | 0.008 | 0.001 | 0.000 | 0.000 | 0.000 |
| | 2 | 0.149 | 0.294 | 0.311 | 0.296 | 0.209 | 0.109 | 0.041 | 0.010 | 0.004 | 0.001 | 0.000 |
| | 3 | 0.033 | 0.147 | 0.208 | 0.254 | 0.279 | 0.219 | 0.124 | 0.047 | 0.023 | 0.009 | 0.000 |
| | 4 | 0.005 | 0.046 | 0.087 | 0.136 | 0.232 | 0.273 | 0.232 | 0.136 | 0.087 | 0.046 | 0.005 |
| | 5 | 0.000 | 0.009 | 0.023 | 0.047 | 0.124 | 0.219 | 0.279 | 0.254 | 0.208 | 0.147 | 0.033 |
| | 6 | 0.000 | 0.001 | 0.004 | 0.010 | 0.041 | 0.109 | 0.209 | 0.296 | 0.311 | 0.294 | 0.149 |
| | 7 | 0.000 | 0.000 | 0.000 | 0.001 | 0.008 | 0.031 | 0.090 | 0.198 | 0.267 | 0.336 | 0.383 |
| | 8 | 0.000 | 0.000 | 0.000 | 0.000 | 0.001 | 0.004 | 0.017 | 0.058 | 0.100 | 0.168 | 0.430 |
| 9 | 0 | 0.387 | 0.134 | 0.075 | 0.040 | 0.010 | 0.002 | 0.000 | 0.000 | 0.000 | 0.000 | 0.000 |
| | 1 | 0.387 | 0.302 | 0.225 | 0.156 | 0.060 | 0.018 | 0.004 | 0.000 | 0.000 | 0.000 | 0.000 |
| | 2 | 0.172 | 0.302 | 0.300 | 0.267 | 0.161 | 0.070 | 0.021 | 0.004 | 0.001 | 0.000 | 0.000 |
| | 3 | 0.045 | 0.176 | 0.234 | 0.267 | 0.251 | 0.164 | 0.074 | 0.021 | 0.009 | 0.003 | 0.000 |
| | 4 | 0.007 | 0.066 | 0.117 | 0.172 | 0.251 | 0.246 | 0.167 | 0.074 | 0.039 | 0.017 | 0.001 |
| | 5 | 0.001 | 0.017 | 0.039 | 0.074 | 0.167 | 0.246 | 0.251 | 0.172 | 0.117 | 0.066 | 0.007 |
| | 6 | 0.000 | 0.003 | 0.009 | 0.021 | 0.074 | 0.164 | 0.251 | 0.267 | 0.234 | 0.176 | 0.045 |
| | 7 | 0.000 | 0.000 | 0.001 | 0.004 | 0.021 | 0.070 | 0.161 | 0.267 | 0.300 | 0.302 | 0.172 |
| | 8 | 0.000 | 0.000 | 0.000 | 0.000 | 0.004 | 0.018 | 0.060 | 0.156 | 0.225 | 0.302 | 0.387 |
| | 9 | 0.000 | 0.000 | 0.000 | 0.000 | 0.000 | 0.002 | 0.010 | 0,040 | 0.075 | 0.134 | 0.387 |
| 10 | 0 | 0.349 | 0.107 | 0.056 | 0.028 | 0.006 | 0.001 | 0.000 | 0.000 | 0.000 | 0.000 | 0.000 |
| | 1 | 0.387 | 0.268 | 0.188 | 0.121 | 0.040 | 0.010 | 0.002 | 0.000 | 0.000 | 0.000 | 0.000 |
| | 2 | 0.194 | 0.302 | 0.282 | 0.233 | 0.121 | 0.044 | 0.011 | 0.001 | 0.000 | 0.000 | 0.000 |
| | 3 | 0.057 | 0.201 | 0.250 | 0.267 | 0.215 | 0.117 | 0.042 | 0.009 | 0.003 | 0.001 | 0.000 |
| | 4 | 0.011 | 0.088 | 0.146 | 0.200 | 0.251 | 0.205 | 0.111 | 0.037 | 0.016 | 0.006 | 0.000 |
| | 5 | 0.001 | 0.026 | 0.058 | 0.103 | 0.201 | 0.246 | 0.201 | 0.103 | 0.058 | 0.026 | 0.001 |
| | 6 | 0.000 | 0.006 | 0.016 | 0.037 | 0.111 | 0.205 | 0.251 | 0.200 | 0.146 | 0.088 | 0.011 |
| | 7 | 0.000 | 0.001 | 0.003 | 0.009 | 0.042 | 0.117 | 0.215 | 0.267 | 0.250 | 0.201 | 0.057 |
| | 8 | 0.000 | 0.000 | 0.000 | 0.001 | 0.011 | 0.044 | 0.121 | 0.233 | 0.282 | 0.302 | 0.194 |
| | 9 | 0.000 | 0.000 | 0.000 | 0.000 | 0.002 | 0.010 | 0.040 | 0.121 | 0.188 | 0.268 | 0.387 |
| | 10 | 0.000 | 0.000 | 0.000 | 0.000 | 0.000 | 0.001 | 0.006 | 0.028 | 0.056 | 0.107 | 0.349 |
| 11 | 0 | 0.314 | 0.086 | 0.042 | 0.020 | 0.004 | 0.000 | 0.000 | 0.000 | 0.000 | 0.000 | 0.000 |
| | 1 | 0.384 | 0.236 | 0.155 | 0.093 | 0.027 | 0.005 | 0.001 | 0.000 | 0.000 | 0.000 | 0.000 |
| | 2 | 0.213 | 0.295 | 0.258 | 0.200 | 0.089 | 0.027 | 0.005 | 0.001 | 0.000 | 0.000 | 0.000 |
| | 3 | 0.071 | 0.221 | 0.258 | 0.257 | 0.177 | 0.081 | 0.023 | 0.004 | 0.001 | 0.000 | 0.000 |
| | 4 | 0.016 | 0.111 | 0.172 | 0.220 | 0.236 | 0.161 | 0.070 | 0.017 | 0.006 | 0.002 | 0.000 |
| | 5 | 0.002 | 0.039 | 0.080 | 0.132 | 0.221 | 0.226 | 0.147 | 0.057 | 0.027 | 0.010 | 0.000 |
| | 6 | 0.000 | 0.010 | 0.027 | 0.057 | 0.147 | 0.226 | 0.221 | 0.132 | 0.080 | 0.039 | 0.002 |
| | 7 | 0.000 | 0.002 | 0.006 | 0.017 | 0.070 | 0.161 | 0.236 | 0.220 | 0.172 | 0.111 | 0.016 |
| | 8 | 0.000 | 0.000 | 0.001 | 0.004 | 0.023 | 0.081 | 0.177 | 0.257 | 0.258 | 0.221 | 0.071 |
| | 9 | 0.000 | 0.000 | 0.000 | 0.001 | 0.005 | 0.027 | 0.089 | 0.200 | 0.258 | 0.295 | 0.213 |
| | 10 | 0.000 | 0.000 | 0.000 | 0.000 | 0.001 | 0.005 | 0.027 | 0.093 | 0.155 | 0.236 | 0.384 |
| | 11 | 0.000 | 0.000 | 0.000 | 0.000 | 0.000 | 0.000 | 0.004 | 0.020 | 0.042 | 0.086 | 0.314 |

*(continued)*

## Table A-2 *(continued)*

Binomial probabilities:

$\binom{n}{x} p^x(1-p)^{n-x}$

| | | p | | | | | | | | | | |
|---|---|---|---|---|---|---|---|---|---|---|---|---|
| n | x | 0.1 | 0.2 | 0.25 | 0.3 | 0.4 | 0.5 | 0.6 | 0.7 | 0.75 | 0.8 | 0.9 |
| 12 | 0 | 0.282 | 0.069 | 0.032 | 0.014 | 0.002 | 0.000 | 0.000 | 0.000 | 0.000 | 0.000 | 0.000 |
| | 1 | 0.377 | 0.206 | 0.127 | 0.071 | 0.017 | 0.003 | 0.000 | 0.000 | 0.000 | 0.000 | 0.000 |
| | 2 | 0.230 | 0.283 | 0.232 | 0.168 | 0.064 | 0.016 | 0.002 | 0.000 | 0.000 | 0.000 | 0.000 |
| | 3 | 0.085 | 0.236 | 0.258 | 0.240 | 0.142 | 0.054 | 0.012 | 0.001 | 0.000 | 0.000 | 0.000 |
| | 4 | 0.021 | 0.133 | 0.194 | 0.231 | 0.213 | 0.121 | 0.042 | 0.008 | 0.002 | 0.001 | 0.000 |
| | 5 | 0.004 | 0.053 | 0.103 | 0.158 | 0.227 | 0.193 | 0.101 | 0.029 | 0.011 | 0.003 | 0.000 |
| | 6 | 0.000 | 0.016 | 0.040 | 0.079 | 0.177 | 0.226 | 0.177 | 0.079 | 0.040 | 0.016 | 0.000 |
| | 7 | 0.000 | 0.003 | 0.011 | 0.029 | 0.101 | 0.193 | 0.227 | 0.158 | 0.103 | 0.053 | 0.004 |
| | 8 | 0.000 | 0.001 | 0.002 | 0.008 | 0.042 | 0.121 | 0.213 | 0.231 | 0.194 | 0.133 | 0.021 |
| | 9 | 0.000 | 0.000 | 0.000 | 0.001 | 0.012 | 0.054 | 0.142 | 0.240 | 0.258 | 0.236 | 0.085 |
| | 10 | 0.000 | 0.000 | 0.000 | 0.000 | 0.002 | 0.016 | 0.064 | 0.168 | 0.232 | 0.283 | 0.230 |
| | 11 | 0.000 | 0.000 | 0.000 | 0.000 | 0.000 | 0.003 | 0.017 | 0.071 | 0.127 | 0.206 | 0.377 |
| | 12 | 0.000 | 0.000 | 0.000 | 0.000 | 0.000 | 0.000 | 0.002 | 0.014 | 0.032 | 0.069 | 0.282 |
| 13 | 0 | 0.254 | 0.055 | 0.024 | 0.010 | 0.001 | 0.000 | 0.000 | 0.000 | 0.000 | 0.000 | 0.000 |
| | 1 | 0.367 | 0.179 | 0.103 | 0.054 | 0.011 | 0.002 | 0.000 | 0.000 | 0.000 | 0.000 | 0.000 |
| | 2 | 0.245 | 0.268 | 0.206 | 0.139 | 0.045 | 0.010 | 0.001 | 0.000 | 0.000 | 0.000 | 0.000 |
| | 3 | 0.100 | 0.246 | 0.252 | 0.218 | 0.111 | 0.035 | 0.006 | 0.001 | 0.000 | 0.000 | 0.000 |
| | 4 | 0.028 | 0.154 | 0.210 | 0.234 | 0.184 | 0.087 | 0.024 | 0.003 | 0.001 | 0.000 | 0.000 |
| | 5 | 0.006 | 0.069 | 0.126 | 0.180 | 0.221 | 0.157 | 0.066 | 0.014 | 0.005 | 0.001 | 0.000 |
| | 6 | 0.001 | 0.023 | 0.056 | 0.103 | 0.197 | 0.209 | 0.131 | 0.044 | 0.019 | 0.006 | 0.000 |
| | 7 | 0.000 | 0.006 | 0.019 | 0.044 | 0.131 | 0.209 | 0.197 | 0.103 | 0.056 | 0.023 | 0.001 |
| | 8 | 0.000 | 0.001 | 0.005 | 0.014 | 0.066 | 0.157 | 0.221 | 0.180 | 0.126 | 0.069 | 0.006 |
| | 9 | 0.000 | 0.000 | 0.001 | 0.003 | 0.024 | 0.087 | 0.184 | 0.234 | 0.210 | 0.154 | 0.028 |
| | 10 | 0.000 | 0.000 | 0.000 | 0.001 | 0.006 | 0.035 | 0.111 | 0.218 | 0.252 | 0.246 | 0.100 |
| | 11 | 0.000 | 0.000 | 0.000 | 0.000 | 0.001 | 0.010 | 0.045 | 0.139 | 0.206 | 0.268 | 0.245 |
| | 12 | 0.000 | 0.000 | 0.000 | 0.000 | 0.000 | 0.002 | 0.011 | 0.054 | 0.103 | 0.179 | 0.367 |
| | 13 | 0.000 | 0.000 | 0.000 | 0.000 | 0.000 | 0.000 | 0.001 | 0.010 | 0.024 | 0.055 | 0.254 |
| 14 | 0 | 0.229 | 0.044 | 0.018 | 0.007 | 0.001 | 0.000 | 0.000 | 0.000 | 0.000 | 0.000 | 0.000 |
| | 1 | 0.356 | 0.154 | 0.083 | 0.041 | 0.007 | 0.001 | 0.000 | 0.000 | 0.000 | 0.000 | 0.000 |
| | 2 | 0.257 | 0.250 | 0.180 | 0.113 | 0.032 | 0.006 | 0.001 | 0.000 | 0.000 | 0.000 | 0.000 |
| | 3 | 0.114 | 0.250 | 0.240 | 0.194 | 0.085 | 0.022 | 0.003 | 0.000 | 0.000 | 0.000 | 0.000 |
| | 4 | 0.035 | 0.172 | 0.220 | 0.229 | 0.155 | 0.061 | 0.014 | 0.001 | 0.000 | 0.000 | 0.000 |
| | 5 | 0.008 | 0.086 | 0.147 | 0.196 | 0.207 | 0.122 | 0.041 | 0.007 | 0.002 | 0.000 | 0.000 |
| | 6 | 0.001 | 0.032 | 0.073 | 0.126 | 0.207 | 0.183 | 0.092 | 0.023 | 0.008 | 0.002 | 0.000 |
| | 7 | 0.000 | 0.009 | 0.028 | 0.062 | 0.157 | 0.209 | 0.157 | 0.062 | 0.028 | 0.009 | 0.000 |
| | 8 | 0.000 | 0.002 | 0.008 | 0.023 | 0.092 | 0.183 | 0.207 | 0.126 | 0.073 | 0.032 | 0.001 |
| | 9 | 0.000 | 0.000 | 0.002 | 0.007 | 0.041 | 0.122 | 0.207 | 0.196 | 0.147 | 0.086 | 0.008 |
| | 10 | 0.000 | 0.000 | 0.000 | 0.001 | 0.014 | 0.061 | 0.155 | 0.229 | 0.220 | 0.172 | 0.035 |
| | 11 | 0.000 | 0.000 | 0.000 | 0.000 | 0.003 | 0.022 | 0.085 | 0.194 | 0.240 | 0.250 | 0.114 |
| | 12 | 0.000 | 0.000 | 0.000 | 0.000 | 0.001 | 0.006 | 0.032 | 0.113 | 0.180 | 0.250 | 0.257 |
| | 13 | 0.000 | 0.000 | 0.000 | 0.000 | 0.000 | 0.001 | 0.007 | 0.041 | 0.083 | 0.154 | 0.356 |
| | 14 | 0.000 | 0.000 | 0.000 | 0.000 | 0.000 | 0.000 | 0.001 | 0.007 | 0.018 | 0.044 | 0.229 |

*(continued)*

## Table A-2 *(continued)*

Binomial probabilities:

$$\binom{n}{x} p^x (1-p)^{n-x}$$

| n | x | p = 0.1 | 0.2 | 0.25 | 0.3 | 0.4 | 0.5 | 0.6 | 0.7 | 0.75 | 0.8 | 0.9 |
|---|---|---|---|---|---|---|---|---|---|---|---|---|
| 15 | 0 | 0.206 | 0.035 | 0.013 | 0.005 | 0.000 | 0.000 | 0.000 | 0.000 | 0.000 | 0.000 | 0.000 |
| | 1 | 0.343 | 0.132 | 0.067 | 0.031 | 0.005 | 0.000 | 0.000 | 0.000 | 0.000 | 0.000 | 0.000 |
| | 2 | 0.267 | 0.231 | 0.156 | 0.092 | 0.022 | 0.003 | 0.000 | 0.000 | 0.000 | 0.000 | 0.000 |
| | 3 | 0.129 | 0.250 | 0.225 | 0.170 | 0.063 | 0.014 | 0.002 | 0.000 | 0.000 | 0.000 | 0.000 |
| | 4 | 0.043 | 0.188 | 0.225 | 0.219 | 0.127 | 0.042 | 0.007 | 0.001 | 0.000 | 0.000 | 0.000 |
| | 5 | 0.010 | 0.103 | 0.165 | 0.206 | 0.186 | 0.092 | 0.024 | 0.003 | 0.001 | 0.000 | 0.000 |
| | 6 | 0.002 | 0.043 | 0.092 | 0.147 | 0.207 | 0.153 | 0.061 | 0.012 | 0.003 | 0.001 | 0.000 |
| | 7 | 0.000 | 0.014 | 0.039 | 0.081 | 0.177 | 0.196 | 0.118 | 0.035 | 0.013 | 0.003 | 0.000 |
| | 8 | 0.000 | 0.003 | 0.013 | 0.035 | 0.118 | 0.196 | 0.177 | 0.081 | 0.039 | 0.014 | 0.000 |
| | 9 | 0.000 | 0.001 | 0.003 | 0.012 | 0.061 | 0.153 | 0.207 | 0.147 | 0.092 | 0.043 | 0.002 |
| | 10 | 0.000 | 0.000 | 0.001 | 0.003 | 0.024 | 0.092 | 0.186 | 0.206 | 0.165 | 0.103 | 0.010 |
| | 11 | 0.000 | 0.000 | 0.000 | 0.001 | 0.007 | 0.042 | 0.127 | 0.219 | 0.225 | 0.188 | 0.043 |
| | 12 | 0.000 | 0.000 | 0.000 | 0.000 | 0.002 | 0.014 | 0.063 | 0.170 | 0.225 | 0.250 | 0.129 |
| | 13 | 0.000 | 0.000 | 0.000 | 0.000 | 0.000 | 0.003 | 0.022 | 0.092 | 0.156 | 0.231 | 0.267 |
| | 14 | 0.000 | 0.000 | 0.000 | 0.000 | 0.000 | 0.000 | 0.005 | 0.031 | 0.067 | 0.132 | 0.343 |
| | 15 | 0.000 | 0.000 | 0.000 | 0.000 | 0.000 | 0.000 | 0.000 | 0.005 | 0.013 | 0.035 | 0.206 |
| 20 | 0 | 0.122 | 0.012 | 0.003 | 0.001 | 0.000 | 0.000 | 0.000 | 0.000 | 0.000 | 0.000 | 0.000 |
| | 1 | 0.270 | 0.058 | 0.021 | 0.007 | 0.000 | 0.000 | 0.000 | 0.000 | 0.000 | 0.000 | 0.000 |
| | 2 | 0.285 | 0.137 | 0.067 | 0.028 | 0.003 | 0.000 | 0.000 | 0.000 | 0.000 | 0.000 | 0.000 |
| | 3 | 0.190 | 0.205 | 0.134 | 0.072 | 0.012 | 0.001 | 0.000 | 0.000 | 0.000 | 0.000 | 0.000 |
| | 4 | 0.090 | 0.218 | 0.190 | 0.130 | 0.035 | 0.005 | 0.000 | 0.000 | 0.000 | 0.000 | 0.000 |
| | 5 | 0.032 | 0.175 | 0.202 | 0.179 | 0.075 | 0.015 | 0.001 | 0.000 | 0.000 | 0.000 | 0.000 |
| | 6 | 0.009 | 0.109 | 0.169 | 0.192 | 0.124 | 0.037 | 0.005 | 0.000 | 0.000 | 0.000 | 0.000 |
| | 7 | 0.002 | 0.055 | 0.112 | 0.164 | 0.166 | 0.074 | 0.015 | 0.001 | 0.000 | 0.000 | 0.000 |
| | 8 | 0.000 | 0.022 | 0.061 | 0.114 | 0.180 | 0.120 | 0.035 | 0.004 | 0.001 | 0.000 | 0.000 |
| | 9 | 0.000 | 0.007 | 0.027 | 0.065 | 0.160 | 0.160 | 0.071 | 0.012 | 0.003 | 0.000 | 0.000 |
| | 10 | 0.000 | 0.002 | 0.010 | 0.031 | 0.117 | 0.176 | 0.117 | 0.031 | 0.010 | 0.002 | 0.000 |
| | 11 | 0.000 | 0.000 | 0.003 | 0.012 | 0.071 | 0.160 | 0.160 | 0.065 | 0.027 | 0.007 | 0.007 |
| | 12 | 0.000 | 0.000 | 0.001 | 0.004 | 0.035 | 0.120 | 0.180 | 0.114 | 0.061 | 0.022 | 0.000 |
| | 13 | 0.000 | 0.000 | 0.000 | 0.001 | 0.015 | 0.074 | 0.166 | 0.164 | 0.112 | 0.055 | 0.002 |
| | 14 | 0.000 | 0.000 | 0.000 | 0.000 | 0.005 | 0.037 | 0.124 | 0.192 | 0.169 | 0.109 | 0.009 |
| | 15 | 0.000 | 0.000 | 0.000 | 0.000 | 0.001 | 0.015 | 0.075 | 0.179 | 0.202 | 0.175 | 0.032 |
| | 16 | 0.000 | 0.000 | 0.000 | 0.000 | 0.000 | 0.005 | 0.035 | 0.130 | 0.190 | 0.218 | 0.090 |
| | 17 | 0.000 | 0.000 | 0.000 | 0.000 | 0.000 | 0.001 | 0.012 | 0.072 | 0.134 | 0.205 | 0.190 |
| | 18 | 0.000 | 0.000 | 0.000 | 0.000 | 0.000 | 0.000 | 0.003 | 0.028 | 0.067 | 0.137 | 0.285 |
| | 19 | 0.000 | 0.000 | 0.000 | 0.000 | 0.000 | 0.000 | 0.000 | 0.007 | 0.021 | 0.058 | 0.270 |
| | 20 | 0.000 | 0.000 | 0.000 | 0.000 | 0.000 | 0.000 | 0.000 | 0.001 | 0.003 | 0.012 | 0.122 |

# Chi-Square Table

Table A-3 shows right-tail probabilities for the Chi-square distribution (you can use Chapter 14 as a reference for the Chi-square test). To use Table A-3, you need three pieces of information from the particular problem you're working on:

- The sample size, *n*.
- The value of $\chi$-squared, for which you want the right-tail probability.
- If you're working with a two-way table, you need $r$ = number of rows and $c$ = number of columns. If you're working with a goodness-of-fit test, you need $k - 1$, where $k$ is the number of categories.

The degrees of freedom for the Chi-square test statistic is $(r - 1) * (c - 1)$ if you're testing for an association between two variables, where $r$ and $c$ are the number of rows and columns in the two-way table, respectively. Or, the degrees of freedom is $k - 1$ in a goodness-of-fit test, where $k$ is the number of categories; see Chapter 15.

Go across the row for your degrees of freedom until you find the value in that row closest to your Chi-square test statistic. Look up at the number at the top of that column. That value is the area to the right (beyond) that particular Chi-square statistic.

## Table A-3 The Chi-Square Table

Numbers in the table represent Chi-square values whose area to the right equals *p*.

| df/p | 0.10 | 0.05 | 0.025 | 0.01 | 0.005 |
|---|---|---|---|---|---|
| 1 | 2.71 | 3.84 | 5.02 | 6.64 | 7.88 |
| 2 | 4.61 | 5.99 | 7.38 | 9.21 | 10.60 |
| 3 | 6.25 | 7.82 | 9.35 | 11.35 | 12.84 |
| 4 | 7.78 | 9.49 | 11.14 | 13.28 | 14.86 |
| 5 | 9.24 | 11.07 | 12.83 | 15.09 | 16.75 |
| 6 | 10.65 | 12.59 | 14.45 | 16.81 | 18.55 |
| 7 | 12.02 | 14.07 | 16.01 | 18.48 | 20.28 |
| 8 | 13.36 | 15.51 | 17.54 | 20.09 | 21.96 |
| 9 | 14.68 | 16.92 | 19.02 | 21.67 | 23.59 |
| 10 | 15.99 | 18.31 | 20.48 | 23.21 | 25.19 |
| 11 | 17.28 | 19.68 | 21.92 | 24.73 | 26.76 |
| 12 | 18.55 | 21.03 | 23.34 | 26.22 | 28.30 |
| 13 | 19.81 | 22.36 | 24.74 | 27.69 | 29.819 |
| 14 | 21.06 | 23.69 | 26.12 | 29.14 | 31.32 |
| 15 | 22.31 | 25.00 | 27.49 | 30.58 | 32.80 |
| 16 | 23.54 | 26.30 | 28.85 | 32.00 | 34.27 |
| 17 | 24.77 | 27.59 | 30.19 | 33.41 | 35.72 |
| 18 | 25.99 | 28.87 | 31.53 | 34.81 | 37.16 |
| 19 | 27.20 | 30.14 | 32.85 | 36.19 | 38.58 |
| 20 | 28.41 | 31.41 | 34.17 | 37.57 | 40.00 |
| 21 | 29.62 | 32.67 | 35.48 | 38.93 | 41.40 |
| 22 | 30.81 | 33.92 | 36.78 | 40.29 | 42.80 |
| 23 | 32.01 | 35.17 | 38.08 | 41.64 | 44.18 |
| 24 | 33.20 | 36.42 | 39.36 | 42.98 | 45.56 |
| 25 | 34.38 | 37.65 | 40.65 | 44.31 | 46.93 |
| 26 | 35.56 | 38.89 | 41.92 | 45.64 | 48.29 |
| 27 | 36.74 | 40.11 | 43.20 | 46.96 | 49.65 |
| 28 | 37.92 | 41.34 | 44.46 | 48.28 | 50.99 |
| 29 | 39.09 | 42.56 | 45.72 | 49.59 | 52.34 |
| 30 | 40.26 | 43.77 | 46.98 | 50.89 | 53.67 |
| 40 | 51.81 | 55.76 | 59.34 | 63.69 | 66.77 |
| 50 | 63.17 | 67.51 | 71.42 | 76.15 | 79.49 |

# Rank Sum Table

Table A-4 shows the critical values for the rank sum test where $\alpha$ is 0.05 for two-sided tests (equivalent to 0.025 for one-sided tests); see Chapter 18 for more on this test. To use Table A-4, you need two pieces of information from the particular problem you're working on:

- The rank sum statistic, $T$
- The sample sizes of the two samples, $n_1$ and $n_2$

To find the critical value for your rank sum statistic using Table A-4, go to the column representing $n_1$ and the row representing $n_2$. Intersect the row and the column on Table A-4, and you find the lower and upper critical values (denoted $T_L$ and $T_U$) for the rank sum test.

**Table A-4 Rank Sum Table**

$\alpha$ = .025 One-Sided; $\alpha$ = .05 Two-Sided

| $n_2$ \ $n_1$ | 3 | | 4 | | 5 | | 6 | | 7 | | 8 | | 9 | | 10 | |
|---|---|---|---|---|---|---|---|---|---|---|---|---|---|---|---|---|
| | $T_L$ | $T_U$ | $T_L$ | $T_U$ | $T_L$ | $T_U$ | $T_L$ | $T_U$ | $T_L$ | $T_U$ | $T_L$ | $T_U$ | $T_L$ | $T_U$ | $T_L$ | $T_U$ |
| 3 | 5 | 16 | 6 | 18 | 6 | 21 | 7 | 23 | 7 | 26 | 8 | 28 | 8 | 31 | 9 | 33 |
| 4 | 6 | 18 | 11 | 25 | 12 | 28 | 12 | 32 | 13 | 35 | 14 | 38 | 15 | 41 | 16 | 44 |
| 5 | 6 | 21 | 12 | 28 | 18 | 37 | 19 | 41 | 20 | 45 | 21 | 49 | 22 | 53 | 24 | 56 |
| 6 | 7 | 23 | 12 | 32 | 19 | 41 | 26 | 52 | 28 | 56 | 29 | 61 | 31 | 65 | 32 | 70 |
| 7 | 7 | 26 | 13 | 35 | 20 | 45 | 28 | 56 | 37 | 68 | 39 | 73 | 41 | 78 | 43 | 83 |
| 8 | 8 | 28 | 14 | 38 | 21 | 49 | 29 | 61 | 39 | 73 | 49 | 87 | 53 | 93 | 54 | 98 |
| 9 | 8 | 31 | 15 | 41 | 22 | 53 | 31 | 65 | 41 | 78 | 51 | 93 | 63 | 108 | 66 | 114 |
| 10 | 9 | 33 | 16 | 44 | 24 | 56 | 32 | 70 | 43 | 83 | 54 | 98 | 66 | 114 | 79 | 131 |

*(continued)*

**Table A-4 *(continued)***

$\alpha$ = .05 One-Sided; $\alpha$ = .10 Two-Sided

| $n_2$ \ $n_1$ | 3 | | 4 | | 5 | | 6 | | 7 | | 8 | | 9 | | 10 | |
|---|---|---|---|---|---|---|---|---|---|---|---|---|---|---|---|---|
| | $T_L$ | $T_U$ | $T_L$ | $T_U$ | $T_L$ | $T_U$ | $T_L$ | $T_U$ | $T_L$ | $T_U$ | $T_L$ | $T_U$ | $T_L$ | $T_U$ | $T_L$ | $T_U$ |
| 3 | 6 | 15 | 7 | 17 | 7 | 20 | 8 | 22 | 9 | 24 | 9 | 27 | 10 | 29 | 11 | 31 |
| 4 | 7 | 17 | 12 | 24 | 13 | 27 | 14 | 30 | 15 | 33 | 16 | 36 | 17 | 39 | 18 | 42 |
| 5 | 7 | 20 | 13 | 37 | 19 | 36 | 20 | 40 | 22 | 43 | 24 | 46 | 25 | 50 | 26 | 54 |
| 6 | 8 | 22 | 14 | 30 | 20 | 40 | 28 | 50 | 30 | 54 | 32 | 58 | 33 | 63 | 35 | 67 |
| 7 | 9 | 24 | 15 | 33 | 22 | 43 | 30 | 54 | 39 | 66 | 41 | 71 | 43 | 76 | 46 | 80 |
| 8 | 9 | 27 | 16 | 36 | 24 | 46 | 32 | 58 | 41 | 71 | 52 | 84 | 54 | 90 | 57 | 95 |
| 9 | 10 | 29 | 17 | 39 | 25 | 50 | 33 | 63 | 43 | 76 | 54 | 90 | 66 | 105 | 69 | 111 |
| 10 | 11 | 31 | 18 | 42 | 26 | 54 | 35 | 67 | 46 | 80 | 57 | 95 | 69 | 111 | 83 | 127 |

# F-Table

Table A-5 shows the critical values on the *F*-distribution where $\alpha$ is equal to 0.05. (Critical values are those values that represent the boundary between rejecting Ho and not rejecting Ho; refer to Chapter 9.) To use Table A-5, you need three pieces of information from the particular problem you're working on:

- The sample size, $n$
- The number of populations (or treatments being compared), $k$
- The value of $F$ for which you want the cumulative probability

To find the critical value for your *F*-test statistic using Table A-5, go to the column representing the degrees of freedom you need ($k - 1, n - k$). Intersect the column degrees of freedom ($k - 1$) with the row degrees of freedom ($n - k$), and you find the critical value on the *F*-distribution. For more on the *F*-test, see Chapter 9.

## Table A-5 The F-Table

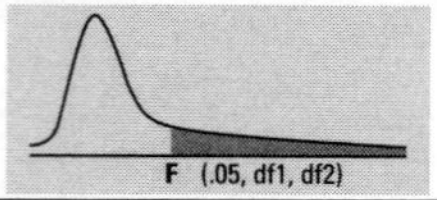


| df2/df1 | 1 | 2 | 3 | 4 | 5 | 6 | 7 | 8 | 9 | 10 | 12 | 15 | 20 | 24 | 30 | 40 | 60 | 120 |
|---|---|---|---|---|---|---|---|---|---|---|---|---|---|---|---|---|---|---|
| 1 | 161.4476 | 199.5000 | 215.7073 | 224.5832 | 230.1619 | 233.9860 | 236.7684 | 238.8827 | 240.5433 | 241.8817 | 243.9060 | 245.9499 | 248.0131 | 249.0518 | 250.0951 | 251.1432 | 252.1957 | 253.252 |
| 2 | 18.5128 | 19.0000 | 19.1643 | 19.2468 | 19.2964 | 19.3295 | 19.3532 | 19.3710 | 19.3848 | 19.3959 | 19.4125 | 19.4291 | 19.4458 | 19.4541 | 19.4624 | 19.4707 | 19.4791 | 19.487 |
| 3 | 10.1280 | 9.5521 | 9.2766 | 9.1172 | 9.0135 | 8.9406 | 8.8867 | 8.8452 | 8.8123 | 8.7855 | 8.7446 | 8.7029 | 8.6602 | 8.6385 | 8.6166 | 8.5944 | 8.5720 | 8.549 |
| 4 | 7.7086 | 6.9443 | 6.5914 | 6.3882 | 6.2561 | 6.1631 | 6.0942 | 6.0410 | 5.9988 | 5.9644 | 5.9117 | 5.8578 | 5.8025 | 5.7744 | 5.7459 | 5.7170 | 5.6877 | 5.658 |
| 5 | 6.6079 | 5.7861 | 5.4095 | 5.1922 | 5.0503 | 4.9503 | 4.8759 | 4.8183 | 4.7725 | 4.7351 | 4.6777 | 4.6188 | 4.5581 | 4.5272 | 4.4957 | 4.4638 | 4.4314 | 4.398 |
| 6 | 5.9874 | 5.1433 | 4.7571 | 4.5337 | 4.3874 | 4.2839 | 4.2067 | 4.1468 | 4.0990 | 4.0600 | 3.9999 | 3.9381 | 3.8742 | 3.8415 | 3.8082 | 3.7743 | 3.7398 | 3.704 |
| 7 | 5.5914 | 4.7374 | 4.3468 | 4.1203 | 3.9715 | 3.8660 | 3.7870 | 3.7257 | 3.6767 | 3.6365 | 3.5747 | 3.5107 | 3.4445 | 3.4105 | 3.3758 | 3.3404 | 3.3043 | 3.267 |
| 8 | 5.3177 | 4.4590 | 4.0662 | 3.8379 | 3.6875 | 3.5806 | 3.5005 | 3.4381 | 3.3881 | 3.3472 | 3.2839 | 3.2184 | 3.1503 | 3.1152 | 3.0794 | 3.0428 | 3.0053 | 2.966 |
| 9 | 5.1174 | 4.2565 | 3.8625 | 3.6331 | 3.4817 | 3.3738 | 3.2927 | 3.2296 | 3.1789 | 3.1373 | 3.0729 | 3.0061 | 2.9365 | 2.9005 | 2.8637 | 2.8259 | 2.7872 | 2.747 |
| 10 | 4.9646 | 4.1028 | 3.7083 | 3.4780 | 3.3258 | 3.2172 | 3.1355 | 3.0717 | 3.0204 | 2.9782 | 2.9130 | 2.8450 | 2.7740 | 2.7372 | 2.6996 | 2.6609 | 2.6211 | 2.580 |
| 11 | 4.8443 | 3.9823 | 3.5874 | 3.3567 | 3.2039 | 3.0946 | 3.0123 | 2.9480 | 2.8962 | 2.8536 | 2.7876 | 2.7186 | 2.6464 | 2.6090 | 2.5705 | 2.5309 | 2.4901 | 2.448 |
| 12 | 4.7472 | 3.8853 | 3.4903 | 3.2592 | 3.1059 | 2.9961 | 2.9134 | 2.8486 | 2.7964 | 2.7534 | 2.6866 | 2.6169 | 2.5436 | 2.5055 | 2.4663 | 2.4259 | 2.3842 | 2.341 |
| 13 | 4.6672 | 3.8056 | 3.4105 | 3.1791 | 3.0254 | 2.9153 | 2.8321 | 2.7669 | 2.7144 | 2.6710 | 2.6037 | 2.5331 | 2.4589 | 2.4202 | 2.3803 | 2.3392 | 2.2966 | 2.252 |
| 14 | 4.6001 | 3.7389 | 3.3439 | 3.1122 | 2.9582 | 2.8477 | 2.7642 | 2.6987 | 2.6458 | 2.6022 | 2.5342 | 2.4630 | 2.3879 | 2.3487 | 2.3082 | 2.2664 | 2.2229 | 2.177 |
| 15 | 4.5431 | 3.6823 | 3.2874 | 3.0556 | 2.9013 | 2.7905 | 2.7066 | 2.6408 | 2.5876 | 2.5437 | 2.4753 | 2.4034 | 2.3275 | 2.2878 | 2.2468 | 2.2043 | 2.1601 | 2.114 |
| 16 | 4.4940 | 3.6337 | 3.2389 | 3.0069 | 2.8524 | 2.7413 | 2.6572 | 2.5911 | 2.5377 | 2.4935 | 2.4247 | 2.3522 | 2.2756 | 2.2354 | 2.1938 | 2.1507 | 2.1058 | 2.058 |
| 17 | 4.4513 | 3.5915 | 3.1968 | 2.9647 | 2.8100 | 2.6987 | 2.6143 | 2.5480 | 2.4943 | 2.4499 | 2.3807 | 2.3077 | 2.2304 | 2.1898 | 2.1477 | 2.1040 | 2.0584 | 2.010 |
| 18 | 4.4139 | 3.5546 | 3.1599 | 2.9277 | 2.7729 | 2.6613 | 2.5767 | 2.5102 | 2.4563 | 2.4117 | 2.3421 | 2.2686 | 2.1906 | 2.1497 | 2.1071 | 2.0629 | 2.0166 | 1.968 |
| 19 | 4.3807 | 3.5219 | 3.1274 | 2.8951 | 2.7401 | 2.6283 | 2.5435 | 2.4768 | 2.4227 | 2.3779 | 2.3080 | 2.2341 | 2.1555 | 2.1141 | 2.0712 | 2.0264 | 1.9795 | 1.930 |
| 20 | 4.3512 | 3.4928 | 3.0984 | 2.8661 | 2.7109 | 2.5990 | 2.5140 | 2.4471 | 2.3928 | 2.3479 | 2.2776 | 2.2033 | 2.1242 | 2.0825 | 2.0391 | 1.9938 | 1.9464 | 1.896 |
| 21 | 4.3248 | 3.4668 | 3.0725 | 2.8401 | 2.6848 | 2.5727 | 2.4876 | 2.4205 | 2.3660 | 2.3210 | 2.2504 | 2.1757 | 2.0960 | 2.0540 | 2.0102 | 1.9645 | 1.9165 | 1.865 |
| 22 | 4.3009 | 3.4434 | 3.0491 | 2.8167 | 2.6613 | 2.5491 | 2.4638 | 2.3965 | 2.3419 | 2.2967 | 2.2258 | 2.1508 | 2.0707 | 2.0283 | 1.9842 | 1.9380 | 1.8894 | 1.838 |
| 23 | 4.2793 | 3.4221 | 3.0280 | 2.7955 | 2.6400 | 2.5277 | 2.4422 | 2.3748 | 2.3201 | 2.2747 | 2.2036 | 2.1282 | 2.0476 | 2.0050 | 1.9605 | 1.9139 | 1.8648 | 1.812 |
| 24 | 4.2597 | 3.4028 | 3.0088 | 2.7763 | 2.6207 | 2.5082 | 2.4226 | 2.3551 | 2.3002 | 2.2547 | 2.1834 | 2.1077 | 2.0267 | 1.9838 | 1.9390 | 1.8920 | 1.8424 | 1.789 |
| 25 | 4.2417 | 3.3852 | 2.9912 | 2.7587 | 2.6030 | 2.4904 | 2.4047 | 2.3371 | 2.2821 | 2.2365 | 2.1649 | 2.0889 | 2.0075 | 1.9643 | 1.9192 | 1.8718 | 1.8217 | 1.768 |
| 26 | 4.2252 | 3.3690 | 2.9752 | 2.7426 | 2.5868 | 2.4741 | 2.3883 | 2.3205 | 2.2655 | 2.2197 | 2.1479 | 2.0716 | 1.9898 | 1.9464 | 1.9010 | 1.8533 | 1.8027 | 1.748 |
| 27 | 4.2100 | 3.3541 | 2.9604 | 2.7278 | 2.5719 | 2.4591 | 2.3732 | 2.3053 | 2.2501 | 2.2043 | 2.1323 | 2.0558 | 1.9736 | 1.9299 | 1.8842 | 1.8361 | 1.7851 | 1.730 |
| 28 | 4.1960 | 3.3404 | 2.9467 | 2.7141 | 2.5581 | 2.4453 | 2.3593 | 2.2913 | 2.2360 | 2.1900 | 2.1179 | 2.0411 | 1.9586 | 1.9147 | 1.8687 | 1.8203 | 1.7689 | 1.713 |
| 29 | 4.1830 | 3.3277 | 2.9340 | 2.7014 | 2.5454 | 2.4324 | 2.3463 | 2.2783 | 2.2229 | 2.1768 | 2.1045 | 2.0275 | 1.9446 | 1.9005 | 1.8543 | 1.8055 | 1.7537 | 1.698 |
| 30 | 4.1709 | 3.3158 | 2.9223 | 2.6896 | 2.5336 | 2.4205 | 2.3343 | 2.2662 | 2.2107 | 2.1646 | 2.0921 | 2.0148 | 1.9317 | 1.8874 | 1.8409 | 1.7918 | 1.7396 | 1.683 |
| 40 | 4.0847 | 3.2317 | 2.8387 | 2.6060 | 2.4495 | 2.3359 | 2.2490 | 2.1802 | 2.1240 | 2.0772 | 2.0035 | 1.9245 | 1.8389 | 1.7929 | 1.7444 | 1.6928 | 1.6373 | 1.576 |
| 60 | 4.0012 | 3.1504 | 2.7581 | 2.5252 | 2.3683 | 2.2541 | 2.1665 | 2.0970 | 2.0401 | 1.9926 | 1.9174 | 1.8364 | 1.7480 | 1.7001 | 1.6491 | 1.5943 | 1.5343 | 1.467 |
| 120 | 3.9201 | 3.0718 | 2.6802 | 2.4472 | 2.2899 | 2.1750 | 2.0868 | 2.0164 | 1.9588 | 1.9105 | 1.8337 | 1.7505 | 1.6587 | 1.6084 | 1.5543 | 1.4952 | 1.4290 | 1.351 |

# Index

## • Numbers & Symbols •



## • A •



## • B •

## • N •

## • O •



## • P •



## • Q •

## • T •



## • U •



## • V •

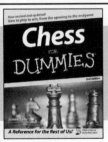

## SPORTS, FITNESS, PARENTING, RELIGION & SPIRITUALITY

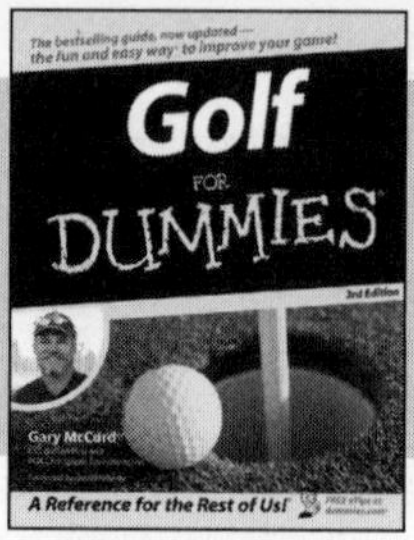


0-471-76871-5

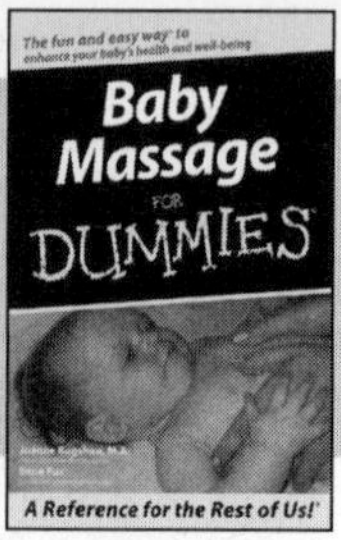


0-7645-7841-3

**Also available:**

- Catholicism For Dummies 0-7645-5391-7
- Exercise Balls For Dummies 0-7645-5623-1
- Fitness For Dummies 0-7645-7851-0
- Football For Dummies 0-7645-3936-1
- Judaism For Dummies 0-7645-5299-6
- Potty Training For Dummies 0-7645-5417-4
- Buddhism For Dummies 0-7645-5359-3
- Pregnancy For Dummies 0-7645-4483-7 †
- Ten Minute Tone-Ups For Dummies 0-7645-7207-5
- NASCAR For Dummies 0-7645-7681-X
- Religion For Dummies 0-7645-5264-3
- Soccer For Dummies 0-7645-5229-5
- Women in the Bible For Dummies 0-7645-8475-8

## TRAVEL

0-7645-7749-2

0-7645-6945-7

**Also available:**

- Alaska For Dummies 0-7645-7746-8
- Cruise Vacations For Dummies 0-7645-6941-4
- England For Dummies 0-7645-4276-1
- Europe For Dummies 0-7645-7529-5
- Germany For Dummies 0-7645-7823-5
- Hawaii For Dummies 0-7645-7402-7
- Italy For Dummies 0-7645-7386-1
- Las Vegas For Dummies 0-7645-7382-9
- London For Dummies 0-7645-4277-X
- Paris For Dummies 0-7645-7630-5
- RV Vacations For Dummies 0-7645-4442-X
- Walt Disney World & Orlando For Dummies 0-7645-9660-8

## GRAPHICS, DESIGN & WEB DEVELOPMENT

0-7645-8815-X

0-7645-9571-7

**Also available:**

- 3D Game Animation For Dummies 0-7645-8789-7
- AutoCAD 2006 For Dummies 0-7645-8925-3
- Building a Web Site For Dummies 0-7645-7144-3
- Creating Web Pages For Dummies 0-470-08030-2
- Creating Web Pages All-in-One Desk Reference For Dummies 0-7645-4345-8
- Dreamweaver 8 For Dummies 0-7645-9649-7
- InDesign CS2 For Dummies 0-7645-9572-5
- Macromedia Flash 8 For Dummies 0-7645-9691-8
- Photoshop CS2 and Digital Photography For Dummies 0-7645-9580-6
- Photoshop Elements 4 For Dummies 0-471-77483-9
- Syndicating Web Sites with RSS Feeds For Dummies 0-7645-8848-6
- Yahoo! SiteBuilder For Dummies 0-7645-9800-7

## NETWORKING, SECURITY, PROGRAMMING & DATABASES

0-7645-7728-X

0-471-74940-0

**Also available:**

- Access 2007 For Dummies 0-470-04612-0
- ASP.NET 2 For Dummies 0-7645-7907-X
- C# 2005 For Dummies 0-7645-9704-3
- Hacking For Dummies 0-470-05235-X
- Hacking Wireless Networks For Dummies 0-7645-9730-2
- Java For Dummies 0-470-08716-1
- Microsoft SQL Server 2005 For Dummie 0-7645-7755-7
- Networking All-in-One Desk Referenc For Dummies 0-7645-9939-9
- Preventing Identity Theft For Dummies 0-7645-7336-5
- Telecom For Dummies 0-471-77085-X
- Visual Studio 2005 All-in-One Desk Reference For Dummies 0-7645-9775-2
- XML For Dummies 0-7645-8845-1

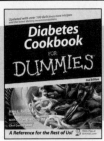

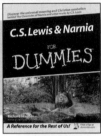